Molecular Farming

Edited by
Rainer Fischer and Stefan Schillberg

Related Titles

Paul Christou and Harry Klee (ed.)

Handbook of Plant Biotechnology

2004
ISBN 0-471-85199-X

Oliver Kayser and Rainer H. Müller (eds.)

Pharmaceutical Biotechnology

Drug Discovery and Clinical Applications

2004
ISBN 3-527-30554-8

Jörg Knäblein and Rainer H. Müller (eds.)

Modern Biopharmaceuticals

Design, Development and Optimization

2005
ISBN 0-527-31184-X

Klaus Dembowsky and Peter Stadler (eds.)

Novel Therapeutic Proteins

Selected Case Studies

2000
ISBN 3-527-30270-0

Rolf D. Schmid and Ruth Hammelehle

Pocket Guide to Biotechnology and Genetic Engineering

2003
ISBN 3-527-30895-4

Knut J. Heller

Genetically Engineered Food

Methods and Detection

2003
ISBN 3-527-30309-X

Shuryo Nakai and H. Wayne Modler (eds.)

Food Proteins

Processing Applications

2000
ISBN 0-471-29785-2

Shuryo Nakai and H. Wayne Modler (eds.)

Food Proteins

Properties and Characterization

1996
ISBN 0-471-18614-7

Molecular Farming

Plant-made Pharmaceuticals and Technical Proteins

Edited by
Rainer Fischer and Stefan Schillberg

WILEY-VCH

WILEY-VCH Verlag GmbH & Co. KGaA

Edited by:

Rainer Fischer
RWTH Aachen
Molecular Biotechnology
Worringerweg 1
52074 Aachen
Germany

Stefan Schillberg
Molecular Farming
Fraunhofer IME
Worringerweg 1
52074 Aachen
Germany

■ This book was carefully produced. Nevertheless, authors, editor and publisher do not warrant the information contained therein to be free of errors. Readers are advised to keep in mind that statements, data, illustrations, procedural details or other items may inadvertently be inaccurate.

Library of Congress Card No.: applied for

British Library Cataloguing-in-Publication Data: A catalogue record for this book is available from the British Library.

Bibliographic information published by Die Deutsche Bibliothek
Die Deutsche Bibliothek lists this publication in the Deutsche Nationalbibliografie; detailed bibliographic data is available in the Internet at http://dnb.ddb.de.

© 2004 Wiley-VCH Verlag GmbH & Co. KGaA, Weinheim

Printed in the Federal Republic of Germany
Printed on acid-free paper

Cover   Grafik-Design Schulz, Fußgönheim
Typesetting   ProSatz Unger, Weinheim
Printing   Strauss GmbH, Mörlenbach
Bookbinding   Litges & Dopf Buchbinderei GmbH, Heppenheim

ISBN   3-527-30786-9

Preface

Mankind has used plants as a source of raw materials and medicines for thousands of years. From the earliest stages of civilization, plant extracts have been used to obtain technical materials and drugs to ease suffering and cure disease. Since the late seventies, many valuable therapeutic and diagnostic proteins have been discovered through molecular biology research and molecular medicine, but widespread use of these molecules has been hampered by production bottlenecks such as low yields, poor and inconsistent product quality and a shortage of production capacity. In the late 1980s, the application of recombinant DNA and protein technology in plants allowed the exploration of plant-based expression systems for the production of safer and cheaper protein medicines (Table 1). Over the last decade, plants have emerged as a convenient, safe and economical alternative to mainstream expression systems which are based on the large-scale culture of microbes or animal cells, or transgenic animals. The production of plant-made pharmaceuticals and technical proteins is known as *Molecular Farming (Molecular Pharming^{TM})*. The objective is to harness the power of agriculture to cultivate and harvest plants or plant cells producing recombinant therapeutics, diagnostics, industrial enzymes and green chemicals.

Molecular Farming has the potential to provide virtually unlimited quantities of recombinant antibodies, vaccines, blood substitutes, growth factors, cytokines, chemokines and enzymes for use as diagnostic and therapeutic tools in health care, the life sciences and the chemical industry. Plants are now gaining widespread acceptance as a general platform for the large-scale production of recombinant proteins. The principle has been demonstrated by the success of a diverse repertoire of proteins, with therapeutic proteins showing the greatest potential for added value and technical enzymes the first to reach commercial status.

We are facing a growing demand for protein diagnostics and therapeutics, but lack the capacity to meet those demands using established facilities. Moreover, recombinant proteins will become more important as high throughput genomics, proteomics, metabolomics and glycomics projects spawn new product candidates, disease targets and eventually new remedies. A shift to plant bioreactors may therefore become necessary within the next few years. However, the production of pharmaceutical proteins in plants will only realize its huge potential if the products achieve consistent highest quality standards, enabling the provision of clinical grade proteins that will gain regulatory approval and can be used routinely in clinical trials and

Molecular Farming. Edited by Rainer Fischer, Stefan Schillberg
Copyright © 2004 WILEY-VCH Verlag GmbH & Co. KGaA, Weinheim
ISBN: 3-527-30786-9

treatments. The achievement of these goals is conditional on the development of technologies for improving yields, ensuring product sustainability and quality, including extraction and processing steps that comply with current good manufacturing practice (cGMP) standards. Moreover, there are several further challenges concerning the environmental impact, biosafety and risk assessment of *Molecular Farming*, which reflect the release of transgenic plants as well the safety of the plant-derived products themselves.

This book covers the most recent achievements and challenges of *Molecular Farming* technology written by experts working in this field. The first few chapters focus on the technological aspects of plant-based protein production, while the second part address the two major target product groups expressed in plant systems: pharmaceutical and technical proteins. Finally, issues concerning the production pipeline are discussed, including production and product safety, quantity and quality control.

We thank all the authors for their contributions and the time and effort they dedicated to compiling this book, which helped to make it a comprehensive and state-of-the-art overview of the technological, economical, commercial and regulatory aspects of *Molecular Farming*. We also gratefully acknowledge the help and support of Dr. Richard Twyman and the team at Wiley. Without all their help, this book would not have been possible.

Aachen, 2004   *Rainer Fischer* and *Stefan Schillberg*

Tab. 1 Key events in the history of *Molecular Farming.*

Year	Highlight	Reference
1986	First plant-derived recombinant therapeutic protein – human growth hormone in tobacco and sunflower [1)]	1
1989	First plant-derived recombinant antibody – full-size IgG in tobacco	2
1990	First native human protein produced in plants – human serum albumin in tobacco and potato	3
1992	First plant-derived vaccine candidate – hepatitis B virus surface antigen in tobacco	4
1992	First plant-derived industrial enzyme – α-amylase in tobacco	5
1995	Secretory IgA produced in tobacco	6
1996	First plant-derived protein polymer – artificial elastin in tobacco	7
1997	First clinical trial using recombinant bacterial antigen delivered in a transgenic potato	8
1997	Commercial production of avidin in maize	9
1999	First glycan analysis of plant-produced recombinant glycoprotein	10
2000	Human growth hormone produced in tobacco chloroplasts	11
2000	Triple helix assembly and processing of human collagen produced in tobacco	12
2001	Highest recombinant protein accumulation achieved in plants so far – 46.1% total soluble protein for *Bacillus thuringiensis* Cry2Aa2 protein	13
2001	First multi-component vaccine candidate expressed in potato – cholera toxin B and A2 subunits, rotavirus enterotoxin and enterotoxigenic *Escherichia coli* fimbrial antigen fusions for protection against several enteric diseases	14
2001	Glycan modification of a foreign protein produced in a plant host using a human glycosyltransferase	15
2003	Expression and assembly of a functional antibody in algae	16
2003	Commercial production of bovine trypsin in maize	17
2004	Genetic modification of the *N*-glycosylation pathway in *Arabidopsis thaliana* resulting in complex *N*-glycans lacking β1,2-linked xylose and core α1,3-linked fucose	18

[1)] Human growth hormone was expressed as fusion with the *Agrobacterium tumefaciens* nopaline synthase enzyme but only transcript was detectable

[1] A. Barta, K. Sommergruber, D. Thompson et al., *Plant Mol. Biol.* **1986**, *6* (5), 347–357.

[2] A. Hiatt, R. Cafferkey, K. Bowdish, *Nature* **1989**, *342* (6245), 76–78.

[3] P.C. Sijmons, B.M. Dekker, B. Schrammeijer et al., *Bio/Technology (N Y)* **1990**, *8* (3), 217–221.

[4] H.S. Mason, D.M. Lam, C.J. Arntzen, *Proc Natl Acad Sci U S A.* **1992**, *89* (24), 11745–11749.

[5] J. Pen, L. Molendijk, W.J. Quax et al. *Bio/Technology* **1992**, *10* (3), 292–296.

[6] J.K. Ma, A. Hiatt, M. Hein et al., *Science* **1995**, *268* (5211), 716–719.

[7] X. Zhang, D.W. Urry, H. Daniell, *Plant Cell Rep.* **1996**, *16* (3–4), 174–179.

[8] C.O. Tacket, H.S. Mason, G. Losonsky et al., *Nat. Med.* **1998**, *4* (5), 607–609.

[9] E.E. Hood, D.R. Witcher, S. Maddock et al., *Mol. Breeding* **1997**, *3* (4), 291–306.

[10] M. Cabanes-Macheteau, A.C. Fitchette-Laine, C. Loutelier-Bourhis et al., *Glycobiology* **1999**, *9* (4), 365–372.

[11] J.M. Staub, B. Garcia, J. Graves, et al., *Nat. Biotechnol.* **2000**, *18* (3), 333–338.

[12] F. Ruggiero, J.Y. Exposito, P. Bournat et al., *FEBS Letter* **2000**, *469* (1), 132–136.

[13] B. De Cosa, W. Moar, S.B. Lee et al., *Nat. Biotechnol.* **2001**, *19* (1), 71–74.

[14] J. Yu, W.H. Langridge, *Nat. Biotechnol.* **2001**, *19* (6), 548–552.
[15] H. Bakker, M. Bardor, J.W. Molthoff et al., *Proc Natl Acad Sci U S A* **2001**, *98* (5), 2899–2904.
[16] S.P. Mayfield, S.E. Franklin, R.A. Lerner, *Proc Natl Acad Sci U S A* **2003**, *100* (2), 438–442.
[17] S.L. Woodard, J.M. Mayor, M.R. Bailey, et al., *Biotechnol. Appl. Biochem.* **2003**, *38* (2), 123–130.
[18] R. Strasser, F. Altmann, L. Mach, et al., *FEBS Letters* **2004**, *561* (1–3), 132–136.

Contents

Molecular Farming. Edited by Rainer Fischer, Stefan Schillberg
Copyright © 2004 WILEY-VCH Verlag GmbH & Co. KGaA, Weinheim
ISBN: 3-527-30786-9

List of Contributors

Klaus Ammann
University of Bern, Botanic Garden
Altenbergrain 21
CH-3013 Bern
Switzerland

Marc-Andre D'Aoust
Medicago Inc.
1020 Route de l'Église
Ste-Foy, Québec, G1V 3V9
Canada

Charles J. Arntzen
Center for Infectious Diseases and
Vaccinology
The Biodesign Institute at Arizona State
University
Box 4501
Tempe, AZ 85287-4501
USA

Pierre Bilodeau
Medicago Inc.
1020 Route de l'Église
Ste-Foy, Québec, G1V 3V9
Canada

Friedrich Bischoff
Drogenpflanzenlabor HPZ 6425-EG-00
Boehringer Ingelheim Pharma GmbH &
Co. KG
D-55216 Ingelheim
Germany

Jim Brandle
Agriculture and Agri-Food Canada
1391 Sand ford Street
London, Ontario, N5V 4T3
Canada

Brittany E. Burns
Department of Molecular Biology and
Microbiology
University of Central Florida, Bio-
molecular Science
Bldg #20, Room 336
Orlando, FL 32816–2360
USA

Ursula Busse
Medicago Inc.
1020 Route de l'Église
Ste-Foy, Québec, G1 V 3V9
Canada

Olga Carmona-Sanchez
Department of Molecular Biology and
Microbiology
University of Central Florida, Bio-
molecular Science
Bldg #20, Room 336
Orlando, FL 32816–2360
USA

Molecular Farming. Edited by Rainer Fischer, Stefan Schillberg
Copyright © 2004 WILEY-VCH Verlag GmbH & Co. KGaA, Weinheim
ISBN: 3-527-30786-9

Daniel Chargelegue
Department of Cellular and Molecular
Medicine
Molecular Immunology Unit
St. George's Hospital Medical School
Cranmer Terrace
London, S W17 ORE
United Kingdom

Paul Christou
Fraunhofer-Institute for Molecular
Biology and Applied Ecology (IME)
Grafschaft, Auf dem Aberg 1
D-57392 Schmallenberg
Germany

Ulrich Commandeur
Institute for Molecular Biotechnology,
Biology VII
RWTH Aachen
Worringerweg 1
D-52074 Aachen
Germany

Udo Conrad
Institute for Plant Genetics and Crop
Plant Research (IPK)
Leibniz Institute
Corrensstr. 3
D-06466 Gatersleben
Germany

Henry Daniell
Department of Molecular Biology and
Microbiology
University of Central Florida, Bio-
molecular Science
Bldg #20, Room 336
Orlando, FL 32816-2360
USA

Pauline M. Doran
School of Biotechnology and Bio-
molecular Sciences
University of New South Wales
Sydney, NSW 2052
Australia

Simone Dorfmüller
Fraunhofer-Institute for Molecular
Biology and Applied Ecology (IME)
Worringerweg 1
D-52074 Aachen
Germany

Pascal M.W. Drake
Department of Cellular and Molecular
Medicine
Molecular Immunology Unit
St. George's Hospital Medical School
Cranmer Terrace
London, SW17 ORE
United Kingdom

Juergen Drossard
Institute for Molecular Biotechnology,
Biology VII
RWTH Aachen
Worringerweg 1
D-52074 Aachen
Germany

Loïc Faye
CNRS UMR 6037
IFRMP 23, GDR 2590
Université de Rouen
F-76821 Mont Saint Aignan
France

Anne-Catherine Fitchette
CNRS UMR 6037
IFRMP 23, GDR 2590
Université de Rouen
F-76821 Mont Saint Aignan
France

Martin Giersberg
Novoplant GmbH
D-06466 Gatersleben
Germany

Veronique Gomord
CNRS UMR 6037
IFRMP 23, GDR 2590
Université de Rouen
F-76821 Mont Saint Aignan
France

Guruatma Khalsa
Center for Infectious Diseases and
Vaccinology
The Biodesign Institute at Arizona State
University
Box 4501
Tempe, AZ 85287-4501
USA

Kimmo Koivu
UniCrop Ltd
Helsinki Business and Science Park
Viikinkaari 4
FIN-00790 Helsinki
Finland

Patrice Lerouge
CNRS UMR 6037
IFRMP 23, GDR 2590
Université de Rouen
F-76821 Mont Saint Aignan
France

Julian K-C. Ma
Department of Cellular and Molecular
Medicine
Molecular Immunology Unit
St. George's Hospital Medical School
Cranmer Terrace
London, SW17 0RE
United Kingdom

Hugh S. Mason
Center for Infectious Diseases and
Vaccinology
The Biodesign Institute at Arizona State
University
Box 4501
Tempe, AZ 85287-4501
USA

Patricia Obregon
Department of Cellular and Molecular
Medicine
Molecular Immunology Unit
St. George's Hospital Medical School
Cranmer Terrace
London, SW17 0RE
United Kingdom

Shailaja Rabindran
Fraunhofer USA Center for Molecular
Biotechnology
9 Innovation Way
Newark, DE 19711
USA

Isolde Saalbach
Helmut Bäumlein
Institute of Plant Genetics and Crop
Plant Research (IPK)
Corrensstr. 3
D-06466 Gatersleben
Germany

Jürgen Scheller
Institute of Biochemistry
Christian Albrecht University of Kiel
Olshausenstr. 40
D-24098 Kiel
Germany

Andreas Schiermeyer
Fraunhofer-Institute for Molecular
Biology and Applied Ecology (IME)
Worringerweg 1
D-52074 Aachen
Germany

Helga Schinkel
Fraunhofer-Institute for Molecular
Biology and Applied Ecology (IME)
Worringerweg 1
D-52074 Aachen
Germany

Fiona S. Shadwick
School of Biotechnology and Bio-
molecular Sciences
University of New South Wales
Sydney, NSW 2052
Australia

Eva Stoger
Institute for Molecular Biotechnology,
Biology VII
RWTH Aachen
Worringerweg 1
D-52074 Aachen
Germany

Sonia Trepanier
Medicago Inc.
1020 Route de I'Église
Ste-Foy, Québec, G1V 3V9
Canada

Richard M. Twyman
Department of Biological Sciences
University of York
Heslington, York, YO10 5DD
United Kingdom

Louis-Philippe Vezina
Medicago Inc.
1020 Route de I'Église
Ste-Foy, Québec, G 1 V 3V9
Canada

Vidadi Yusibov
Fraunhofer USA Center for Molecular
Biotechnology
9 Innovation Way
Newark, DE 19711
USA

1
Efficient and Reliable Production of Pharmaceuticals in Alfalfa

Marc-André D'Aoust, Patrice Lerouge, Ursula Busse, Pierre Bilodeau,
Sonia Trépanier, Véronique Gomord, Loïc Faye and Louis-Philippe Vézina

1.1
Introduction

In 1986, it was shown that tobacco plants and sunflower calluses could express recombinant human growth hormone as a fusion protein [1]. Since then, a diverse range of plant systems has been used for the production of pharmaceuticals [2, 3]. We have developed a production system based on the leaves of alfalfa (*Medicago sativa* L.), a choice made originally because of the plant's many favorable agronomic characteristics. Alfalfa is a perennial plant, so vegetative growth can be maintained for many years. For molecular farming, this characteristic, combined with the ease of clonal propagation through stem cutting, makes alfalfa a robust bioreactor with regard to batch-to-batch reproducibility. Among perennial plants, legume forage crops such as alfalfa have the advantage of fixing atmospheric nitrogen, thus reducing the need for fertilizers. Moreover, as a feed fodder crop, alfalfa has benefited from important research aiming to increase leaf protein content, so that today's varieties produce as much as 30 mg total protein per gram fresh weight.

In addition to these appealing agronomic characteristics, biotechnological research has revealed additional benefits for the production of pharmaceuticals in alfalfa. Expression cassettes have been optimized for protein expression in alfalfa leaves. Methods for transient protein expression have been developed so that it is now possible to use agroinfiltration or the transformation of protoplasts for early-stage demonstration and validation steps. In addition, glycosylation studies have shown that alfalfa is capable of producing recombinant glycoproteins with homogenous (uniform) glycosylation patterns.

This chapter provides an overview of the tools that have been developed and optimized specifically for the production of pharmaceuticals in alfalfa, with the emphasis on recent technological breakthroughs. The ability of alfalfa leaves to produce complex recombinant proteins of pharmaceutical interest is discussed and illustrated with recent data obtained in our laboratories. Data are presented concerning the production and characterization of alfalfa-derived C5-1, a diagnostic anti-human

Molecular Farming. Edited by Rainer Fischer, Stefan Schillberg
Copyright © 2004 WILEY-VCH Verlag GmbH & Co. KGaA, Weinheim
ISBN: 3-527-30786-9

IgG developed by Héma-Québec (Québec, Canada) for phenotyping and cross matching red blood cells from donors and recipients in blood banks [4].

1.2
Alfalfa-specific Expression Cassettes

The first hurdle encountered during the development of alfalfa as a recombinant protein production system was the relative inefficiency of the available expression cassettes. A study in which a tomato proteinase inhibitor I transgene was expressed in tobacco and alfalfa under the control of the cauliflower mosaic virus (CaMV) 35S promoter showed that 3–4 times more protein accumulated in tobacco leaves compared to alfalfa leaves [5]. Despite the low efficiency of the CaMV 35S promoter in alfalfa, biopharmaceutical production using this system has been reported in the scientific literature. Such reports include expression of the foot and mouth disease virus antigen [6], an enzyme to improve phosphorus utilization [7] and the anti-human IgG C5-1 [8]. In this last work, the C5-1 antibody accumulated to 1% total soluble protein [8].

Given the relatively high level of C5-1 antibody detected in alfalfa leaves using the weak CaMV 35S promoter, it was expected that expression cassette optimization would lead to significantly higher yields. The first family of expression cassettes we developed was thus designed to achieve strong expression in the aerial parts of alfalfa plants. The MED-2000 series (patent pending) consists of strong, leaf-specific expression cassettes, and is based on regulatory sequences from the alfalfa plastocyanin gene. Using cassettes of this family to drive the *gusA* reporter gene, it was possible to achieve up to 14-fold the level of expression obtained in alfalfa leaves with the 35S promoter (Fig. 1.1). Interestingly, although the MED-2000 promoters were derived from alfalfa genomic sequences, they also produced up to 25-fold higher β-glucuronidase (GUS) activity than the 35S-*gusA-nos* construct in the leaves of transgenic tobacco plants.

Because pharmaceuticals are bioactive molecules, their accumulation in plant cells could have a deleterious effect on the growth and development of the host plant. Therefore, we have developed a second series of expression cassettes incorporating inducible promoter elements. The regulatory elements of the MED-1000 series expression cassettes are derived from the alfalfa nitrite reductase (NiR) gene [9]. The induction strategy used with these expression cassettes exploits the ability of alfalfa to grow abundantly in the absence of mineral nitrogen while fixing atmospheric nitrogen through its symbiosis with rhizobium, but also takes into account the fact that NiR genes are highly inducible by nitrate fertilizers [10, 11, 12].

We have demonstrated that the alfalfa NiR promoter is an excellent candidate for the inducible control of transgene expression in alfalfa leaves. As an example, a 3-kb genomic fragment corresponding to an alfalfa NiR promoter was isolated and fused to the *gusA* gene for analysis. We have shown that the promoter remains silent in nodulated plants grown in a nitrate-free medium. Upon the addition of nitrate, however, *gusA* gene expression is induced, and the reporter enzyme accumulates to a similar level to that observed in the leaves of 35S-*gusA* alfalfa plants (Fig. 1.1).

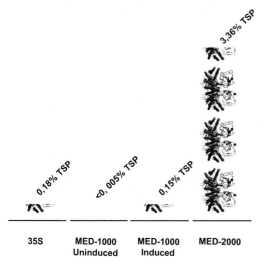

Fig. 1.1 Promoter activity in alfalfa leaves. Accumulation of β-glucuronidase achieved in transgenic alfalfa leaves expressing the *gusA* gene under the control of CaMV 35S and alfalfa promoters. %TSP, percentage of total soluble proteins.

3,36% TSP

0,18% TSP   <0,005% TSP   0,15% TSP

| 35S | MED-1000 Uninduced | MED-1000 Induced | MED-2000 |

1.3
Alfalfa Transformation Methods

Genetic transformation, which results in the stable integration of foreign DNA into the genome, is one of the key technologies underpinning the production of pharmaceuticals in alfalfa. Plant transformation at the industrial level must be optimized for efficiency, predictability and reproducibility in all aspects ranging from explant preparation to the physical conditions of DNA intake and the recovery of transgenic plants. This is an interesting challenge because plant transformation efficiency depends on many factors, including DNA conformation, explant type, plant species, plant genotype and the culture medium. In addition, the development of a plant-based expression platform to produce pharmaceuticals, nutraceuticals and industrial enzymes adds further requirements in terms of plant transformation. For example, a key issue in prototype development is the rapidity with which the ability of the system to produce a selected molecule can be tested, and this reflects the identification of optimal regulatory sequences to drive transgene expression. In order to address these various issues, we have adapted documented transformation methods and developed an alfalfa transformation portfolio ranging from proof-of-concept technology that allows rapid screening of target proteins, to stable expression in transgenic plants or cell cultures for sustainable commercial-scale production. Table 1.1 lists the characteristics of different transformation methods used with alfalfa.

As for many plants, alfalfa is amenable to transformation by various methods including *Agrobacterium*-mediated transfer, direct DNA transfer to protoplasts using polyethylene glycol, and particle bombardment (reviewed in [13]). In recent years, we have developed a medium-throughput system to manage the various activities related to plant transformation, from plant preparation through to transformation and regeneration. This allows us to maintain a continuous production schedule. The sys-

Tab. 1.1 Characteristics of alfalfa transformation methods

	Agrobacterium-mediated stable transformation – Plant	Agrobacterium-mediated stable transformation – Cell culture	Transient protoplast transformation	Particle bombardment-based transient expression	Agrobacterium-mediated transient expression
Plasmid type	Binary	Binary	pUC-based	pUC-based	Binary
Tissue	Leaves	Isolated cells	Protoplasts	Leaves	Leaves
Working conditions	Sterile	Sterile	Sterile	Sterile	Non-sterile
Integration in the genome	Yes	Yes	No	No	No
Timing	6 months minimum	5 weeks	2 days	2 days	5 days
Amount of protein produced	Micrograms or greater	Nanograms	Nanograms	Minimal	Micrograms
Complex protein assembly	Yes	Yes	Yes	No	Yes
Possibility to purify	Yes	Limited	Limited	No	Yes

tem allows the introduction of up to six constructs per week, which represents approximately 600 explants, and this capacity can easily be scaled up by increasing the number of staff and the availability of appropriate equipment. Thus far, more than 180 constructs have been integrated into alfalfa, and several thousand transgenic plants have been generated in our facilities. Given that 98% of the regenerated plants are PCR positive for the gene of interest, our medium-throughput system appears to work very efficiently.

In order to reduce the time required to confirm the accumulation of a given recombinant protein, we have developed a cell culture system in which transgenic alfalfa callus material produced at the proliferation step of *Agrobacterium*-based transformation is used to initiate cell cultures. These cell suspensions can be subcultured to sustain batch production of modest protein amounts. The protein blot shown in Fig. 1.2 demonstrates our ability to detect a recombinant protein in total

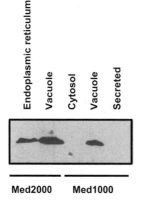

Fig. 1.2 Protein blot analysis of human therapeutic protease inhibitor (HTPI) produced in alfalfa cell cultures using different promoters and subcellular targeting peptides as shown. Equal amounts of total soluble proteins from cell cultures were separated by sodium dodecylsulfate polyacrylamide gel electrophoresis (SDS-PAGE) and blotted onto a polyvinyldifluoride (PVDF) membrane. Monoclonal anti-HTPI IgGs were used for detection.

soluble protein extracts from alfalfa cell cultures. It must be emphasized at this point that the recovered protein is most likely derived from several transformation events involving the same gene construct. This technique allows the detection of recombinant proteins 6–8 weeks after transformation, which is three times faster than the 20 weeks required to regenerate and screen transgenic plants following *Agrobacterium*-mediated transformation. This development has also shown that our alfalfa expression cassettes, although more adapted for leaf expression, provide adequate expression in cell cultures.

Although cell culture considerably reduces the time required to achieve proof-of-concept for new molecules, this time frame still needs to be reduced. In addition, there is some concern that the cell culture system might not correctly predict the ability of alfalfa to assemble complex proteins, and might not be a suitable guide for the selection of subcellular targeting strategies. We have therefore adapted several transient transformation methods to work with the alfalfa platform, including PEG-based protoplast transformation, particle bombardment and *Agrobacterium*-mediated transient transformation of leaves (agroinfiltration). The last method turned out to be particularly successful for the selection of optimal targeting strategies for a given candidate protein. Figure 1.3 shows that, for a given recombinant protein expressed in alfalfa leaves, the level of accumulation is dependent on the subcellular destination of the protein. More importantly, the figure shows that the relative protein accumulation in the different subcellular compartments is similar in leaves from agroinfiltrated and transgenic plants. In the case presented here, chloroplast targeting led to the highest accumulation both in agroinfiltrated leaves and transgenic plants, followed by targeting to the cytosol and mitochondria.

Agrobacterium-mediated transient gene expression has become the method of choice for rapid validation of gene constructs and targeting strategies in alfalfa leaves. It was adapted for alfalfa from a method published by Kapila *et al.* (1997) [14]. In this system, an *Agrobacterium* culture carrying the T-DNA of interest is forced to enter into the intercellular spaces of the leaves under high vacuum. Once the physi-

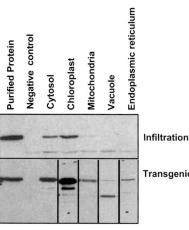

Fig. 1.3 Prediction of the most appropriate subcellular targeting strategies by agroinfiltration. The levels of an industrial enzyme (IE) are shown in agroinfiltrated and transgenic alfalfa leaves using different subcellular targeting peptides. Equal amounts of total soluble leaf proteins were separated by SDS-PAGE and blotted onto a PVDF membrane. Polyclonal anti-IE IgGs were used for detection.

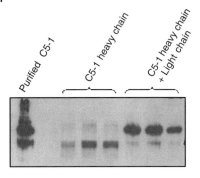

Fig. 1.4 Protein blot analysis of C5-1 assembly in agroinfiltrated alfalfa leaves. Total leaf soluble proteins, extracted 4 days after infiltration were separated by SDS-PAGE under non-reducing conditions and blotted onto a PVDF membrane. Polyclonal anti-mouse IgGs were used for detection. Purified C5-1 was mixed with total soluble proteins from control infiltrated alfalfa leaves and loaded as a standard.

cal barrier of the epidermis is crossed, the bacteria infect neighboring cells, transferring T-DNA copies into the nucleus. Although the T-DNA exists inside the nucleus only transiently, the genes present on the T-DNA are transcribed, leading to the production of the recombinant protein in each infected cell. The efficiency of this method is thus highly dependent on the ability to distribute the bacterial culture evenly inside the leaf tissue.

As well as its short time frame, agroinfiltration has several further advantages for recombinant protein production. The method allows the expression of multiple genes by infiltrating cells with a mixture of two or more *Agrobacterium* cultures (co-infiltration), thus eliminating the need to clone several genes within the same T-DNA. Agroinfiltration is also readily scalable. Routinely, 25 leaves are infiltrated for immunological verification of expression or the comparison of targeting strategies. However, after the selection of an ideal transgene construct, infiltration of 7500 leaves per week can be carried out by a limited number of staff, in a continuous process, for the production of micrograms of recombinant protein.

The production of C5-1 by co-infiltration illustrates the impressive capacity of this method. Results presented in Fig. 1.4 show that the production of C5-1 in detached alfalfa leaves was validated within 5 days from infiltration. In these experiments, different bacteria bearing the light- and the heavy-chain constructs were used to infect the cells. Most of the infected cells were occupied by both strains, and a protein corresponding to fully assembled C5-1 was detected in the infiltrated leaf extract. This result demonstrates the potential of agroinfiltration for testing the adequate expression and assembly of complex proteins in alfalfa leaves using different *Agrobacterium* strains.

1.4
Characteristics of Alfalfa-derived Pharmaceuticals

When recombinant proteins are produced in a heterologous system, there may potentially be differences between the final product and the natural molecule. Hence, for each new protein produced in alfalfa, a thorough analysis of the processing, folding, assembly and post-translational modification is conducted to ensure the conformity of the purified molecules. This section describes the analysis of alfalfa-derived

C5-1 antibodies to demonstrate the ability of alfalfa plants to produce large amounts of high-quality molecules for therapeutic or diagnostic applications.

Purified C5-1 has been obtained from alfalfa leaf extracts by affinity chromatography on either a human IgG-Sepharose column or a Streamline rProtein A-Sepharose column. Interestingly, the purified product obtained with these two methods differed significantly. As shown in Fig. 1.5a, the antibody fraction obtained from the human IgG column contained a mixture of different intermediate assembly forms of the heavy (H) and light (L) chains, ranging from H2 to the fully assembled H2L2 form. In comparison, purification on rProtein A-Sepharose resulted in the isolation of H2L2 form alone (Fig. 1.5b). This situation emphasizes the major impact that purification methods can have on the characteristics of the end product.

In some heterologous production systems, improper removal of the signal peptide may occur during the expression of secreted proteins, which would result in the addition or removal of amino acids at the N-terminal end. In most cases, these modifi-

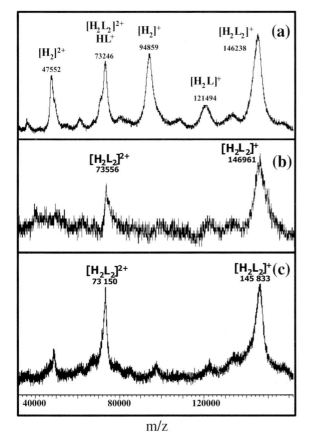

Fig. 1.5 MALDI-TOF mass spectra of purified alfalfa-derived C5-1 using **(a)** human IgG or **(b)** protein A. **(c)** Hybridoma-derived C5-1 as control. Used with permission from Ref 18.

cations are undesirable in a therapeutic context. For C5-1 expression in alfalfa, the natural sequence encoding the signal peptide was retained during the assembly of the expression cassettes. Although most examples show that mammalian signal peptides are correctly processed in plants, N-terminal amino acid sequencing was performed on the heavy chain of alfalfa-derived C5-1 in order to confirm the N-terminal integrity of the antibody. The N-terminal sequence of the heavy chain was confirmed as EIQLV, which is identical to that of the hybridoma-derived C5-1 and indicates the correct processing of the signal peptide in alfalfa.

N-glycosylation is another important issue when considering the conformity of therapeutic proteins produced in heterologous systems. Although every eukaryotic expression system N-glycosylates proteins targeted to the secretory pathway, each system links a different form of N-glycan to the recombinant protein. The glycans synthesized in a heterologous production system only rarely correspond to those found in the natural source of the protein. In this context, the ability of plants to perform complex glycosylation [15] represents an advantage over yeast and insect cells, and places the plant system in the group of Chinese hamster ovary cells (CHO) and murine myeloma cell lines (NSO). Importantly, however, the analysis of recombinant IgGs produced in tobacco indicates heterogeneity in the structure of N-glycans [16, 17].

In contrast, glycosylation analysis of alfalfa-derived C5-1 showed that a single, unique N-linked glycan form is found on the antibody (Fig. 1.5). The glycoform is representative of plant complex N-glycans, and includes core $\beta(1,2)$-xylose and $\alpha(1,3)$-fucose. Figure 1.6 shows a comparison of N-glycan structures found on alfalfa- and mouse-derived C5-1. Homogenous N-glycosylation of a recombinant protein ensures batch-to-batch reproducibility, but also provides an ideal substrate for in vitro modification of the N-glycan. For example, it has been shown that incubating the purified alfalfa-derived C5-1 with $\beta(1,4)$-galactosyltransferase in the presence of UDP-galactose resulted in an efficient addition of $\beta(1,4)$-galactose to the terminal GlcNAc residues of the N-linked glycans [18].

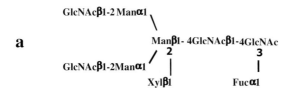

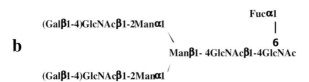

Fig. 1.6 Structure of N-glycans isolated from
(a) alfalfa-derived C5-1 and **(b)** murine C5-1.
Used with permission from Ref 18.

Tab. 1.2 Activity of alfalfa- and hybridoma-derived C5-1. Used with permission from Ref 8.

Extract	Specific activity (OD/100 ng)	True affinity (K_Ds)
C5-1 from hybridoma	$0.235 \pm 0,020$	4.6×10^{-10} M
C5-1 from alfalfa	$0.267 \pm 0,080$	4.7×10^{-10} M

In order to compare the specific activity of plant-derived C5-1 to that of the hybridoma-derived antibody, the antigen-binding capacity of antibodies produced in each system was assayed by enzyme-linked immunosorbent assay (ELISA). As shown in Table 1.2, antibodies from both sources demonstrated similar binding characteristics against human IgGs [8]. Furthermore, the stability of alfalfa-derived C5-1 in the blood stream of Balb/c mice was comparable to that of the hybridoma-derived IgG [8].

In the light of the results presented above, we conclude that alfalfa offers a suitable system for the high-yield production of correctly assembled complex proteins, including multimeric glycoproteins. The post-translational capacities of alfalfa indicate that this system is one of the best-suited for the production of molecules for therapeutic and diagnostic applications.

1.5
Industrial Production of Recombinant Proteins in Alfalfa

1.5.1
Ramping Up Alfalfa Biomass

In addition to yielding large amounts of high quality protein, an efficient system for large-scale recombinant protein production should include a rapid biomass amplification method. In alfalfa, different propagation methods can be applied including stem cutting, somatic embryogenesis and seed production. The stem cutting method offers the possibility of rapid biomass amplification by quickly creating a clonal population from an elite plant. This method is very reliable with respect to maintaining the capacity and batch-to-batch consistency of crucial aspects such as expression level and product uniformity.

Alfalfa stem propagation can be achieved without the addition of hormones so long as the cuttings are maintained in a humid environment. One transgenic alfalfa plant can generate a clonal population filling a 1000-m^2 greenhouse within 14 months. Stem cutting is also the method of choice when small quantities (milligrams) of recombinant protein are required within a limited time frame, as it is the case for product testing and pre-clinical studies.

1.5.2
Alfalfa Harvest, and Recovery of Recombinant Molecules

In greenhouses, alfalfa can be harvested 8–10 times per year using minimal equipment. A 5-ha greenhouse containing mature alfalfa plants will yield about 130 tons of fresh biomass at each harvest, if 75% of the surface is cultivated. Harvested 10 times, such a greenhouse will generate 900 tons of alfalfa annually, although for practical reasons the harvests are distributed throughout the year. For example, within the same greenhouse, the plants are distributed into 10 plots of 0.375 ha each. If each plot is harvested every 5 weeks, this means that 100 harvests of 13 tons are performed per year. This biomass of fresh alfalfa tissue can easily be handled by medium-scale processing machinery.

Extraction of soluble proteins from alfalfa tissue begins with a maceration step in a hammer mill. This step is intended to break the cells to facilitate extraction of confined water and soluble content. From the resulting mash, the green juice is extracted using a screw press. Four presses, each with a 400 kg h^{-1} biomass intake capacity will produce 800 L h^{-1}, making a total of 6500 L of green juice per harvest. This capacity is essential to minimize the delay between harvest and extraction. The soluble protein content of alfalfa green juice produced in this manner is about 2%, so a 5-ha greenhouse generates 130 kg of soluble proteins twice weekly. Finally, with current recombinant protein expression levels of 0.1% to 1% of soluble proteins, the yield from the 5-ha greenhouse is estimated to be between 13 kg and 130 kg of recombinant protein per year.

Purification of recombinant proteins from alfalfa extracts can be performed using several strategies, depending on the required purity and the intended application of the protein. Sophisticated and powerful methods are under development in our laboratories. In the case of C5-1, two purification methods have been investigated. The first method is an adaptation of the method used at Héma-Québec for the purification of C5-1 from hybridoma cells. It involves affinity chromatography using a human IgG1 (the antigen recognized by C5-1) to purify the protein. Alternatively, an expanded bed affinity chromatography method using staphylococcal protein A (Streamline rProteinA, Amersham Biosciences, Piscataway, NJ) has better potential for larger scale preparation of C5-1. These large-capacity columns, in which the liquid flows upward, can be loaded with unclarified green juice, and can bind up to 20 mg of human IgG per mL of medium. Unfortunately, protein A has a low affinity for mouse IgGs. In our hands, at pH9, we have obtained up to 5 mg C5-1 per mL of medium. Current improvements of the purification step will include coupling the expanded bed column with streptococcal protein G, which shows a higher affinity for mouse IgGs than protein A.

Although greenhouses can supply enough biomass to produce kilograms of recombinant proteins, field production would be necessary if tons of recombinant proteins were required. At this scale, molecular farming will benefit from the current knowledge developed for the animal feed industry. Among the most significant developments impacting on large-scale alfalfa processing, the wet fractionation process, currently used by Sativa 2000 in Champagne (France), treats up to 750,000 tons of fresh alfalfa per year. In the wet fractionation process, temperature and pH are used to separate alfalfa proteins and isolate protein-rich fractions, which are dried into pellets for the animal feed industry. A refined version of this large-scale protein separation process is used at Viridis (Aulnay-aux-Planches, France) to produce purified Rubisco from alfalfa as a food additive for human consumption.

1.6
Conclusions

The intrinsic qualities of alfalfa justify its selection as a platform for the production of heterologous proteins. Alfalfa plants are easily propagated by stem cutting to create large populations. In greenhouses, these populations can be harvested 10 times per year, and the plants can be maintained for more than 5 years. Alfalfa is also capable of producing and processing complex proteins, and adds homogenous N-glycan chains to secreted glycoproteins.

The expression cassettes developed in our laboratories facilitate the strong expression of recombinant proteins in alfalfa leaves. High levels of the candidate molecules accumulate when they are targeted to the most appropriate subcellular compartment. A rapid capacity to evaluate expression strategies has been developed based on the accumulation of proteins in agroinfiltrated leaves. By combining this rapid selection method with the efficient genetic transformation of alfalfa, the solid foundation of an effective heterologous expression platform has been secured.

References

[1] A. Barta, K. Sommergruber, D. Thompson et al., *Plant Mol. Biol.* 1986, 6 (5), 347–357.

[2] R. Fischer, N. Emans, *Transgenic Res.* 2000, 9 (4–5), 279–299.

[3] E. Stoger, M. Sack, Y. Perrin et al., *Mol. Breeding* 2002, 9 (3), 149–158.

[4] M. St Laurent, A. Marcil, S. Verrette et al., *Vox. Sang.* 1993, 64 (2), 99–105.

[5] J. Narvàez-Vàsquez, M. L. Orozco-Càrdenas, C. A. Ryan, *Plant Mol. Biol.* 1992, 20 (6), 1149–1157.

[6] A. Wigdorovitz, C. Carrillo, M. J. Dus Santos et al., *Virology* 1999, 255 (2), 347–353.

[7] S. Austin-Phillips, R. G. Koegel, R. J. Straub et al., US patent application 5900525, 1999.

[8] H. Khoudi, S. Laberge, J. M. Ferullo et al., *Biotechnol. Bioeng* 1999, 64 (2), 135–143.

[9] L. P Vézina, M. A. D'Aoust, US patent application 6420548, 2002.

[10] E. Back, W. Burkhart, M. Moyer et al., *Mol. Gen. Genet.* 1988, 212 (1), 20–26.

[11] E. Back, W. Dunne, A. Schneider-Bauer et al., *Plant Mol. Biol.* 1991, 17 (1), 9–18.

[12] L. Sander, P. E. Jensen, L. F. Back et al., *Plant Mol. Biol.* 1995, 27 (1), 165–177.

[13] M.-A. D'Aoust, U. Busse, M. Martel et al., Alfalfa: An efficient bioreactor for continuous recombinant protein production, in *Molecular Farming of Plants and Animals for Human and Veterinary Medicine.* 2003, eds. L. Erickson, W.-J. Yu, J. Brandle and R. Rymerson, Kluwer Academic Publishers, Amsterdam, 33–43.

[14] J. Kapila, R. De Rycke, M. Van Montagu et al., *Plant Sci.* 1997, 122 (1), 101–108.

[15] P. Lerouge, M. Cabanes-Macheteau, C. Rayon et al., *Plant Mol. Biol.* **1998**, *38* (1–2), 31–48.

[16] M. Cabanes-Macheteau, A. C. Fichette-Lainé, C. Loutellier-Bourhis et al., *Glycobiology* **1999**, *9* (4), 365–372.

[17] H. Bakker, M. Bardor, J. W. Molthoff et al., *Proc. Natl Acad. Sci. USA* **2001**, *98* (5) 2899–2904.

[18] M. Bardor, C. Loutelier-Bourhis, T. Paccalet et al., *Plant Biotech. J.* **2003**, *1* (6) 451–462.

2

Foreign Protein Expression Using Plant Cell Suspension and Hairy Root Cultures

Fiona S. Shadwick and Pauline M. Doran

2.1
Foreign Protein Production Systems

The demand of the pharmaceutical industry for large quantities of mammalian proteins has led to the development of heterologous expression systems for the production of proteins and peptides of varying complexity. The host organisms used range from bacteria to eukaryotic systems such as yeast, insect, mammalian and plant cell cultures and transgenic animals and plants. Bacteria can produce relatively high levels of foreign protein but develop insoluble inclusion bodies and offer only limited post-translational modification. In contrast, eukaryotic expression systems are able to glycosylate proteins and carry out post-translational processing, although different post-translational effects can result in the formation of products that are not identical in all respects to the native protein. The cost of protein production using different host organisms and expression systems varies widely. The complexity of the protein, its end use, the scale of production and the degree of similarity required between the transgenic and native proteins are important factors to consider when determining the type of production system to apply.

Plant-based production systems are now being used commercially for the synthesis of foreign proteins [1–3]. Post-translational modification in plant cells is similar to that carried out by animal cells; plant cells are also able to fold multimeric proteins correctly. The sites of glycosylation on plant-produced mammalian proteins are the same as on the native protein; however, processing of N-linked glycans in the secretory pathway of plant cells results in a more diverse array of glycoforms than is produced in animal expression systems [4]. Glycoprotein activity is retained in plant-derived mammalian proteins.

Agricultural production of foreign proteins in crop plants can deliver large quantities of product at low cost [5]. This is possible even if protein expression levels are low relative to other heterologous systems, as agriculture is a cheap technology and production targets can be achieved using additional plantings or cropping larger areas of land [6]. The ability to scale-up economically and to utilize production and extraction technologies already developed for the food industry significantly reduces the cost of high-volume protein production using agriculture compared with alterna-

Molecular Farming. Edited by Rainer Fischer, Stefan Schillberg
Copyright © 2004 WILEY-VCH Verlag GmbH & Co. KGaA, Weinheim
ISBN: 3-527-30786-9

tive methods such as cell culture. The cost of agricultural production is also reduced substantially if the foreign protein does not require purification from the plant biomass, as is the case for edible vaccines.

Despite the advantages of agriculture, and even if the product can be shown to have appropriate biological activity, agricultural production of foreign proteins may not provide adequate assurance of product safety and quality. For example, foreign proteins produced in the field are subject to contamination with pesticides, herbicides and mycotoxins; field-grown plants also experience variable weather conditions, non-uniform soil compositions and infestation by pests and diseases. Any or all of these factors may result in unpredictable product yield and quality. The inability to control production conditions could mean that whole-plant systems fail to comply with good manufacturing practice in many countries around the world, particularly if the transgenic protein is to be used as a therapeutic. Issues of environmental crop safety also arise, especially if the foreign protein is toxic, e. g. to soil microorganisms or to wild-life capable of consuming the plants [7]. In situations where the plant-derived product has suitable activity but where contamination, inconsistent quality and/or regulatory issues prohibit the use of agricultural methods, large-scale plant tissue culture offers an alternative route for foreign protein production [8]. This is particularly so if the volume of protein required is low.

As well as overcoming many of the inherent problems associated with agriculture, plant tissue culture also offers a number of advantages over conventional animal cell culture methods currently being applied to produce biopharmaceutical proteins commercially [8]. As plant culture media are relatively simple in composition and do not contain proteins, the cost of the process raw materials is reduced and protein recovery from the medium is easier and cheaper compared with animal cell culture. In addition, as most plant pathogens are unable to infect humans, the risk of pathogenic infections being transferred from the cell culture via the product is also substantially reduced.

2.2
Production of Foreign Proteins Using Plant Tissue Culture

In this review, we focus on the use of plant tissue culture to produce foreign proteins that have direct commercial or medical applications. The development of large-scale plant tissue culture systems for the production of biopharmaceutical proteins requires efficient, high-level expression of stable, biologically active products. To minimize the cost of protein recovery and purification, it is preferable that the expression system releases the product in a form that can be harvested from the culture medium. In addition, the relevant bioprocessing issues associated with bioreactor culture of plant cells and tissues must be addressed.

Extensive research has been carried out into the molecular aspects of foreign protein production in whole plants to enhance the yield, quality and stability of the product and to facilitate protein separation and purification from the biomass [3, 6, 9]. In contrast, comparatively little research has been undertaken to investigate the specific issues associated with producing foreign proteins in plant tissue culture. Many of the problems that need to be addressed, such as low protein yields, are similar *in vivo* and *in vitro*. However, the differences between these systems means that different solutions may be required.

Some developments that have proven useful for enhancing foreign protein yields in whole plants, such as chloroplast and organ-specific expression systems, have little or no practical application in tissue culture. On the other hand, although not yet demonstrated *in vitro*, the use of viral vectors to increase protein production, and novel approaches for simplifying product purification such as targeting of proteins to oil bodies, may have significant implications for the production of foreign proteins in tissue culture systems.

Foreign proteins that have been produced using plant tissue culture are listed in Table 2.1. Several of these proteins, e.g. antibodies, interleukins, erythropoietin, human granulocyte-macrophage colony stimulating factor (hGM-CSF) and hepatitis B antigen, have pharmaceutical or therapeutic uses and would be suitable for further commercial development if the production levels could be increased. However, to date, few plant cell or organ cultures have been shown to accumulate or secrete foreign proteins at concentrations sufficient for commercial viability.

As indicated in Table 2.1, tobacco (*Nicotiana tabacum*) has been the host species in most studies of foreign protein expression in plant tissue culture. Rice (*Oryza sativa*) has also been used by several groups. Accumulation of hGM-CSF was found to be significantly higher in rice suspensions with an inducible promoter than in tobacco suspensions with constitutive transgene expression [10]. The predominance of tobacco-based expression systems in tissue culture studies differs from the situation with whole plants, where advantages associated with producing foreign proteins in edible species or in storage organs such as seeds have resulted in a variety of plant species being transformed.

Most research into *in vitro* foreign protein production has been undertaken using cell suspensions. However, other forms of plant tissue culture such as hairy roots and shooty teratomas have also been tested in a number of studies (Table 2.1). The characteristics of different types of plant tissue culture and their utility for large-scale foreign protein production are outlined in the following sections.

2.2.1
Suspended Cell Cultures

Plant cell suspensions comprise small clumps of dedifferentiated plant cells in liquid nutrient medium. Dedifferentiation of the cells occurs under the influence of plant growth regulators, which must be provided in the medium to promote rapid growth and maintain the culture morphology. Transgenic cell suspensions can be developed from callus initiated using explants from transformed plants; alternatively, wild-type suspensions may be transformed directly using *Agrobacterium tumefaciens*-mediated transfection or biolistic delivery of plasmid DNA into the cells (Table 2.1).

Plant suspensions are being used to produce an increasing number of foreign proteins. These include complete antibodies, antibody fragments, hGM-CSF, interleu-

Tab. 2.1 Expression of foreign proteins in plant tissue culture

Foreign protein	Culture type	Plant species	Transformation method	Promoter	Leader sequence	Production level (maximum)	Reference
Alkaline phosphatase, human placental	Whole plant, hydroponic	Nicotiana tabacum (tobacco)	A. tumefaciens transformation of leaf explant	Mannopine synthase (mas2')	Not reported	20 µg day^{-1} g^{-1} root dry weight (e) 3% of total medium protein (e)	72
Antibody, scFv, against phytochrome	Suspension	Nicotiana tabacum (tobacco)	A. tumefaciens transformation of leaf explant	CaMV 35S	Tobacco pathogenesis related protein (PR1a)	0.5 mg L^{-1} (e) 5.0% of total medium protein (e)	73
Antibody, scFv, against carcinoembryonic antigen	Callus	Oryza sativa (rice)	Microparticle bombardment of callus	Maize ubiquitin-1	Murine heavy- and light-chain IgG	0.45 µg g^{-1} fresh weight without KDEL (i) 3.8 µg g^{-1} fresh weight with KDEL (i)	31
Antibody, heavy chain, against p-azophenyl-arsonate	Suspension	Nicotiana tabacum (tobacco)	A. tumefaciens transformation of suspension	CaMV 35S	Native murine Native murine	150 µg L^{-1} (i) 430 µg L^{-1} (t) with DMSO 170 µg L^{-1} (i) 10 µg L^{-1} (e) 360 µg L^{-1} (e) with PVP 80 µg mL^{-1} (i) 300 µg mL^{-1} (e)	64 11 67
Antibody, murine IgG-2b/κ, against tobacco mosaic virus	Suspension	Nicotiana tabacum (tobacco)	A. tumefaciens transformation of leaf explant	CaMV 35S	Murine	15 µg g^{-1} wet weight (i) 45 µg g^{-1} wet weight (i) with amino acids	62
Antibody, murine Ig	Suspension	Nicotiana tabacum (tobacco)	A. tumefaciens transformation of leaf explant, regenerated plants sexually crossed	CaMV 35S	Murine Ig	1.2 mg g^{-1} dry weight (t) 7.5 mg L^{-1} (t) 3.6 mg L^{-1} (e) 6.5% TSP (t) 12% TSP (t) with PVP	17

Tab. 2.1 (continued)

Foreign protein	Culture type	Plant species	Transformation method	Promoter	Leader sequence	Production level (maximum)	Reference
Antibody, murine IgG$_1$, against *Streptococcus mutans* surface antigen	Hairy root	*Nicotiana tabacum* (tobacco)	*A. tumefaciens* transformation of leaf explant, regenerated plants sexually crossed	CaMV 35S	Murine Ig	18 mg L^{-1} (t) 1.8% TSP (i) 3.2 mg L^{-1} (e) 10.8 mg L^{-1} (e) with PVP 1.1 mg g^{-1} dry weight (t) 7.0 mg L^{-1} (t) 1.4 mg L^{-1} (e) 3.0% TSP (t) 4.0% TSP (t) with PVP	19 17
	Shooty teratoma	*Nicotiana tabacum* (tobacco)	*A. tumefaciens* transformation of leaf explant, regenerated plants sexually crossed	CaMV 35S	Murine Ig	0.28 mg g^{-1} dry weight (t) 3.2 mg L^{-1} (t)	17
α_1-antitrypsin, human	Suspension	*Oryza sativa* (rice)	Microparticle bombardment of callus	Rice α-amy-lase, inducible	Rice α-amylase	85 mg L^{-1} (e) 5.7 mg g^{-1} dry weight (e) 51 mg L^{-1} (e)	29 39
Bryodin 1	Suspension	*Nicotiana tabacum* (tobacco)	Microparticle bombardment of suspension	CaMV 35S	Extensin	30 mg L^{-1} (e)	74
Cytochrome P450 2E1, rabbit	Hairy root	*Atropa belladonna*	*A. rhizogenes* transformation of leaf explant	CaMV 35S	Not reported	Not reported	22
Erythropoietin, human	Suspension	*Nicotiana tabacum* (tobacco)	*A. tumefaciens* transformation of suspension	CaMV 35S	Native human erythropoietin	0.8 µg L^{-1} (t) 0.0026% TSP	61

Tab. 2.1 (continued)

Foreign protein	Culture type	Plant species	Transformation method	Promoter	Leader sequence	Production level (maximum)	Reference
Granulocyte-macrophage colony stimulating factor, human (hGM-CSF)	Suspension	*Nicotiana tabacum* (tobacco)	*A. tumefaciens* transformation of callus	CaMV 35S	Tobacco etch virus	150 μg L^{-1} (i) 240 μg L^{-1} (e)	40
			A. tumefaciens transformation of explant	CaMV 35S	Natural mammalian	180 μg L^{-1} (e) 783 μg L^{-1} (e) with gelatin	65
		Oryza sativa (rice)	Microparticle bombardment of callus	Rice α-amylase, inducible	Rice α-amylase	129 mg L^{-1} (e) 25% total medium protein (e)	10
Green fluorescent protein	Hairy root	*Hyoscyamus muticus*	*A. tumefaciens* transformation of hairy root	CaMV 35S	–	Not reported	23
		Catharanthus roseus	*A. rhizogenes* transformation of plant	Glucocorticoid-inducible	–	–	34
	Whole plant, hydroponic	*Nicotiana tabacum* (tobacco)	*A. tumefaciens* transformation of plant	Mannopine synthase (mas2′)	*N. plumbaginifolia* calreticulin	296–923 ng day^{-1} g^{-1} root dry weight (e)	72
Hepatitis B surface antigen	Suspension	*Glycine max* (soybean)	Microparticle bombardment of suspension	Chimeric ocs-mas	Not reported	1.7 mg g^{-1} dry weight (i) 20–22 mg L^{-1} (i)	14
		Nicotiana tabacum (tobacco)	*A. tumefaciens* transformation of suspension	Chimeric ocs-mas	Not reported	0.31 mg g^{-1} dry weight (i)	14
Interleukin-2, human	Suspension	*Nicotiana tabacum* (tobacco)	*A. tumefaciens* transformation of suspension	CaMV 35S	Natural mammalian	75 μg L^{-1} (i) 10 μg L^{-1} (e)	75

Tab. 2.1 (continued)

Foreign protein	Culture type	Plant species	Transformation method	Promoter	Leader sequence	Production level (maximum)	Reference
Interleukin-4, human	Suspension	*Nicotiana tabacum* (tobacco)	*A. tumefaciens* transformation of suspension	CaMV 35S	Natural mammalian	275 µg L^{-1} (i) 180 µg L^{-1} (e)	75
Interleukin-12, human	Suspension	*Nicotiana tabacum* (tobacco)	*A. tumefaciens* transformation of leaf explant, regenerated plants sexually crossed	CaMV 35S	Native human interleukin-12 subunit	60 µg L^{-1} (i) 175 µg L^{-1} (e) 700 µg L^{-1} (e) with gelatin	60
Invertase, carrot	Suspension	*Nicotiana tabacum* (tobacco)	*A. tumefaciens* transformation of suspension	CaMV 35S	Native carrot invertase	1400 U L^{-1} (i) 40 U g^{-1} dry weight (i) 150 U L^{-1} (e)	15
Lysozyme, human	Suspension	*Oryza sativa* (rice)	Microparticle bombardment of callus	Rice α-amylase, inducible	Rice α-amylase	4% TSP	30

i = protein in the biomass; e = protein in the medium; t = total protein
CaMV = cauliflower mosaic virus; DMSO = dimethylsulfoxide; PVP = polyvinylpyrrolidone; TSP = total soluble protein; U = unit

kin-2, interleukin-4 and interleukin-12, erythropoietin, hepatitis B surface antigen, α_1-antitrypsin, human lysozyme and carrot invertase. Although, in some instances, stable production of foreign proteins has been found to occur over extended periods [11–15], suspension cultures are subject to various types of genetic instability through the effects of somaclonal variation [16]. Significant reductions in the yield of foreign proteins over time, possibly caused by genetic instability, have been reported in plant cell suspensions [12–14, 17]. In some cases, cell lines with stable production characteristics were isolated by screening and selection from a large number of cultures [12–14].

2.2.2
Hairy Root Cultures

Hairy roots are neoplastic roots produced by transformation of plant cells with *Agrobacterium rhizogenes*. When cultured in liquid medium, hairy roots often exhibit rapid growth relative to untransformed roots. Hairy roots can be propagated indefinitely in liquid medium and retain their morphological integrity and stability in the absence of exogenous plant growth regulators. Hairy root cultures have been found to have significantly greater long-term stability than suspended plant cells for the production of foreign proteins [17].

Transgenic hairy root cultures can be initiated by infecting transgene-containing plants or explants with *A. rhizogenes* (Table 2.1). Using this approach, it is relatively easy to generate hairy roots expressing multiple foreign genes, as plants containing multiple transgenes (produced by sexually crossing transgenic plants carrying single transgenes) may be used for hairy root initiation. For example, transgenic tobacco plants developed by crossing antibody-heavy-chain-expressing plants with antibody-light-chain-expressing plants [18] were used subsequently to generate hairy roots capable of synthesising complete IgG_1 antibody [19–21]. Transgene-expressing hairy roots can also be obtained by performing root initiation and transformation at the same time using genetically-modified *A. rhizogenes* with the transgene inserted into plasmid constructs [22]. Alternatively, established hairy root cultures can be induced to produce foreign proteins by direct *A. tumefaciens*-mediated transformation [23].

2.2.3
Shooty Teratoma Cultures

Shooty teratomas are a form of differentiated organ culture produced by transformation of plants with particular strains of *Agrobacterium tumefaciens* [24]. Foreign protein production in transgenic shooty teratomas has been reported by only one group [17, 21]. In this system, shooty teratomas of tobacco were used to produce an IgG_1 antibody. Antibody yields in the teratoma cultures were lower than in suspended cell and hairy root cultures [17]. The growth characteristics of shooty teratomas were also not conducive to liquid culture as the shoots tended to callus and were very susceptible to hyperhydricity (vitrification).

2.2.4
Scale-up Considerations for Different Forms of Plant Tissue Culture

As indicated in Table 2.2, several studies of foreign protein production have been carried out using plant cell suspensions or hairy roots in bioreactors. Bioprocess development for the large-scale culture of suspended plant cells is relatively well established, as this type of culture has been examined extensively for the production of plant secondary metabolites such as paclitaxel (taxol), ginseng and shikonin [25–27]. Accordingly, if suspension cultures suitable for the commercial production of foreign proteins were developed, the basic technology for large-scale operations is already available. Research into bioreactor systems for more complex forms of tissue culture such as roots and shoots is not as well developed. The principal reactor types trialed for large-scale hairy root culture have been reviewed by Giri and Lakshmi Narasu [28]. Difficulties associated with providing a low-shear environment while maintaining adequate mixing and oxygen transfer present significant problems for

Tab. 2.2 Production of foreign proteins using plant tissue culture in bioreactors

Foreign protein	Culture type	Plant species	Bioreactor system	Production level (maximum)	Reference
Antibody, heavy chain, against *p*-azophenylarsonate	Suspension	*Nicotiana tabacum* (tobacco)	3.5 L working volume, stirred, 6-blade disc impeller, batch operation, 8 days	$110\ \mu g\ L^{-1}$ (i) $190\ \mu g\ L^{-1}$ (e)	67
Antibody, murine IgG$_1$, against *Streptococcus mutans* surface antigen	Hairy roots	*Nicotiana tabacum* (tobacco)	2 L, magnetic stirrer, vertical cylindrical wire-mesh cage for biomass support, batch operation, 30 days	$1.9\ mg\ L^{-1}$ (e) $0.45\ mg\ g^{-1}$ dry weight (i)	19
Antibodies, antibody fragments, antibody fusion proteins (unspecified)	Suspension	*Nicotiana tabacum* (tobacco)	40 L working volume, stirred, 3-blade impeller, batch operation, 150 h	Not reported	76
α_1-antitrypsin, human	Suspension	*Oryza sativa* (rice)	5 L working volume, stirred, single pitched-blade impeller, 2-stage batch operation, 6–8 days	$51\ mg\ L^{-1}$ (e) $7.3\ mg\ day^{-1}\ L^{-1}$ (e)	39
			2 L, stirred, 2-stage batch operation, 13 days	$25.6\ mg\ L^{-1}$ (e) $4.6\ mg\ g^{-1}$ dry weight (e)	29
Invertase, carrot	Suspension	*Nicotiana tabacum* (tobacco)	10 L, stirred, dual 6-blade turbines, continuous operation, 75 days	$20.8\ U\ h^{-1}\ L^{-1}$ (i) 0.8–$1.0\ U\ mg^{-1}$ protein (i)	15

i = intracellular protein; e = protein in the medium

the scale-up of root reactors. However, despite these engineering challenges, if advantages such as enhanced culture stability are associated with hairy roots compared with suspended cells [17], further technical development of root cultures for foreign protein production would be worthwhile.

2.3
Strategies for Improving Foreign Protein Accumulation and Product Recovery in Plant Tissue Culture

As indicated by the protein accumulation levels in Table 2.1, it is currently possible using plant tissue culture to achieve moderate levels of foreign protein expression in some systems. To take advantage of the cost benefits associated with recovering products from the culture medium rather than from homogenized biomass, expression systems for protein secretion have also been developed. Yet, relative to the production levels attained in animal cell cultures, foreign protein concentrations in plant cultures are typically very low. This is a major hurdle preventing plant systems being utilized more widely for commercial protein production. Therefore, a key research objective has been to increase the accumulation of active foreign proteins in plant tissue culture. Although the reasons for the low yields in plant systems are not yet fully understood, two different approaches have been taken to increase product levels *in vitro*. These are: (i) to increase the level of gene expression in the cells by altering the transgene constructs and methods of expression, and (ii) to increase the retention and stability of foreign protein in the cultures after the protein is produced. Efforts have also been made to enhance the availability of product in the culture medium to facilitate subsequent recovery and purification.

2.3.1
Expression Systems

Compared with whole plants, there has been limited development of foreign protein expression systems specifically for use in tissue culture. Some modifications of expression constructs have resulted in improved protein accumulation or have allowed simplified protein recovery. However, in general, modified expression systems have been tested only in a restricted number of cases and have not resulted in the large increases in product yield required for plant cultures to compete with other foreign protein production vehicles. Transient expression techniques, for example using viral vectors, that have been developed for use in whole plants have not yet been applied in plant tissue culture.

2.3.1.1 Modifications to Existing Expression Constructs
Several molecular strategies have been successful in increasing foreign protein production in cultured plant cells. These include using promoters for inducible expression [10, 12, 29], optimizing codon usage [30] and adding the KDEL sequence to ensure protein retention in the endoplasmic reticulum [31]. Application of different

promoters has been the most common approach to the modification of expression constructs.

As indicated in Table 2.1, most of the promoters used in plant tissue culture have been based on the constitutive cauliflower mosaic virus (CaMV) 35S promoter. In contrast, inducible promoters have the advantage of allowing foreign proteins to be expressed at a time that is most conducive to protein accumulation and stability. Although a considerable number of inducible promoters has been developed and used in plant culture applications, e.g. [32–37], the only one to be applied thus far for the production of biopharmaceutical proteins is the rice α-amylase promoter. This promoter controls the production of an α-amylase isozyme that is one of the most abundant proteins secreted from cultured rice cells after sucrose starvation. The rice α-amylase promoter has been used for expression of hGM-CSF [10], α_1-anti-trypsin [12, 29, 38, 39] and human lysozyme [30].

Alterations to the proteins and pre-proteins expressed by cultured plant cells have been used to facilitate product recovery. A leader sequence is required for foreign protein secretion from plant cells into the apoplast and then into the culture medium. As indicated in Table 2.1, plant, mammalian and viral sequences have been employed to achieve the entry of transgenic proteins into the bulk-flow pathway in plant cultures.

To facilitate product recovery and purification, molecular tags may be added to foreign proteins. Attachment of a functional His_6 tag to a secreted therapeutic protein, hGM-CSF, has been examined in tobacco suspension cultures [40]. The His_6 tag consisted of six histidine residues attached to the protein terminus and gave the protein the ability to bind strongly to metal ions. The presence of the His_6 tag allowed the specific removal of product from the culture medium using iminodiacetic acid metal affinity resin [41].

2.3.1.2 Transient Expression Using Viral Vectors

Genetically modified viral vectors have been applied in many whole-plant systems for the production of therapeutic proteins and epitope vaccines, e.g. [42–46]. Foreign proteins produced using viral vectors can be in the form of free cytosolic proteins or fusions to viral proteins. Viral expression systems exploit the ability of viruses to propagate rapidly and achieve high concentrations in plant tissues. For example, tobacco mosaic virus (TMV) can accumulate in infected tobacco leaves to levels greater than $60 \ mg \ g^{-1}$ dry weight [47] and produce amounts of TMV coat protein accounting for 10–40% of the total protein content of the leaves [42]. Provided the movement proteins on recombinant viruses remain functional, viral vectors are able to spread throughout the entire plant from a single infection point via the plasmodesmata between individual cells and the vascular system. Therefore, in principle, when foreign protein is co-expressed with the virus, large amounts of product can be formed.

To date, application of transgenic viruses in whole plants has not resulted in the production of foreign proteins to the same high levels as viral proteins from non-transgenic virus infections. This is probably because the genetic construct carried by the virus interferes to some extent with the normal folding, packaging, transmission or replication processes [48]. Nevertheless, foreign protein yields achieved using viral vectors can be substantial. For example, transgenic viruses with coat protein fusions have been reported to accumulate to levels of 1–3 mg g^{-1} of plant tissue [49, 50].

A vector that facilitates high-level protein expression in plant tissue culture, particularly a transient expression system that could be applied to existing wild-type cultures, would be advantageous for *in vitro* foreign protein production. However, such a system has not yet been developed. The success of this approach depends in part on whether appropriate levels of viral infection, replication and transmission can be established within tissue culture systems.

Previous work has shown that mechanical inoculation techniques, for example, rubbing cells with abrasive powder [51] or vibrating cell suspensions in a vortex mixer [52], can be used to infect plant cell suspensions with viruses. Other methods that have been tested include microinjection of viruses into plant cells [53] and inoculation of callus by pricking the tissues with needles dipped in a virus suspension [51]. Viral infection has been reported to occur to some extent even without special mechanical treatment of cultured plant cells [52]. It is thought that viral agents are able to enter cells via the plasmodesmata observed to be present in dedifferentiated cultures [54]. Infection of suspended cells was found to be most successful when friable cell clumps were freshly dispersed from callus into liquid medium containing the virus [55]. The reason given for this was that the protoplasmic connections between the cells were broken in this procedure so that the plasmodesmata were exposed allowing viral entry into the cells [52]. Additional non-intentional injury may also occur to cells in agitated culture, thus providing other routes for virus infection.

In previous work, levels of viral accumulation in plant cell suspensions have been significantly lower than those achieved in whole plants. For example, suspended tobacco cells have been reported to accumulate only one-thirtieth to one-fortieth the concentration of TMV attainable in tobacco leaves [56]. In other experiments, maximum TMV coat-protein levels of only about 250 µg g^{-1} fresh weight were measured in tobacco cell suspensions [57]. Even though plant cells in suspension tend to aggregate so that individual cells in clumps are connected by plasmodesmata, the spread of virus between infected and uninfected cells that are not in direct contact is likely to be very limited. Therefore, compared with the recombinant systems already available for tissue culture applications, suspended plant cells may offer no significant advantage for improving foreign protein yields using virus-based expression. However, other forms of plant tissue in which the cells are in close and constant contact with each other, such as differentiated organs, may prove feasible hosts for high-level viral expression of foreign proteins *in vitro*.

We have examined the characteristics of virus infection of hairy roots to determine if root cultures would be suitable for foreign protein production using viral vectors. Extensive cell-to-cell contact occurs in hairy roots and some viral transport may also be possible through the vascular tissue. As shown in Figure 2.1, TMV was produced in significant quantities in *Nicotiana benthamiana* hairy roots, mostly during the period of active root growth. The concentration of virus in replicate cultures was found to vary considerably. The average concentration of virus between days 21 and 36 was approximately 2 mg g^{-1} dry weight; however, levels of virus in individual cultures were as high as 5 mg g^{-1}. These results demonstrate the potential of hairy roots for

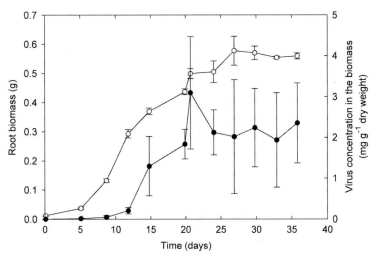

Fig. 2.1 Root growth (○) and accumulation of tobacco mosaic virus
(TMV) (●) in hairy roots of *N. benthamiana*. TMV concentrations
were measured by ELISA. The error bars indicate standard errors
from four replicate shake-flask cultures.

the propagation of plant viruses. Further work is underway to test the production of
co-expressed foreign proteins in hairy root cultures using a modified TMV vector.

2.3.2
Secretion of Foreign Proteins

Proteins produced in plant cells can remain within the cell or are secreted into the
apoplast via the bulk transport (secretory) pathway. In whole plants, because levels of
protein accumulated intracellularly, e.g. using the KDEL sequence to ensure retention in the endoplasmic reticulum, are often higher than when the product is
secreted [58], foreign proteins are generally not directed for secretion. However, as
protein purification from plant biomass is potentially much more difficult and expensive than protein recovery from culture medium, protein secretion is considered
an advantage in tissue culture systems. For economic harvesting from the medium,
the protein should be stable once secreted and should accumulate to high levels in
the extracellular environment.

Secretion of foreign proteins into the medium requires that the protein molecules
move through the cell walls. The pores in plant cell walls are thought to allow passage of globular proteins of maximum size around 20 kDa; however, a small number
of wider pores may serve as channels for relatively slow permeation of larger molecules [59]. Foreign proteins with molecular weights significantly greater than 20 kDa
have been recovered in substantial quantities from plant culture media [17, 29, 60].
In other cases, despite having signal sequences that allow the protein to reach and
traverse the plasma membrane, recombinant proteins such as erythropoietin [61]

and IgG-2b/κ antibody [62] remain associated with the plant cell wall and fail to be released from the biomass. These results suggest that protein composition and structure may affect the extracellular availability of secreted foreign proteins.

The presence of foreign protein in the medium of plant cultures does not necessarily mean that all or even most of the product can be recovered from the medium. In many expression systems where an appropriate signal sequence has been used, considerable amounts of foreign protein remain within the plant cells and/or tissues. For example, in a comparison of IgG_1 antibody production in tobacco cell suspension and hairy root cultures, a maximum of 72% of the total antibody was found in the medium of the suspension cultures whereas only 26% was found in the medium of the hairy root cultures [17]. This result could indicate that secretion and/or transport across the cell wall was slower in the hairy roots; alternatively, it could indicate poorer stability of the secreted protein in the hairy root medium. If foreign proteins are to be purified from the medium, improved secretion and extracellular product stability are desirable.

2.3.3
Foreign Protein Stability

There is considerable evidence that foreign proteins are subject to a significant degree of degradation and instability in plant expression systems, both inside and outside of the cells.

2.3.3.1 Stability Inside the Cells
Foreign protein fragments often appear in addition to the intact protein in western blots of extracts from transgenic plants and plant cells [21]. This phenomenon is not confined to plant tissue cultures or particular host species, and occurs in seeds as well as vegetative tissues. A detailed investigation of IgG_1 antibody fragments in tobacco cell suspension and hairy root cultures has been carried out by Sharp and Doran [21]. Although various explanations have been offered for the presence of foreign protein fragments, such as protease release during sample homogenization, the presence of assembly intermediates, and variations in the extent of protein glycosylation, in the case of IgG_1 antibody these explanations could not adequately account for all the observed molecular properties of the fragments. Instead, with the aid of a range of affinity probes, glycosylation and secretion inhibitors and glycan-reactive agents, proteolytic degradation in the apoplasm of tissues such as hairy roots, and between the endoplasmic reticulum and Golgi apparatus in both hairy roots and suspended cells, was identified as the most likely mechanism of fragment formation.

2.3.3.2 Stability Outside the Cells
The simplicity of plant culture media is considered an advantage for foreign protein production in tissue culture systems. However, as a mixture of salts and sugar containing several heavy metals but negligible protein (except for any protein secreted

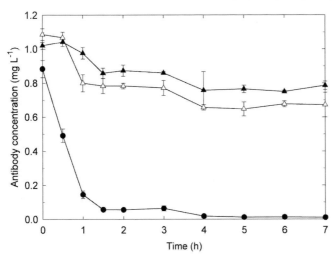

Fig. 2.2 Stability of IgG$_1$ monoclonal antibody added to sterile plant and animal cell culture media. (●) Murashige and Skoog (MS) medium; (▲) Dulbecco's minimal essential medium (DMEM) with 10% serum; and (△) serum-free Ex-cell 302 medium. The error bars indicate standard errors from triplicate flasks. (Reproduced with permission, from B. M. -Y. Tsoi and P. M. Doran, *Biotechnol. Appl. Biochem.* **2002**, *35*, 171–180. © Portland Press on behalf of the IUBMB.)

from the plant cells), plant culture medium provides an environment for foreign proteins that is very different from the physiological conditions inside the cells.

The effect of medium composition on the concentration of IgG$_1$ antibody in solution is illustrated in Figure 2.2. In this experiment, antibody was added at a concentration of 1.0 mg L^{-1} to fresh, sterile media in shake flasks. The flasks were then incubated on an orbital shaker at 25°C and the antibody concentration was measured as a function of time using an enzyme-linked immunosorbent assay (ELISA) [63]. As shown in Figure 2.2, there is a significant difference between the extent of antibody retention in Murashige and Skoog (MS) plant culture medium and in media designed to support the growth of animal cells. After 7 hours, about 80% of the added antibody was retained in Dulbecco's minimal essential medium (DMEM) containing 10% fetal bovine serum and about 70% was present in serum-free Ex-cell 302 medium. In contrast, in MS medium, less than 10% of the added antibody could be detected after only 1.5 h. These results indicate that fresh, sterile plant culture medium does not support the retention and stability of proteins in solution.

There have been many reports from several groups that plant culture medium is not conducive to protein stability, and that the retention of secreted proteins in culture media can be very poor [10, 11, 17, 40, 60, 63–66]. The mechanisms responsible for protein loss from plant culture media are not completely understood; however, current indications are that multiple factors may be involved. Processes that have been proposed to affect foreign proteins in plant media include protein degradation due to protease activity [10, 17, 20, 38, 60, 65], protein instability due to defined or undefined conditions or components in the medium [40, 63, 64, 66], surface adsorption of proteins onto the culture vessel [11, 17] and protein aggregation or insolubility [17].

As declining levels of foreign protein in plant tissue culture have been associated in a number of studies with an increase in the concentration of extracellular proteases [10, 60, 65], minimizing protease levels in the medium and/or reducing the susceptibility of heterologous proteins to protease degradation have been investigated as methods for improving foreign protein accumulation. Several approaches have been tested to achieve this objective with varying levels of success. These include adding the broad-spectrum protease inhibitor bacitracin to the medium of cell suspension and hairy root cultures [20], adjusting the osmolarity of the medium to minimize cell disruption and protease release [38], adding gelatin as a possible alternative substrate for protease activity [60], reducing protease accumulation by using inducible promoters to allow separation of the growth and production phases [10] and using host species such as rice that are considered to secrete lower levels of proteases than the more commonly applied tobacco [10].

2.3.3.3 Medium Additives
Protein stabilizing agents such as polyvinylpyrrolidone (PVP) [11, 17, 19, 60, 65, 66], gelatin [19, 60, 65], bovine serum albumin (BSA) [40] and salt (NaCl) [40] have been demonstrated to improve the retention of several types of foreign protein in plant tissue culture media. The precise mode of action of these additives in protecting proteins is unclear; however, they may prevent protein aggregation, conformational change and/or adsorption onto the internal surfaces of the holding vessel. The appropriate stabilizing polymer must be identified for each foreign protein production system, as their effects appear to vary depending on the specific culture and its protein product. The addition of biopolymers, particularly proteins such as BSA, to tissue culture media has the potential to complicate downstream processing operations for product recovery. However, when the resulting increase in protein yield is large, the additional cost of product purification may be acceptable.

Several medium additives other than the protein stabilizing agents mentioned above have also been tested to improve foreign protein accumulation in plant tissue cultures. These include dimethylsulfoxide (DMSO) [64], polyethylene glycol [60, 65], nitrate [19], amino acids [62], heamin [63], gibberellic acid [63] and glutamine [67]. The mechanisms by which these components might affect intra- and extracellular foreign protein levels include improving protein expression and synthesis, increasing protein secretion, reducing the extent of intracellular protein degradation and improving protein stability in the medium.

The results of empirical studies carried out to test the effects of medium additives on foreign protein accumulation in plant tissue culture are summarized below.

Polyvinylpyrrolidone (PVP)
PVP is a metabolically inert, water-soluble polymer with excellent protein stabilizing properties. As an example of the beneficial effect of PVP on foreign protein accumulation in plant tissue culture, data for growth and IgG_1 antibody levels in transgenic

tobacco hairy root cultures with and without PVP are shown in Figure 2.3. In these experiments, PVP with a relative molecular mass of 360,000 was added to the medium at a concentration of 1.5 g L^{-1}. PVP had no significant effect on root growth (Fig. 2.3a); in a number of studies, PVP at concentrations up to 3 g L^{-1} has been found to have little effect on growth of plant cell and organ cultures [11, 17, 19, 66]. The primary effect of PVP was a substantial increase in the amount of foreign protein in the culture medium (Fig. 2.3b); the maximum level of antibody in the medium with PVP was about four-fold greater than that without PVP. As indicated in Figure 2.3c, on average throughout the culture period the effect of PVP on antibody levels in the root biomass was relatively small. Similar results have also been reported for PVP-treated plant cell suspensions producing foreign protein [66].

The addition of PVP 360,000 at a concentration of 0.75 g L^{-1} has been reported to yield a 35-fold increase in the level of extracellular foreign protein in suspended plant cell cultures [66]. The effectiveness of PVP in stabilizing secreted proteins depends on both the polymer molecular weight and its concentration. Low-molecular-weight (10,000 and 40,000) PVP was found to be less effective than PVP 360,000

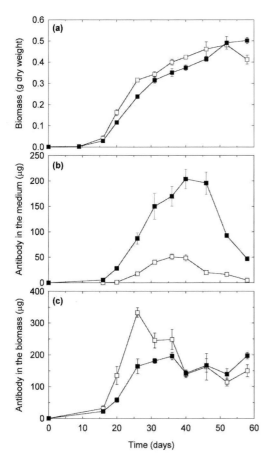

Fig. 2.3 Effect of PVP on **(a)** growth; **(b)** antibody in the medium; and **(c)** antibody in the biomass, for *N. tabacum* hairy roots expressing IgG₁ antibody. (■) Cultures with added PVP; (□) cultures without PVP. The error bars indicate standard errors from triplicate cultures; the initial culture volume was 50 mL. Much higher amounts of antibody were retained in the medium with PVP. (Reproduced with permission, from J. M. Sharp and P. M. Doran, *Biotechnol. Prog.* **2001**, *17*, 979–992. Copyright 2001 Am. Chem. Soc.)

[66]. Increasing concentrations of PVP 360,000 up to 1.0 g L^{-1} improved antibody accumulation in hairy root culture medium; however, above this concentration there was no further increase in antibody levels [19]. Addition of PVP after extracellular foreign protein levels had decreased during plant suspension culture did not result in a recovery of the protein [66].

Although PVP has proven successful as a foreign protein stabilizer in several systems and has yielded substantial increases in product concentrations as discussed above, it has been found relatively ineffective in other plant tissue cultures producing foreign proteins [60, 65].

Gelatin

Gelatin has been shown to enhance foreign protein levels in the medium of transgenic plant tissue cultures [19, 60, 65]. However, growth in both suspended cell and hairy root cultures was reduced, with the negative effect increasing with gelatin concentration [19]. It has been suggested that gelatin could act either as a protein stabilizing agent or as an alternative substrate for protease activity [60].

The addition of gelatin has been associated with significant increases in extracellular foreign protein levels in plant cultures. The concentration of interleukin-12 in suspension culture medium increased four-fold in the presence of 2% gelatin [60], while 0.1–0.9% gelatin increased the concentration of IgG_1 antibody in hairy root medium by factors of between four and eight [19]. The addition of 5% gelatin to cell suspensions expressing hGM-CSF resulted in a 4.6-fold improvement in the yield of extracellular product [65].

As gelatin is a common food additive with applications in the pharmaceutical industry, its introduction into foreign protein production systems may generate fewer regulatory concerns than other biopolymers.

Bovine Serum Albumin (BSA)

The addition of BSA to tobacco suspensions producing hGM-CSF increased the maximum protein concentration in the medium by a factor of two without affecting intracellular antibody levels or cell growth [40]. However, as an animal-derived protein, BSA has the potential to introduce mammalian pathogens into plant tissue cultures, thus negating an important advantage associated with using plant systems for the synthesis of pharmaceutical and therapeutic proteins. The introduction of exogenous proteins into plant cultures may also complicate the downstream processing of foreign proteins.

Salt (NaCl)

The addition of 50–100 mM NaCl to tobacco cell suspensions reduced cell growth but resulted in a 50% increase in the maximum level of hGM-CSF secreted into the culture medium [40]. Intracellular hGM-CSF was relatively unaffected by this treatment.

Dimethylsulfoxide (DMSO)

When added to plant suspension cultures, DMSO functions as a cell permeabilizing agent facilitating the release of intracellular products into the culture medium [68]. Use of DMSO as an effector for antibody secretion and/or a protein-stabilizing agent has been tested using tobacco cell suspensions and hairy roots [64, 69]. The addition of DMSO to suspended cells expressing an antibody heavy chain resulted in a significant increase in both intracellular and extracellular product levels [64]. Whereas untreated cultures accumulated a maximum of 150 μg L^{-1} of total antibody, treated cultures accumulated 210 μg L^{-1} in the presence of 2.8% DMSO and 430 μg L^{-1} in the presence of 4% DMSO. In contrast, when DMSO at a concentration of 2.8% was added to transgenic hairy roots at various times during the culture, there was no significant effect on IgG$_1$ antibody accumulation in either the biomass or medium [69].

Nitrate

Supplementation of Gamborg's B5 medium with 0.1% KNO$_3$ (in addition to the 0.25% KNO$_3$ usually present in B5 medium) resulted in a significant increase in antibody levels in tobacco hairy root cultures [19]. The antibody concentration in the medium increased 2.8-fold and the total antibody accumulation increased by 90% relative to cultures without added nitrate.

Amino Acids

Transgenic tobacco suspensions supplemented with a cocktail of essential and non-essential amino acids produced three-fold more IgG-2b/κ antibody than untreated cultures [62]. The amino acids were added 10 h prior to harvesting of the cultures in the presence of 1 mM CaCl$_2$ to facilitate amino acid uptake.

Other Additives

Further additives have been tested for their influence on foreign protein accumulation in plant tissue cultures. These include heamin ($C_{34}H_{32}O_4N_4FeCl$), an effective inhibitor of ubiquitin-dependent proteolysis, the plant hormone gibberellic acid, which has been shown to enhance the activity of the endoplasmic reticulum and secretory pathway in plant cells, and the amino acid glutamine [63, 67]. However, these compounds had relatively little effect on foreign protein levels. Addition of polyethylene glycol to plant culture medium also did not significantly improve the accumulation of foreign protein in the systems tested [60, 65].

2.3.3.4 Medium Properties

Modifying the properties of plant culture media, including increasing the osmolarity, reducing the effective concentration of selected heavy metals and altering the pH, has resulted in enhanced foreign protein accumulation or stability in some systems.

Osmolarity

Hyperosmolar medium is known to be advantageous for the production of proteins such as antibodies in animal cell cultures [70, 71]. Likewise, raising the osmolarity of MS medium using mannitol was found to improve the accumulation of foreign protein in both the biomass and medium in plant cell suspensions [63]. When IgG_1-secreting tobacco cell suspensions were grown in hyperosmolar medium, maximum antibody levels in the medium increased by up to 2.4-fold compared with standard MS medium. This result was possibly due to greater amounts of antibody being produced or secreted from the cells, as IgG_1 antibody added to fresh, sterile hyperosmolar medium did not show increased stability compared with unmodified medium [63].

The influence of medium osmolarity on foreign protein accumulation and activity was also demonstrated by Terashima *et al.* [38] using rice suspensions. In this system, α_1-antitrypsin was expressed using an inducible promoter that allowed protein production to occur when the sucrose content of the medium was low. Sucrose starvation was achieved by replacing the plant growth medium with sugar-free production medium. However, the removal of sucrose decreased the osmolarity of the medium and the activity of the α_1-antitrypsin produced was relatively low. Increasing the osmolarity of the production medium using mannitol or N-(2-hydroxyethyl) piperazine-N'-(2-ethanesulfonic acid) (HEPES)/NaCl resulted in a significant improvement in α_1-antitrypsin activity [38].

Heavy Metal Concentration

The heavy metals copper, manganese, cobalt and zinc were omitted individually and in combination from MS and B5 media to determine the effect on antibody stability in solution [63]. When IgG_1 antibody was added to these modified media in experiments similar to the one represented in Figure 2.2, only the B5 medium without Mn showed a significant improvement in antibody retention relative to normal culture media. Nevertheless, protein losses were considerable as only about 30% of the added antibody could be detected in the Mn-free medium after about 5 h. The beneficial effect of removing Mn was lost when all four heavy metals, Cu, Mn, Co and Zn, were omitted simultaneously. The reason for these results is unclear. Addition of the metal chelating agent ethylenediaminetetraacetate (EDTA) had a negligible effect on antibody retention in both MS and B5 media [63].

pH

Solution pH has a strong influence on the structure, conformation and solubility of globular proteins. However, when IgG_1 antibody was added to fresh, sterile B5 medium adjusted to pH values between 4.0 and 8.0, there was no significant difference in the rate at which the antibody disappeared from the different solutions [63]. In other work, antibody-expressing tobacco hairy roots were cultured in B5 medium with initial pH between 3.0 and 11.0 [69]. Root growth was affected severely at the lowest and highest pH values and total antibody levels declined as the initial pH was increased above 5.0–6.0.

2.3.4
Bioprocess Developments

The development of novel bioreactors and bioreactor operating strategies has the potential to improve the performance of cell culture systems. Some progress has been made in this area to enhance foreign protein production in plant tissue culture.

2.3.4.1 Product Recovery from the Medium

Continuous or periodic harvesting of secreted foreign proteins from plant culture media could be used to increase product yields by removing active protein before it can be degraded or otherwise lost from the culture.

Shake-flask-based affinity-chromatography bioreactors have been used for simultaneous production and purification of foreign proteins *in vitro* [41]. Heavy chain antibody secreted by suspended plant cells was recovered by continuously recycling the culture medium through a column containing Protein G resin. The harvested antibody was eluted from the column at the end of the culture period. The extent to which the antibody bound to the resin depended on the recirculation flow rate and the column pH. Although extracellular antibody concentrations were similar in the affinity-chromatography and control flasks, protein harvesting increased the total production of secreted protein more than eight-fold [41]. In similar experiments, hGM-CSF with a His_6 tag [40] was harvested semi-continuously by recycling the medium through a metal affinity column [41]. Cell growth was reduced in this system; however, the total amount of hGM-CSF accumulated was more than twice that obtained without product removal.

In other work, periodic foreign protein recovery from cell suspension and hairy root cultures was achieved using hydroxyapatite resin [17]. Beginning 13 days after inoculation of the hairy roots and 7 days after inoculation of the suspensions, medium was periodically separated from the biomass, exposed to the resin and then returned to the cultures. Recovered IgG_1 antibody was eluted from the resin and the total amount of antibody produced by the cultures determined by ELISA. Even though growth was unaffected by the hydroxyapatite, the overall effect of this process on antibody accumulation was relatively small, with maximum total antibody levels only 20–21% higher with periodic harvesting than without. Considering the rapidity with which antibodies may be lost from plant culture media (Figure 2.2), continuous rather than periodic removal of product may be required to achieve more substantial improvements in yield [17].

2.3.4.2 Oxygen Transfer and Dissolved Oxygen Concentration

Adequate aeration of plant tissue cultures is crucial for achieving maximum production of foreign proteins [17, 67]. Poor oxygen transfer has been found to limit cell growth and reduce antibody heavy chain production in genetically modified tobacco cell suspensions in shake flasks [67]. However, although raising the air flow rate in a stirred 5-L bioreactor increased foreign protein levels to a certain extent, excessive aeration resulted in foaming and reduced both growth and antibody production [67].

In transgenic hairy root cultures, the concentration of IgG_1 antibody in the biomass increased by 52% relative to the levels obtained under air when the roots were cultured at a dissolved oxygen tension of 150% air saturation [17]. Compared with root cultures grown at 50% air saturation, which is a realistic operating dissolved oxygen tension in poorly mixed, large-scale root reactors, oxygen enrichment to 150% air saturation improved total antibody accumulation 2.9-fold.

2.4
Conclusions

Plant tissue culture offers an alternative to agriculture as a plant-based system for producing low-to-medium volumes of foreign proteins. Secretion of foreign proteins into the culture medium has significant potential benefits for reducing the cost and complexity of product recovery and purification. A range of molecular, culture and bioprocessing techniques has been employed to enhance the accumulation and stability of foreign proteins in plant cell and organ cultures. Further improvements in product yield are required to raise the economic competitiveness of plant tissue culture as a viable commercial method for the controlled, safe and reliable manufacture of pharmaceutical and therapeutic proteins.

References

[1] J. W. LARRICK, L. YU, J. CHEN et al., *Res. Immunol.* **1998**, *149* (6), 603–608.

[2] E. E. HOOD, J. M. JILKA, *Curr. Opin. Biotechnol.* **1999**, *10* (4), 382–386.

[3] C. L. CRAMER, J. G. BOOTHE, K. K. OISHI, in *Plant Biotechnology: New Products and Applications*, J. Hammond, P. McGarvey, V. Yusibov (eds), Springer, Berlin, **1999**, 95–118.

[4] M. CABANES-MACHETEAU, A.-C. FITCHETTE-LAINÉ, C. LOUTELIER-BOURHIS et al., *Glycobiology* **1999**, *9* (4), 365–372.

[5] R. L. EVANGELISTA, A. R. KUSNADI, J. A. HOWARD et al., *Biotechnol. Prog.* **1998**, *14* (4), 607–614.

[6] A. R. KUSNADI, Z. L. NIKOLOV, J. A. HOWARD, *Biotechnol. Bioeng.* **1997**, *56* (5), 473–484.

[7] L. MIELE, *Trends Biotechnol.* **1997**, *15* (2), 45–50.

[8] P. M. DORAN, *Curr. Opin. Biotechnol.* **2000**, *11* (2), 199–204.

[9] H. DANIELL, S. J. STREATFIELD, K. WYCOFF, *Trends Plant Sci.* **2001**, *6* (5), 219–226.

[10] Y.-J. SHIN, S.-Y. HONG, T.-H. KWON et al., *Biotechnol. Bioeng.* **2003**, *82* (7), 778–783.

[11] N. S. MAGNUSON, P. M. LINZMAIER, J.-W. GAO et al., *Protein Express. Purif.* **1996**, *7* (2), 220–228.

[12] J. HUANG, T. D. SUTLIFF, L. WU et al., *Biotechnol. Prog.* **2001**, *17* (1), 126–133.

[13] J. GAO, J. M. LEE, G. AN, *Plant Cell Rep.* **1991**, *10* (10), 533–536.

[14] M. L. SMITH, H. S. MASON, M. L. SHULER, *Biotechnol. Bioeng.* **2002**, *80* (7), 812–822.

[15] D. VERDELHAN DES MOLLES, V. GOMORD, M. BASTIN et al., *J. Biosci. Bioeng.* **1999**, *87* (3), 302–306.

[16] P. J. LARKIN, W. R. SCOWCROFT, *Theor. Appl. Genet.* **1981**, *60* (4), 197–214.

[17] J. M. SHARP, P. M. DORAN, *Biotechnol. Prog.* **2001**, *17* (6), 979–992.

[18] J. K.-C. MA, T. LEHNER, P. STABILA et al., *Eur. J. Immunol.* **1994**, *24* (1), 131–138.

[19] R. WONGSAMUTH, P. M. DORAN, *Biotechnol. Bioeng.* **1997**, *54* (5), 401–415.

[20] J. M. SHARP, P. M. DORAN, *Biotechnol. Bioprocess Eng.* **1999**, *4* (4), 253–258.

[21] J. M. SHARP, P. M. DORAN, *Biotechnol. Bioeng.* **2001**, *73* (5), 338–346.

[22] S. BANERJEE, T. Q. SHANG, A. M. WILSON et al., *Biotechnol. Bioeng.* **2002**, *77* (4), 462–466.

[23] C. D. MERRITT, S. RAINA, N. FEDOROFF et al., *Biotechnol Prog.* **1999**, *15* (2), 278–282.

[24] M. A. SUBROTO, J. D. HAMILL, P. M. DORAN, *J. Biotechnol.* **1996**, *45* (1), 45–57.

[25] J.-J. ZHONG, *J. Biosci. Bioeng.* **2002**, *94* (6), 591–599.

[26] F. BOURGAUD, A. GRAVOT, S. MILESI et al., *Plant Sci.* **2001**, *161* (5), 839–851.

[27] A. H. SCRAGG, *Curr. Opin. Biotechnol.* **1992**, *3* (2) 105–109.

[28] A. GIRI, M. LAKSHMI NARASU, *Biotechnol. Adv.* **2000**, *18* (1), 1–22.

[29] M. TERASHIMA, Y. MURAI, M. KAWAMURA et al., *Appl. Microbiol. Biotechnol.* **1999**, *52* (4), 516–523.

[30] J. HUANG, L. WU, D. YALDA et al., *Transgenic Res.* **2002**, *11* (3), 229–239.

[31] E. TORRES, C. VAQUERO, L. NICHOLSON et al., *Transgenic Res.* **1999**, *8* (6), 441–449.

[32] K.-I. SUEHARA, S. TAKAO, K. NAKAMURA et al., *J. Ferment. Bioeng.* **1996**, *82* (1), 51–55.

[33] S. SOMMER, M. SIEBERT, A. BECHTHOLD et al., *Plant Cell Rep.* **1998**, *17* (11), 891–896.

[34] E. H. HUGHES, S.-B. HONG, J. V. SHANKS et al., *Biotechnol. Prog.* **2002**, *18* (6), 1183–1186.

[35] K. YOSHIDA, T. KASAI, M. R. C. GARCIA et al., *Appl. Microbiol. Biotechnol.* **1995**, *44* (3–4), 466–472.

[36] C. XIANG, Z.-H. MIAO, E. LAM, *Plant Mol. Biol.* **1996**, *32* (3), 415–426.

[37] H. KURATA, T. TAKEMURA, S. FURUSAKI et al., *J. Ferment. Bioeng.* **1998**, *86* (3), 317–323.

[38] M. TERASHIMA, Y. EJIRI, N. HASHIKAWA et al., *Biochem. Eng. J.* **1999**, *4* (1), 31–36.

[39] M. M. TREXLER, K. A. MCDONALD, A. P. JACKMAN, *Biotechnol. Prog.* **2002**, *18* (3), 501–508.

[40] E. A. JAMES, C. WANG, Z. WANG et al., *Protein Express. Purif.* **2000**, *19* (1), 131–138.

[41] E. JAMES, D. R. MILLS, J. M. LEE, *Biochem. Eng. J.* **2002**, *12* (3), 205–213.

[42] R. N. BEACHY, J. H. FITCHEN, M. B. HEIN, *Ann. N. Y. Acad. Sci.* **1996**, *792*, 43–49.

[43] C. PORTA, V. E. SPALL, T. LIN et al., *Intervirology* **1996**, *39* (1–2), 79–84.

[44] H. B. SCHOLTHOF, K.-B. G. SCHOLTHOF, A. O. JACKSON, *Annu. Rev. Phytopathol.* **1996**, *34*, 299–323.

[45] T. VERCH, V. YUSIBOV, H. KOPROWSKI, *J. Immunol. Meth.* **1998**, *220* (1–2), 69–75.

[46] A. WIGDOROVITZ, D. M. PÉREZ FILGUEIRA, N. ROBERTSON et al., *Virology* **1999**, *264* (1), 85–91.

[47] R. J. COPEMAN, J. R. HARTMAN, J. C. WATTERSON, *Phytopathology* **1969**, *59* (7), 1012–1013.

[48] S. RABINDRAN, W. O. DAWSON, *Virology* **2001**, *284* (2), 182–189.

[49] R. FISCHER, C. VAQUERO-MARTIN, M. SACK et al., *Biotechnol. Appl. Biochem.* **1999**, *30* (2), 113–116.

[50] K. DALSGAARD, Å. UTTENTHAL, T. D. JONES et al., *Nature Biotechnol.* **1997**, *15* (3), 248–252.

[51] B. KASSANIS, T. W. TINSLEY, F. QUAK, *Ann. Appl. Biol.* **1958**, *46* (1), 11–19.

[52] H. H. MURAKISHI, J. N. HARTMANN, R. N. BEACHY et al., *Virology* **1971**, *43* (1), 62–68.

[53] T. E. RUSSELL, R. S. HALLIWELL, *Phytopathology* **1974**, *64* (12), 1520–1526.

[54] D. F. SPENCER, W. C. KIMMINS, *Can. J. Bot.* **1969**, *47* (12), 2049–2050.

[55] H. H. MURAKISHI, *Phytopathology* **1968**, *58* (7), 993–996.

[56] B. KASSANIS, *Methods Virol.* **1967**, *1*, 537–566.

[57] P. THOMAS, G. S. WARREN, *J. Exp. Bot.* **1994**, *45* (276), 987–994.

[58] U. FIEDLER, J. PHILLIPS, O. ARTSAENKO et al., *Immunotechnology* **1997**, *3* (3), 205–216.

[59] N. CARPITA, D. SABULARSE, D. MONTEZINOS et al., *Science* **1979**, *205* (4411), 1144–1147.

[60] T. H. KWON, J. E. SEO, J. KIM et al., *Biotechnol. Bioeng.* **2003**, *81* (7), 870–875.

[61] S. MATSUMOTO, K. IKURA, M. UEDA et al., *Plant Mol. Biol.* **1995**, *27* (6), 1163–1172.

[62] R. Fischer, Y.-C. Liao, J. Drossard, *J. Immunol. Methods* **1999**, *226* (1–2), 1–10.

[63] B. M.-Y. Tsoi, P. M. Doran, *Biotechnol. Appl. Biochem.* **2002**, *35* (3), 171–180.

[64] M. F. Wahl, G. An, J. M. Lee, *Biotechnol. Lett.* **1995**, *17* (5), 463–468.

[65] J.-H. Lee, N.-S Kim, T.-H. Kwon et al., *J. Biotechnol.* **2002**, *96* (3), 205–211.

[66] W. LaCount, G. An, J. M. Lee, *Biotechnol. Lett.* **1997**, *19* (1), 93–96.

[67] F. Liu, J. M. Lee, *Biotechnol. Bioprocess Eng.* **1999**, *4* (4), 259–263.

[68] D. P. Delmer, *Plant Physiol.* **1979**, *64* (4), 623–629.

[69] R. Wongsamuth, P. M. Doran, Hairy roots as an expression system for production of antibodies, in *Hairy Roots: Culture and Applications*, ed. P. M. Doran, Harwood Academic, Amsterdam, **1997**, 89–97.

[70] S. K. W. Oh, P. Vig, F. Chua et al., *Biotechnol. Bioeng.* **1993**, *42* (5) 601–610.

[71] M. S. Lee, G. M. Lee, *Biotechnol. Bioeng.* **2000**, *68* (3), 260–268.

[72] N. V. Borisjuk, L. G. Borisjuk, S. Logendra et al., *Nature Biotechnol.* **1999**, *17* (5), 466–469.

[73] S. Firek, J. Draper, M. R. L. Owen et al., *Plant Mol. Biol.* **1993**, *23* (4), 861–870.

[74] J. A. Francisco, S. L. Gawlak, M. Miller et al., *Bioconjugate Chem.* **1997**, *8* (5), 708–713.

[75] N. S. Magnuson, P. M. Linzmaier, R. Reeves et al., *Protein Express. Purif.* **1998**, *13* (1), 45–52.

[76] R. Fischer, N. Emans, F. Schuster et al., *Biotechnol. Appl. Biochem.* **1999**, *30* (2), 109–112.

3
Novel Sprouting Technology for Recombinant Protein Production
Kimmo Koivu

3.1
Introduction

Sprouted seeds or sprouts are renowned for their excellent nutritional properties. They are enjoyed as a protein-rich food, and also contain beneficial phytochemicals such as vitamins A, E and C, which are antioxidants [1]. We have developed a novel application for sprouting dicotyledonous seeds, namely the production of recombinant proteins. In a contained production system, transgenic seeds carrying the gene encoding a specific protein of interest are first produced and harvested in a greenhouse and then sprouted in an airlift tank. The recombinant protein is produced during sprouting and extracted from the sprouts. Alternatively, protein can be removed directly from the growth medium.

When they are harvested, the seeds are at a developmental stage where all resources required for sprouting are stored and only water is required to initiate germination. We therefore use water as the growth medium, although this may be supplemented with a soluble nitrogen source and antibiotics to prevent protein expression in proplastids (see below). During sprouting, stored resources are mobilized. Seed proteins are broken down into amino acids by the action of a series of specific proteases, forming a free amino acid pool. This is used for the *de novo* synthesis of new proteins, such as the Rubisco enzyme, which can accumulate to represent 50% of the total protein in the cotyledons. In our production system, part of this amino acid pool is diverted to synthesize the recombinant protein of interest.

We have isolated a group of Rubisco small subunit (SSU) promoters that are very active in cotyledons. These are linked to the desired transgene to regulate its expression. As stated above, antibiotics such as streptomycin are added to the growth medium to inhibit expression of the large Rubisco subunit in proplastids. SSU genes are expressed from the nuclear genome and therefore remain unaffected by antibiotic treatment. Thus far we have achieved yields of 0.5 g human serum albumin (HSA) and 1.5 g β-glucuronidase (GUS) per kg of oilseed rape (*Brassica napus* L.) seeds after 72 h sprouting.

Oilseed rape has been selected as the model plant for our production system because of its high protein content, rapid sprouting, large number of seeds, efficient transformation, well-characterized biology and long cultivation history. However, we have also developed an alternative platform based on falseflax (*Camelina sativa* L.). This is a self-pollinating oilseed crop, rarely used for food, which originates from the Fertile Crescent.

3.2
Biology of Sprouting

Our protein production technology exploits the physiology of sprouting dicotyledonous seeds to produce large amounts of any recombinant protein of interest. This section summarizes the major structural and physiological changes occurring during seed germination and seedling growth, showing how these processes can be utilized for molecular farming. Seed germination physiology is reviewed extensively in [2] and subsequent seedling development is discussed in [3].

3.2.1
Structure and Content of Dicotyledonous and Monocotyledonous Seeds

The seeds of dicotyledonous plants have two cotyledons, or seed leaves, which are part of the embryo. The cotyledons usually are the main storage tissue, although in some plants (such as castor bean) the endosperm also has a storage function. During development in the field, seeds gradually accumulate storage oils, proteins and carbohydrates (Table 3.1). In the seed, the cotyledon structure is relatively simple. The remainder of the embryo, the embryonic axis, consists mostly of undifferentiated cells, but provascular tissue can be detected that develops into vascular tissue in the seedling.

Storage proteins are a group of proteins found mainly in seeds, but also in many other tissue types, such as roots and tubers. In seeds they serve as a nitrogen source for the developing plant during germination and sprouting. Storage proteins have no enzymatic activity. They exist naturally in an aggregated state within membrane-surrounded vesicles. The membrane that delimits the protein body is derived from the endoplasmic reticulum (ER) and Golgi network. Storage proteins often comprise a number of different polypeptide chains [4].

In monocotyledonous plants, including all the cereals, storage proteins are found mainly in the endosperm. The major storage compounds are carbohydrates rather

Tab. 3.1 Main seed storage compounds (percentages) of different crop species

	Oil	Protein	Carbohydrate
Oilseed rape (*Brassica napus*)	45–48	22–24	18–20
Soybean (*Glycine max*)	18–20	38–40	25–27
Alfalfa (*Medicago sativa*)	8– 9	38–40	40
Barley (*Hordeum vulgare*)	3	12	76
Maize (*Zea mays*)	4	8	74

than proteins. The first emerging leaf like structure during sprouting is a coleoptile, not a cotyledon. The single cotyledon is reduced and modified to form the scutellum, a tissue layer adjacent to the endosperm. The scutellum secretes digestive enzymes and passes nutrients to the seedling from the endosperm [5].

Structure and Content of Rapeseeds

The cotyledons make up about 90% of the mass of a rapeseed, filling the seed coat and forming a hemisphere around the embryonic axis. The total protein content of rapeseeds is 22–25% depending on cultivar, growing conditions and crop management. The proteins are present as distinct protein bodies. We use cultivar Westar because the tissue culture and transformation methods are well established [6, 7]. One plant can produce up to 4500 seeds with a seed weight of 4–6 mg under our cultivation conditions. The average yield in Europe is about 2500 kg ha^{-1} [8].

The major rapeseed storage proteins are a 12S type globulin, called cruciferin, which makes up 60% of the total protein, and a 2S type albumin, called napin, which makes up 20% of the total protein [9, 10]. Both of these proteins are formed in embryonic cells and stored in specialized vacuoles known as protein bodies.

3.2.2
Germination

Seed germination occurs rapidly (in oilseed rape it takes 6–8 hours) and is defined as the emergence of the radicle, the embryonic root of the seed, through the seed coat. The first phase of germination is characterized by rapid water uptake. After a plateau phase when germination is completed, more water is imbibed. Protein synthesis, using newly synthesized RNAs, begins at the plateau phase within hours after initial water uptake [2]. Enzymes needed for the mobilization of storage reserves are synthesized at the post-germination stage. These break down the seed's insoluble high-molecular-weight compounds into soluble, easily transportable, low-molecular-weight molecules.

Several different types of proteases hydrolyze intact storage proteins first into large fragments and then into smaller peptides and amino acids within the protein body. The peptides are transported to the cytosol where other enzymes, e.g. aminopeptidases, carboxypeptidases, dipeptidases and tripeptidases, cleave them and eventually form a pool of free amino acids [11].

The amino acid composition of storage proteins differs from that of the complete sprout [12, 13]. At least in the case of oilseed rape, alfalfa (*Medicago sativa* L.) and *Camelina sativa*, amino acids in the sprout are used mainly, either directly or indirectly, for the synthesis of the Rubisco proteins. Computer analysis shows that the amino acid composition of cruciferin and napin is completely different to the amino acid composition of Rubisco. This indicates that amino acids released from the seed storage proteins must be converted into other amino acids prior to Rubisco synthesis.

3.2.3
The Sprout

A sprout is a seedling at the stage following seed germination. Seeds generally sprout in 2 to 10 days but this depends on conditions such as temperature and light, which also have a major effect on sprout morphology. Oilseed rape sprouts growing on water agar plates at 24 °C for four days (with 16 h of light and 8 h of darkness) reach approximately 30–40 mm in height and 50–60 mg in weight (Figure 3.1). As the sprout begins to develop, the stored reserves are consumed. Later, when the seedling starts to photosynthesize, it needs light and carbon dioxide to continue its growth.

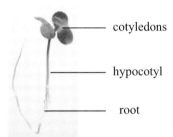

cotyledons

hypocotyl

root

Fig. 3.1 A four-day-old rapeseed seedling, showing the cotyledons, hypocotyl and root.

3.2.4
Rubisco Synthesis

Ribulose-1,5-bisphosphate carboxylase/oxygenase (EC 4.1.1.39; Rubisco) is the key enzyme in the photosynthetic fixation of CO_2. It is localized in the chloroplasts and is composed of eight small subunits (SSU) and eight large subunits (LSU) encoded by nuclear *rbcS* and chloroplast *rbcL* genes, respectively. Rubisco accounts for 30–50% of total *de novo* protein synthesis in the oilseed sprout. It is likely that part of the Rubisco enzyme is not available for carboxylation and instead acts as a source of nitrogen [14]. The dual role of Rubisco as a key photosynthetic enzyme and a dynamic source of nitrogen is very important in the case of short-lived cotyledons. A very rapid conversion of the major protein pool from storage proteins to Rubisco proteins is clearly seen when investigating the soluble protein pool in developing sprouts. Storage proteins in oilseed sprouts are broken down and Rubisco is synthesized within 3 days (Figure 3.2).

The Rubisco SSU genes in the nucleus and the LSU genes in the chloroplast are developmentally regulated and expressed in a tissue-specific manner. The expression of each gene must be coordinated. Light increases the general transcriptional activity of the chloroplast genome including the Rubisco LSU gene. Light, sensed by phytochromes and blue light receptors, has also been shown to induce Rubisco SSU gene transcription through regulatory DNA elements, although high levels of sugars can repress transcription of this gene [15, 16]. Multimeric complex composition requires

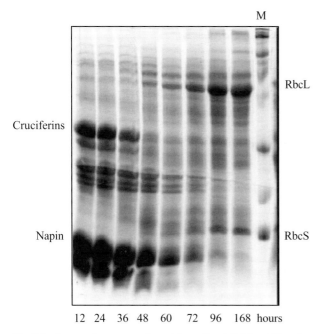

Fig. 3.2 Rapeseeds were germinated from 12 to 168 h in airlift tank. The total soluble proteins were extracted and separated by 15% SDS-PAGE. The gel was stained with Coomassie blue. Between 36 and 60 h, the degradation of storage proteins and the de novo synthesis of Rubisco is clearly visible.

an abundance of unassembled subunits. In antisense *rbcS* plants, where SSU accumulation is limited, LSU levels are adjusted to those of the SSU. This is regulated at the level of translation [17].

3.2.5
Rubisco Promoters

The Rubisco promoter is probably the most efficient promoter in sprouting rapeseed cotyledons. Based on GenBank data and the cloning of novel sequences, we have tested several Rubisco promoters for expression in sprouts. One of the strongest was a promoter isolated from *Brassica rapa*, a close relative of oilseed rape. In some plants, e.g. *Arabidopsis* and tomato [18, 19], the genetic structure of Rubisco genes has been extensively studied and both the Rubisco genes and their regulatory elements have been cloned. In the case of *Brassica*, the sequences of three Rubisco SSU genes can be found in GenBank. Several approaches have been used for promoter isolation, including the construction of cDNA libraries in combination with genome walking techniques. A set of unknown promoter sequences has been obtained. Table 3.2 summarizes some examples of Rubisco genes in different plant species. Rubisco mRNA is highly abundant in cotyledons (Figure 3.3). The untranslated re-

Tab. 3.2 Rubisco SSU genes in different plant species

	No. of Rubisco SSU genes	References
Wheat (*Triticum aestivum*)	12	33
Tobacco (*Nicotiana tabacum*)	7	34
Potato (*Solanum tuberosum*)	5	35
Pea (*Pisum sativum*)	5	36
Arabidopsis thaliana	4	18
Oilseed rape (*Brassica napus*)	>5	37, 38, our unpublished data
Alfalfa (*Medicago sativa*)	>1	Gene Bank accession. no. AF056313; McKersie *et al.*, unpublished data
Sucarcane (*Saccharum officinarum*)	16	39
Tomato (*Lycopersicon esculentum*)	5	19

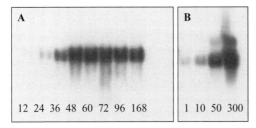

Fig. 3.3   A. Northern blot showing the synthesis of Rubisco SSU mRNA after 24–36 h of sprouting in an airlift tank. Total RNA was isolated from sprouts germinated from 12 to 168 hours. **B.** On the same filter, unlabelled Rubisco RNA produced by in vitro transcription was loaded as a control. The amount of control RNA is indicated in pg.

gions of Rubisco mRNA are responsible for targeting the synthesized polypeptide to the polyribosome system, which is very stable and has a high rate of translation.

3.2.6
Inhibition of Endogenous Gene Expression

To increase the yield of recombinant protein, it is beneficial to downregulate the expression of endogenous genes. Because about 50% of the proteins in the cotyledons are derived from the chloroplast genome, we have focused on downregulating protein expression in chloroplast. This does not interfere with transgene expression since transgenes are introduced into the nuclear genome.

Transcription and protein synthesis in the chloroplast can be inhibited by antibiotics because chloroplast ribosomes are similar in structure to those of bacteria. Chloroplast ribosomes are 70S in size, comprising a large 50S subunit (containing

23S, 5S and 4.5S rRNAs) and a small 30S subunit (containing 16S rRNA). Chloroplast ribosomal proteins are encoded by both nuclear and chloroplast genes.

Chloroplast protein synthesis is controlled largely at the post-transcriptional level [20,21] and can be repressed by the inclusion of antibiotics such as streptomycin in the sprouting medium. Streptomycin binds to the 16S rRNA and causes the ribosome to misread the mRNA sequence, producing incorrect and non-functional proteins [22].

Transcriptional inhibitors could be used simultaneously. Rifampicin blocks chloroplast and mitocondrian RNA synthesis [23, 24], while tagetitoxin is a very specific inhibitor of chloroplast RNA polymerase [25]. Treatment with these antibiotics does not inhibit Rubisco SSU synthesis since the promoter is part of the nuclear genome, while the cytosolic ribosomes are not affected by streptomycin. Therefore SSU promoters can be used to drive transgene expression and facilitate the accumulation of recombinant proteins. Expressed proteins are targeted to a suitable cellular compartment, such as the cytoplasm, apoplastic space or chloroplast, depending on the nature of the protein.

We are also developing an RNA-level silencing strategy, which can be used instead of or in combination with regulation at the level of translation. The advantage of RNA silencing is that the resources of the protein expression machinery are released at an earlier stage for the production of the recombinant protein. For the silencing of specific genes, such as the Rubisco genes, it is possible to use either antisense RNA or RNA interference (RNAi) techniques.

As discussed above, Rubisco levels have been reduced by expressing antisense RNA in transgenic tobacco plants [26]. Plants expressing antisense *rbcS* RNA showed reduced levels of *rbcS* mRNA, normal levels of *rbcL* mRNA, and coordinately reduced levels of LSU and SSU proteins.

Guo and colleagues [27] have been able to silence endogenous gene expression using a chemically regulated, inducible RNAi system. Upon induction at the seed germination and post-germination stages, the phytoene desaturase gene was silenced. A stable and reproducibly inducible RNAi phenotype was obtained in subsequent generations of transgenic plants.

3.3
Expression Cassette Design

We use *Agrobacterium*-mediated transformation for the transfer and integration of expression cassettes into the nuclear genome of oil seed rape. More specifically, we use *Agrobacterium tumefaciens* strain LBA4404 with a binary vector containing kanamycin or hygromycin resistance selectable markers. After regeneration, the transgenic shoots are grown *in vitro* until small seedlings are obtained. The seedlings are planted in soil in the greenhouse and grown until seeds are formed.

We have isolated Rubisco small subunit promoters from several plant species and tested their strength with *gusA* and *ALB* (human serum albumin) transgenes in sprouts. The highest level of expression in *Brassica napus* sprouts has been obtained with a Rubisco promoter isolated from *Brassica rapa*. Testing of protease-, heat shock- 35S and salicylate-inducible promoters resulted only in moderate GUS activity when compared to the Rubisco promoter. A basic version of the expression cassette contains the Rubisco small subunit promoter and 5′ UTR region, the recombinant protein-coding region and a transcription terminator element.

For promoter isolation, a *rbcS*-specific mRNA library was constructed from cotyledons grown in the light basically according to the CloneAmp procedure (Clontech). Sequence analysis of 100 Rubisco cDNAs indicated that 56% of the mRNAs originated from one gene and 29% from another. Promoters for these most abundant Rubisco cDNAs were isolated using the Clontech Universal GenomeWalker kit. The *gusA* gene, or a cDNA encoding mature HSA (which had been codon-optimized for plants and joined to an ER-targeting signal), were linked to the above promoters and corresponding untranslated regions.

There are many commercially important proteins that have multiple subunits and are therefore encoded by several genes. For example, in order to form full size IgG antibodies, two genes must be expressed simultaneously, and for secretory IgAs, four genes are required. Although gene silencing has been extensively studied in the last few years, it is not clear if one promoter can be used repeatedly in a multigene construct or whether each gene should have different promoter. Gene expression levels obtained with multigene constructs where each gene has a different promoter must be tested experimentally. For that reason, several promoters functioning in sprouts have been isolated.

3.4
Sprouting Equipment

In our experimental set-up, the volume of a laboratory-scale sprouting vessel is 10 l (Figure 3.4). It is made of two glass cylinders one within the other. The inner cylinder forms a growth chamber where seeds are sprouted in a medium consisting of tap water. Heated water is circulated through the space between the cylinders and this acts as a heating jacket. The temperature can be regulated accurately using a thermally controlled water bath to heat the circulating water.

The water in the growth chamber is aerated by pumping pressured air through a gas dispersal sinter at the bottom of the cylinder. Air can be sterilized using a 0.2-μm filter. For efficient aeration, 15% of medium volume consists of small air bubbles. Air stirs the medium effectively and enhances the movement of the sprouts. The growth medium can be circulated optionally through the growth chamber. Normally, two to four chamber volumes are circulated through the growth chamber per day.

In the 10-l sprouting vessel, 400–800 g of seeds can be sprouted, depending on the cultivation time. In the case of overload, sprouts stick together and are not uniformly lit. Light is needed to regulate sprout development and for the induction of the promoter driving the transgene of interest. However, sprouts do not need light as an energy source because about 48% of the dry weight of rapeseeds is storage oil that is used for the initial growth of the sprout.

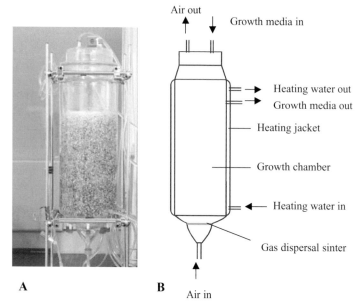

Air out   Growth media in

Heating water out
Growth media out

Heating jacket

Growth chamber

Heating water in

Gas dispersal sinter

A   **B**   Air in

Fig. 3.4 **A**. Laboratory scale 10-L sprouting equipment. The apparatus is made from glass. Oilseed rape sprouts have been grown for
three days under continuous aeration. The sprouting medium is tap
water. **B**. A schematic drawing of the major parts of the apparatus.

3.5
Sprouting Conditions

Several methods for seed sprouting have been investigated. Microbial contamination
has been a problem in solid-state systems under humid conditions in which wet matrix, like sand, glass wool or peat, have been used. Also, the initiation of germination
has not been fully synchronized. Continuous spraying with water reduces microbial
growth.

Other systems, in which the sprouts are grown in a tank and submerged in water
at intervals of several hours, are used often for commercial production of sprouts for
food. Light and aeration conditions are not fully uniform in such systems, because
inside the sprout mass less light is available. The use of light-inducible promoters
clearly demands another kind of sprouting system.

In our case, sprouting in an effectively aerated water medium was selected as the
sprouting method of choice. In our system the sprouts are moving and circulating
vigorously with the water flow. Effective aeration is necessary; seeds submerged in
water without aeration do not sprout and eventually die. Normal tap water or reverse
osmosis purified water (RO-water) can be used.

External factors have a major effect on sprout growth and development. By altering the growth conditions, growth can be accelerated or inhibited. For example, if

properties of the expressed protein require the use of lower production temperatures, rapid growth for 2 days can be achieved at a higher temperature, e.g. 32 °C, and at the time of protein expression, the temperature can be lowered to 16 °C for 4 days.

3.5.1
Sterilization

Microbial contamination, especially by salmonellas, is a risk when sprouts are produced commercially for human consumption. For recombinant protein production, seeds can be washed with water and surface-sterilized using hypochlorite solution. Sprouts can also be surface-sterilized during sprouting, by the addition of mild hypochlorite solution directly into the growth medium. Eventually, the hypochlorite is diluted out with pure water or growth medium. In our experiment on plate count agar [28], the sprouts showed no bacterial growth after sterilization with 1% sodium hypochlorite.

3.5.2
Sprouting Time and Temperature

Imbibition of the dry seeds is initiated by placing them in the airlift tank filled with water. Later, nutrients and other substances can also be added. For example, KNO_3 cannot be added any earlier than 6 hours after the start of germination, or it will inhibit sprout growth.

Germination and subsequent sprouting is faster in an airlift tank than on agar plates or in soil. In an airlift tank, the sprout reaches the same developmental stage after 3 days as it does after 4 days cultivation on agar. The sprouting process is rapid: storage proteins are mobilized within 48 h and the Rubisco protein is synthesized within 96 h. Recombinant protein expression occurs preferably between 48 and 96 h, after most of the storage resources have been used and the cotyledon begins to turn into a more leaf-like structure. The Rubisco promoters we are using for recombinant protein expression are activated after 36 h.

Sprouting can be accelerated by increasing the growth temperature (Figure 3.5) and by using an appropriate nitrogen fertilizer. Shorter sprouting times allow more batches per airlift tank. One batch of sprouting takes four days. In one year, up to 90 batches can be produced. A short expression time in appropriate conditions is beneficial when an unstable recombinant protein is produced. The potential for large-scale accumulation of stable proteins in a short period also exists. For example, natural Rubisco protein accumulates to high levels within 96 h after the onset of germination. The cycle time in our sprouting system is comparable to that of *Agrobacterium*-mediated transient expression in infiltrated tobacco leaves materials like in leaves [29].

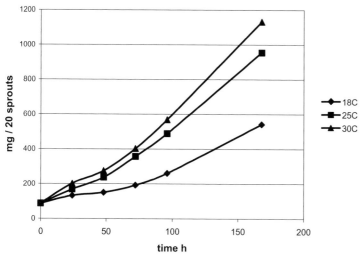

Fig. 3.5 The sprouting of rapeseeds at various temperatures (18 °C, 25 °C and 30 °C) shows that increasing the growth temperature up to 30 °C has a positive effect on sprout growth. Further temperature increments do not increase the growth rate and temperatures over 38 °C inhibit sprouting totally (not shown).

3.5.3
Light

Light is one of the most important environmental factors affecting the sprout development, and the effects of light can differ from one organ or cell type to another. The illumination of growing sprouts reduces cell elongation in the hypocotyl but induces expansion and division in the cotyledon and shoot apex. In cotyledons, proplastids differentiate into chloroplasts that initiate photosynthesis [3]. Light also induces the Rubisco promoters that we used to express HSA in our system (Figure 3.6).

3.5.4
Inhibition of Endogenous Gene Expression

Endogenous Rubisco genes can be inhibited during sprouting through the use of the antibiotic streptomycin. Timing is critical to achieve strong inhibition. If streptomycin is added too early, sprout development is delayed, whereas if it is added too late, the storage reserves will already have been used for endogenous protein expression and no increase in recombinant protein expression is detectable.

The best results are obtained when 100 mg L^{-1} streptomycin is added 48–50 h after the initiation of germination. With streptomycin treatment, 100–400% increases in recombinant protein expression have been obtained. The accumulation of both Rubisco subunits is prevented (Figure 3.7). The specific activity of GUS increases 2.5-fold when streptomycin is used (Figure 3.8).

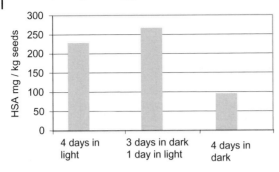

Fig. 3.6 Human serum albumin (HSA) was produced during the sprouting of transgenic rapeseeds. A light-inducible Rubisco promoter was used to control transgene expression. The highest yield was obtained with light induction after three days of continuous darkness. Sprouting was carried out in an airlift fermentor at room temperature. The sprouting medium was water.

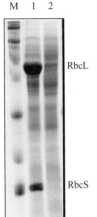

Fig. 3.7 Transgenic rapeseeds were sprouted in an airlift tank with (lane 1) and without (lane 2) of streptomycin at 100 mg L^{-1}. Total proteins were extracted, separated by SDS-PAGE and stained with Coomassie blue. The synthesis of Rubisco large and small subunits was inhibited as clearly shown in lane 2.

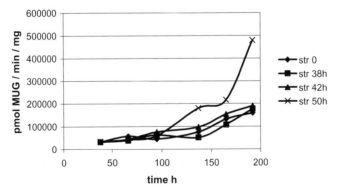

Fig. 3.8 Transgenic rapeseeds expressing the *gusA* reporter gene were germinated in an airlift tank with streptomycin added to the medium (100 mg L^{-1}). Streptomycin was added 0, 38, 42 or 50 h after germination. When streptomycin is added after 50 h, a 2.5-fold increase in GUS activity can be seen. This indicates the importance of correct timing when streptomycin is added to inhibit endogenous Rubisco gene expression.

3.5.5
Growth Regulators

Sprout size can be regulated with auxins and cytokinins. By adding such growth regulators to the sprouting medium, root and hypocotyl extension can be decreased
without affecting cotyledon growth. The advantage of this strategy is that the sprouts
are smaller and more can be grown in same volume. Gibberilic acid (GA) is used in
malting to synchronize the initiation of germination. Seed germination in our system is well synchronized even without GA. There were no significant benefits observed when GA was added to the medium.

3.5.6
Nitrogen Fertilizer

Potassium nitrate (KNO_3) increases the growth rate of sprouts and leads to higher levels of recombinant HSA (Figure 3.9). Germination becomes uneven and is slowed
down if 20 mM KNO_3 is added at the beginning of the process, but when nitrogen is
added 6 h after germination begins there are no negative effects on the sprouting
rate. The effect on sprout size is clear. Differences are visible after 50 h of treatment.
After 96 h of treatment, the average weight of a sprout treated with KNO_3 is more
than twice that of a sprout grown in unsubstituted water. It is known that the availability of nitrogen and carbon during *Arabidopsis* seedling growth regulates photosynthetic gene expression and storage lipid mobilization [30].

3.5.7
Seed Production

Seed quality can be controlled much more easily under greenhouse conditions than
in the field. Environmental factors, such as insects, drought and thunderstorms, do
not affect the harvest. Also the growth conditions are easier to optimize. Fertilizers,
water availability, temperature, light and soil quality all need to be optimized for the
maximum yield and quality of seeds. To increase the seed protein content it is beneficial to use high-nitrogen fertilizers early in development and initiate early ripening
using dry conditions in the greenhouse.

The use of a high-containment-level greenhouse for seed production is an effective
way to prevent pollen escape and outcrossing with natural relatives or food producing plant lines. Inside the greenhouse unit, air is circulated through pollen im-
permeable filters and wastewater is collected for heat sterilization before disposal
into a communal sewer system. Separate transgenic lines are grown in separate
greenhouse units to prevent unwanted cross-pollination.

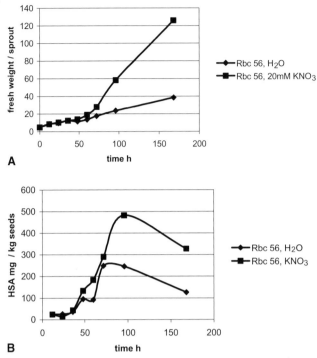

Fig. 3.9 The positive effect of nitrogen fertilizer (potassium nitrate) on the growth and productivity of transgenic sprouts (**A.**) The growth rate, measured as the increase in fresh weight during sprouting. (**B.**) The yield of recombinant HSA, as determined using an enzyme-linked immunosorbent assay (ELISA). The sprouts were germinated in an airlift bioreactor tank for 175 hours in the presence or absence of 20 mM KNO_3. Recombinant HSA was expressed under the control of the Rbc56 promoter isolated in our laboratory.

3.6
Yield Estimates and Benefits of Sprouting Technology in Protein Production

3.6.1
Yield Estimates

According to FAO statistics, the average field harvest of oilseed rape in Europe is 2500 kg ha^{-1} [8] and based on our own experiments, 3000 kg ha^{-1} yields can be obtained in the greenhouse. This translates into an annual harvest potential of 9000 kg ha^{-1}, based on the fact that three harvests each of 3000 kg ha^{-1} can be obtained in a year. The desired protein is produced predominantly in the sprout. The developing seeds do not accumulate recombinant protein.

Because the total protein content of a rapeseed is 25% dry weight, the total protein production level would be 2250 kg ha^{-1} in 1 ha of greenhouse space per year. If the recombinant protein of interest was expressed at the level of 5% total protein, the overall production yield would be 112.5 kg per year.

Currently, we can produce 1.5 g of GUS protein and 0.5 g of HSA per kg of seeds. These levels correspond to 13.5 kg and 4.5 kg production yields per year, respectively. Expression levels are indicated as a weight per weight of starting material (dry seeds) because the sprout fresh weight and total soluble protein content varies depending on cultivation time and conditions.

An airlift tank of 1000–2000 L would be sufficient to sprout the 9000 kg of seeds produced per year in a 1-ha greenhouse. Experimentally, a 70-L airlift tank has been used and a few hundred kg of sprouts have been produced. In the food industry, sprouts are currently produced at the rate of 500 tons per year in the USA and 400 tons per year in Japan. The seeds are harvested easily and the volume of sprout material is manageable because harvesting whole plants is not necessary. This avoids the processing of large volumes of transgenic material. Characteristic benefits of our system in terms of large-scale production can be summarized as follows:

- *Short production cycle time*: Seeds can be sprouted at a relatively high temperature, which reduces the production cycle time to only 2–5 days. This compares favorably to the growth of transgenic plants in open fields (months) or to the production of pharmaceuticals in mammalian cells (weeks).
- *Batch production*: Sprouting is carried out in batches, so the product quality is more consistent than that obtained during the continuous processing of biomass as is necessary with whole plant material. One batch in the bioreactor is a production lot of the protein.
- *Easy process control and low costs*: The use of contained sprouting technology provides good process control. Costs are kept down by the use of water and simple nitrogen compounds as a growth medium. The seeds have natural intracellular storage reserves for initial growth so that complex and expensive media, as used in mammalian cell bioreactors, are unnecessary. Sprouting is also a hygienic process as the seeds are surface sterilized before sprouting.
- *Efficient seed production and high yields*: Oilseed plants are efficient seed producers. The number of seeds per plant is large, and the seeds are small. The high seed protein content results in high yields of recombinant proteins.
- *Easy handling and storage of seeds*: Seeds are small, easy to handle and remain stable during long-term storage. Therefore, in contrast to leaves (which decompose rapidly leading to protein degradation), rapeseeds can be handled and transported without specialized equipment. Transgenic seeds can be stored at a cool temperature and normal humidity and they can be used on demand.
- *Camelina sativa*, a rare and unexploited crop plant: *Camelina sativa* is grown as a crop plant only in Finland and Ireland. Because it is a self-pollinating plant the risk of inadvertently transferring the new trait to naturally occurring plant relatives in the environment is low. *Camelina sativa* has not been extensively used in plant breeding, which means that there are only few varieties of the plant.

3.6.2
Quality and Environmental Aspects

Sprouting technology is designed for the contained production of proteins. Safety and product quality are the main advantages of the system. Three complementary approaches have been adopted to ensure environmental safety during production.

- Containment: seeds are produced in greenhouse and sprouting takes place in an airlift tank.
- Choice of host plant: the plant is not widely cultivated in the area where the transgenic plants are grown, it has no wild relatives with which to cross and it is preferentially self-pollinating.
- Molecular control of gene escape: the use of specific genetic systems to prevent transgene escape into environment.

For environmental safety reasons, the potential of *Camelina sativa* as a producer plant has been studied. *Camelina sativa* is a self-pollinating cruciferous oil-producing plant. It originates from the area around the Mediterranean and Central Asia. *Camelina sativa* was used in Europe as a crop plant during Iron and Bronze Ages when it was an important complement to poppy and flax. Since that time, *Camelina sativa* has not been used extensively. The interest in *Camelina sativa* was renewed in the 1980s as an alternative crop plant due to its early growing season and high yields. Today, *Camelina sativa* is grown mainly for its oil, which has a fatty acid composition favorable to human health. UniCrop has patented a transformation method for *Camelina sativa*.

In addition to the prevention of pollen escape, greenhouse cultivation allows better control and management of crop growth. Seed quality is the first aspect of product quality that needs to be addressed. Variation between sprouting batches is minimized through the use of high-quality seeds. Synchronized sprouting times and uniform sprout quality is achievable only when high-quality seeds are used. The yields of recombinant proteins expressed in sprouts has been high, and we are now focusing on the functional properties of the products, using antibodies and certain enzymes as models. Sprouts are known to have a very simple tissue structure, and the uniformity of glycosylation patterns for proteins produced in sprouts is an essential target for future analysis.

To further ensure the environmental safety UniCrop's GM production plants, the company has developed a specific Transgene Escape Prevention System (patent pending) based on molecular techniques [31, 32].

References

[1] www.isga-sprouts.org.

[2] J.D. BEWLEY, M. BLACK, *Seeds: Physiology of Development and Germination*, Plenum Press, New York and London, **1994**.

[3] A. VON ARNIM, X.W. DENG, *Annu. Rev. Plant Physiol. Plant Mol. Biol.* **1996**, *47*, 215–243.

[4] K. MÜNTZ, *Plant Mol. Biol.* **1998**, *38* (1–2), 77–99.

[5] G. FINCHER, *Annu. Rev. Plant Physiol. Plant Mol. Biol.* **1989**, *40*, 305–346.

[6] M. DeBLOCK, D. DeBROUWER, P. TENNING, *Plant Physiol.* **1989**, *91*(2), 694–701.

[7] M.M. MOLONEY, J.M. WALKER, K.K. SHARMA, *Plant Cell Rep.* **1989**, *8* (4), 238–242.

[8] www.fao.org.

[9] M.L. CROUCH, I. SUSSEX, *Planta* **1981**, *153* (1), 64–74.

[10] J. BARCISZEWSKI, M. SZYMANSKI, T. HAERTLÉ, *J. Protein Chem.* **2000**, *19* (4), 249–254.

[11] K. MÜNTZ, *J. Exp. Botany* **1996**, *47* (298), 605–622.

[12] P. ROZAN, Y. KUO, F. LAMBEIN, *Phytochemistry* **2001**, *58* (2), 281–289.

[13] W. RATAJCZAK, T. LEHMANN, W. POLCYN et al., *Acta Physiol. Plantarum* **1996**, *18* (1), 13–18.

[14] H. EICHELMANN, A. LAISK, *Plant Physiol.* **1999**, *119* (1), 179–189.

[15] R. TANG, J. JIA, L. LI, *Acta Phytohys. Sinica* **1997**, *23* (4), 337–341.

[16] C.L. CHENG, G.N. ACEDO, M. CRISTINSIN et al., *Proc. Natl Acad. Sci. USA* **1992**, *89* (5), 1861–1864.

[17] S. RODERMEL, J. HALEY, C.-Z. JIANG et al., *Proc. Natl Acad. Sci. USA* **1996**, *93* (9), 3881–3885.

[18] A. DEDONDER, R. RETHY, H. FREDERICQ et al., *Plant Physiol.* **1993**, *101* (3), 801–808.

[19] M. SUGITA, T. MANZARA, E. PICHERSKY et al., *Mol. Gen. Genet.* **1997**, *209* (2), 247–256.

[20] X.-W. DENG, W. GRUISSEM, *Cell* **1987**, *49* (3), 379–387.

[21] J.E. MULLET, R.R. KLEIN, *EMBO J.* **1987**, *6* (6), 1571–1579.

[22] D. MOAZED, H.F. NOLLER, *Nature* **1987**, *327* (6121), 389–394.

[23] G. HARTMANN, K.O. HONIKEL, F. KNÜFSEL et al., *Biochim. Biophys. Acta* **1967**, *145* (3), 843–849.

[24] T. PFANNSCHMIDT, K. OGREZEWALLA, S. BAGINSKY et al., *Eur. J. Biochem.* **2000**, *267* (1), 253–261.

[25] D.E. MATHEWS, R.D. DURBIN, *J. Biol. Chem.* **1990**, *265* (1), 493–498.

[26] S. RODERMEL, M. ABBOTT, L. BOGORAD, *Cell* **1988**, *55* (4), 673–681.

[27] H.-S. GUO, J.-F. FEI, Q. XIE et al., *Plant J.* **2003**, *34* (3), 383–392.

[28] F.G. MARTELEY, S.R. JAYASHANKAR, R.C. LAWRENCE, *J. Appl. Bacteriol.* **1970**, *33* (2), 363–370.

[29] M. SACK, C. VAQUERO, R. SRIRAMAN et al., *Bio Tech. International* **2002**, *14* (4), 12–15.

[30] T. MARTIN, O. OSWALD, I.A. GRAHAM, *Plant Physiol.* **2002**, *128* (2), 472–481.

[31] V. KUVSHINOV, K. KOIVU, A. KANERVA et al., *Plant Sci.* **2001**, *160* (3), 517–522.

[32] V. KUVSHINOV, K. KOIVU, A. KANERVA et al., US patent application 20020007500, **2001**.

[33] T. SASANUMA, *Mol. Genet. Genomics* **2001**, *265* (1), 161–171.

[34] N.O. POLANS, N.F. WEEDEN, W.F. THOMPSON, *Proc. Natl Acad. Sci. USA* **1985**, *82* (15), 5083–5087.

[35] F.P. WOLTER, C.C. FRITZ, L. WILLMITZER et al., *Proc. Natl Acad. Sci. USA* **1988**, *85* (3), 846–850.

[36] M.A. GRANDBASTIEN, S. BERRY-LOWE, B.W. SHIRLEY et al., *Plant Mol. Biol.* **1986**, *7* (6), 451–465.

[37] A.M. NANTEL, F. LAFLEUR, R. BOIVIN et al., *Plant Mol. Biol.* **1991**, *16* (6), 955–966.

[38] I. BECK, C. WITTIG, G. LINK, *Bot. Acta* **1995**, 108 (4), 327–333.

[39] W.D. TANG, S.S.C. SUN, *Plant Mol. Biol.* **1993**, *21* (5), 949–951.

4
Monocot Expression Systems for Molecular Farming

PAUL CHRISTOU, EVA STOGER and RICHARD M. TWYMAN

4.1
Introduction

Among the many different agricultural expression systems that can be used for the large-scale production of recombinant proteins, field-grown monocot crops, and more specifically cereals, represent one of the most attractive options [1,2]. The major advantage of cereals is that recombinant proteins can be targeted to accumulate specifically in the seed, e.g. in the embryo, the aleurone layer or the endosperm. The endosperm is particularly suitable because this tissue has evolved for the accretion of storage products, including proteins [3]. Therefore, it contains a well-developed endomembrane system, within which molecular chaperones and disulfide isomerases help to fold proteins correctly, while the desiccated environment in the mature seed protects the stored proteins from degradation [3,4]. In the best cases, recombinant proteins expressed in seeds have remained stable and active after storage at room temperature for more than two years [1,2,4]. The recombinant protein reaches high concentrations in a small volume of tissue, which facilitates extraction and downstream processing. In terms of processing, other advantages of cereal seeds include the lack of phenolic compounds and the relatively simple proteome. Phenolic compounds, such as the alkaloids present in tobacco and the oxalic acid found in alfalfa, can interfere with processing steps, while the presence of many competing proteins can result in the co-purification of non-target endogenous polypeptides with similar properties to the target recombinant protein. A final advantage of seeds is that proteins restricted to seeds do not interfere with the growth of vegetative organs, so even if they could adversely affect the plant, then normal growth is possible. In terms of biosafety, seed expression limits the exposure of non-target organisms, such as pollinating insects, microbes in the rhizosphere or herbivores feeding on leaves, to biologically active recombinant proteins. Furthermore, three of the cereal crops considered below are self-pollinating, reducing the likelihood of outcrossing to non-transgenic crops and wild relatives [5].

Four cereal crops have thus far been utilized for the production of recombinant proteins: maize, rice, wheat and barley. It is notable that, despite the attention given to tobacco, oilseed rape and potatoes as major expression systems, the only cultivated

Molecular Farming. Edited by Rainer Fischer, Stefan Schillberg
Copyright © 2004 WILEY-VCH Verlag GmbH & Co. KGaA, Weinheim
ISBN: 3-527-30786-9

crop currently used to produce commercial recombinant proteins is maize [6,7]. In this chapter, we discuss the relative merits of the four cereal expression systems, consider some of the technical issues surrounding recombinant protein expression in cereal grains, and finally discuss examples of recombinant proteins that have been produced in cereals. Several factors need to be weighed up when choosing the most appropriate cereal expression host for any given protein. These include geographical considerations for crop growth and harvest, the ease of transformation and regeneration, the annual yield of seed per hectare, the yield of recombinant protein per kilogram of seed, the producer price of the crop, the percentage of the seed that is made up of protein and, inevitably, intellectual property issues. Together, these determine the overall cost of production.

4.2
Cereal Production Crops

All four of the cereal production crops have advantages for molecular farming, but maize was the first to be developed into a platform expression system thanks largely to the efforts of scientists at Prodigene Inc., College Station, TX. Maize was chosen over the other cereals because it has the largest annual grain yield (approximately 8300 kg ha^{-1}) but also a relatively high seed protein content (10%), and these properties together offer the highest potential recombinant protein yields per hectare. Maize is also relatively easy to transform and manipulate in the laboratory, while its short generation interval (normally about 17 weeks) facilitates rapid production scale-up in the field. Maize is the most widely cultivated crop in North America, so the complex infrastructure for growing, harvesting, processing, storing and transporting large volumes of corn seed is already in place. Prodigene currently has two commercial products that are expressed in maize [6,7] and is developing further lines for the production of a range of pharmaceutical and technical proteins, including recombinant antibodies, vaccine candidates and enzymes [8–10].

Rice has many advantages in common with maize including the high grain yields, the ease of transformation and manipulation in the laboratory, and the capacity for rapid scale up [11]. At 6600 kg ha^{-1}, the annual grain yield of rice is slightly lower than that of maize, and the seed also has a slightly lower protein content (8%). Importantly, however, rice has emerged as the model cereal species, and is one of the few terrestrial plants to benefit from a completed genome sequence [12,13]. Therefore, many useful expression cassettes have been developed based on endogenous rice genes, including constitutive actin promoters, glutelin promoters for seed-specific expression and the α-amylase inducible promoter system which is often employed in rice suspension cell cultures [14]. The major disadvantage of rice compared to maize is that the producer price is significantly higher, so it may be excluded as a mainstream expression host in the West simply on economic grounds. However, the story may be different in Asia and Africa, where rice is traditionally grown. Rice, along with barley (see below) has been adopted as a platform production technology by Ventria Bioscience, Sacramento, CA.

Wheat has been used only rarely for molecular farming, probably because the technology for gene transfer and regeneration is not so well advanced as in other cereals [4]. The major advantage of wheat for the commercial production of recombinant proteins is its very low producer price compared to maize and rice. The wheat grain also has a higher protein content than these other cereals (>12%). Unfortunately, wheat also has a much lower grain yield than maize and rice (2800 kg ha^{-1} yr^{-1}) which means that more land is required to produce the same amount of protein. While some recombinant proteins, including pea legumin, have been produced at high levels in wheat endosperm, only low yields have been achieved with antibodies, although this situation may change in the future if better expression cassettes can be developed.

Barley has much in common with wheat as a host for molecular farming. It has a low producer price, a high seed protein content (about 13%) and the technology for gene transfer and regeneration is not widely disseminated. Unlike wheat, however, there have been some very encouraging reports of high recombinant protein yields in transgenic barley, including a diagnostic antibody that accumulated to 150 μg g^{-1} seed weight [15] and a recombinant cellulase that accumulated to 1.5% total seed protein [16]. Further recombinant proteins will need to be expressed in barley before its performance as an expression platform can be judged against maize and rice.

4.3
Technical Aspects of Molecular Farming in Cereals

4.3.1
Cereal Transformation

Until the 1990s, cereals were thought to be outside the host range of *Agrobacterium tumefaciens*, so researchers concentrated on alternative transformation methods. Initially, attempts were made to regenerate cereal plants from transformed protoplasts following PEG-mediated DNA transfer or electroporation. After successful experiments using model dicots, protoplast transformation was achieved in wheat [17] and the Italian ryegrass *Lolium multiflorum* [18]. In each case, transgenic callus obtained but it was not possible to recover transgenic plants, probably because monocot protoplasts loose their competence to respond to tissue culture conditions as the cells differentiate. In cereals and grasses, this has been addressed to a certain extent by using embryogenic suspension cultures as a source of protoplasts. Additionally, since many monocot species are naturally tolerant towards kanamycin, the *npt*II marker used in the initial experiments was replaced with alternative markers conferring resistance to hygromycin or phosphinothricin. With these modifications, it has been possible to regenerate transgenic rice and maize of certain varieties with reasonable efficiency [19–21]. However, the extended tissue culture step is unfavorable, often resulting in sterility and other phenotypic abnormalities in the regenerated plants.

An alternative procedure for plant transformation was introduced in 1987, involving the use of a modified shotgun to accelerate small (1–4 μm) metal particles into plant cells at a velocity sufficient to penetrate the cell wall (~250 m s^{-1}) [22]. The original device was gunpowder-driven and rather inaccurate. This has been replaced by a commercially-available pressurized helium gun, resulting in greater control over particle velocity and hence greater reproducibility of transformation conditions. In addition, an apparatus based on electric discharge [23] has been useful for the development of variety-independent gene transfer methods for the more recalcitrant cereals. There appears to be no intrinsic limitation to the scope of particle bombardment since DNA-delivery is governed entirely by physical parameters. Many different types of plant material have been used as transformation targets, including callus pieces, cell suspension cultures and organized tissues such as immature embryos and meristems. Almost all of the commercially-important cereals have been transformed by particle bombardment, including maize [24], rice [25], wheat [26] and barley [27].

During the 1980s, it was shown that asparagus [28] and yam [29] could be infected with *A. tumefaciens* and that tumors could be induced. In the latter case, an important factor in the success of the experiment was pre-treatment of the *A. tumefaciens* suspension with wound exudate from potato tubers (*A. tumefaciens* infection of monocots is inefficient because wounded monocot tissues do not produce phenolic compounds such as acetosyringone at sufficient levels to induce *vir* gene expression). Attention then turned towards the cereals. The first species to be transformed was rice. The *npt*II gene was used as a selectable marker, and successful transformation was demonstrated both by the resistance of transgenic callus to kanamycin or G418, and the presence of T-DNA in the genome [30]. However, these antibiotics interfere with the regeneration of rice plants, so only four transgenic plants were produced. The use of an alternative marker conferring resistance to hygromycin allowed the regeneration of large numbers of transgenic japonica rice plants [31], and the same selection strategy has been used to produce transgenic rice plants representing the remaining subspecies, indica and javanica [32,33]. More recently, efficient *Agrobacterium*-mediated transformation has become possible for other cereals discussed in this chapter: maize [34], wheat [35] and barley [36].

The breakthrough in cereal transformation using *A. tumefaciens* reflected the recognition of several key factors required for efficient infection and gene transfer. The use of explants containing a high proportion of actively dividing cells, such as embryos or apical meristems, was found to increase transformation efficiency greatly, probably because DNA synthesis and cell division favor the integration of exogenous DNA. In dicots, cell division is induced by wounding, whereas wound sites in monocots tend to become lignified. Hiei *et al.* [31] showed that the co-cultivation of *A. tumefaciens* and rice embryos in the presence of 100 mM acetosyringone was a critical factor for successful transformation. Transformation efficiency can be increased further by the use of vectors with enhanced virulence functions. The modification of *A. tumefaciens* to boost virulence has been achieved by increasing the expression of *vir*G (which in turn upregulates the expression of the other *vir* genes) and/or the expression of *vir*E1, which is a major limiting factor in T-DNA transfer [37]. Komari *et al.* [38] used a different strategy, in which a portion of the virulence region from the Ti-plasmid of supervirulent strain A281 was transferred to the T-DNA-carrying plasmid to generate a so-called superbinary vector. The advantage of the latter technique is that the superbinary vector can be used in any *Agrobacterium* strain.

4.3.2
Expression Construct Design

The most important aspect of construct design for molecular farming in cereals is the promoter used to drive transgene expression. In dicots, the cauliflower mosaic virus (CaMV) 35S promoter is the most popular choice because it is strong and constitutive, and therefore drives high-level transgene expression in any relevant organ, including seeds [39,40]. The promoter can be made even more active by various modifications, such as duplicating the enhancer region [41]. Even so, this promoter shows a very low activity in cereals.

Although some plant promoters appear to be just as active in both dicots and monocots [42–44], most promoters that are active in dicots need to be modified in some way before they work efficiently in cereals. One general type of modification that appears to work well with a range of promoters is the addition of an intron, usually in the untranslated region between the promoter and the initiation codon of the transgene open reading frame. Several different introns have been used to modify the CaMV 35S promoter and this has led to improved promoter activity in all four of the cereal host species discussed above, as well as in some grasses. For example, four monocot introns were tested with the CaMV 35S promoter in transgenic maize and bluegrass by Vain et al. [45], specifically those from the *Adh1*, *Sh1*, *Ubi1* and *Act1* genes, as well as a dicot intron (*chsA*). In this comparison, the *Ubi1* intron provided the highest level of enhancement in both species on average, but the dicot *chsA* intron perhaps surprisingly achieved a nearly 100-fold enhancement. Other monocot and dicot introns that have been tested with the CaMV 35S promoter include those from the *Bz1*, *cat1* and *Wx* genes [46–49]. The activity of another dicot promoter, the potato *pin2* promoter, was greatly enhanced in transgenic rice by inserting the first intron from the rice *Act1* gene [50]. More recently, Waterhouse and colleagues [51] adapted their novel pPLEX series of expression constructs, which are based on regulatory elements from subterranean clover stunt virus (SCSV), to be used in monocots. This was achieved by adding either the *Ubi1* or *Act1* introns, as well as GC-rich enhancer sequences from banana bunchy top virus (BBTV) or maize streak virus (MSV). An alternative to dicot promoters enhanced with introns is to use constitutive monocot promoters, of which the maize *Ubi1* promoter (with first intron) is the most widely utilized [52].

Various seed-specific promoters, many derived from seed storage protein genes, have been employed to restrict recombinant protein expression to different parts of the seed. Early examples include the maize *zmZ27* zein promoter, a maize *Waxy* promoter and the rice small subunit ADP-glucose pyrophosphorylase promoter, all of which are endosperm-specific in maize, and the rice glutelin 1 (*Gt1*) promoter, which is endosperm-specific in rice and maize [53–55]. More recently, recombinant proteins have been expressed in maize using the embryo-preferred globulin-1 promoter [56], in rice using an endosperm-specific globulin promoter [57] and two aleurone-specific promoters from barley [58], and in barley using the wheat *Bx17* high-molecular-weight glutenin promoter [59]. Care must be taken, however, because although these promoters are often described as 'seed-specific', a low level of activity may be present in other tissues. One pertinent example is the maize *Waxy* promoter, which is endosperm-preferred but shows a low level of activity in pollen [53].

In addition to an active promoter, a strong polyadenylation signal is required for transcript stability [1,2]. For molecular farming in cereals, terminators derived from the CaMV 35S transcript, the *A. tumefaciens nos* gene, and the potato *PinII* (protease inhibitor II) gene have been popular choices. The structure of the 5′ and 3′ untranslated regions should be inspected for AU-rich sequences, and these should be removed where possible [60]. Such sequences can act as cryptic splice sites, cryptic polyadenylation sites and mRNA instability elements. Some sequences, such as the 5′ leader of the petunia chalcone synthase gene and the 5′ leader of tobacco mosaic virus RNA (also known as the omega sequence) have been identified as translational enhancers although they may not always work effectively in cereals. Further important factors that influence translation include the presence of a single AUG codon within a consensus Kozak sequence, since multiple AUG codons, where present, often result in pausing and inefficient translational initiation. Different species also have very different codon preferences when specifying degenerate amino acids. Taking the amino acid arginine as an example, the codon CGU is preferred in alfalfa, and is 50 times more likely to occur than the rarest codon, CGG. In contrast, both of these codons are equally prevalent in maize, but the preferred choice is CGC. The expression of foreign transgenes in cerelas can therefore be very inefficient if infrequently-used codons predominate [60].

One critical consideration for the improvement of protein yields is subcellular protein targeting, because the compartment in which a recombinant protein accumulates strongly influences the interrelated processes of folding, assembly and post-translational modification [61]. For protein accumulation in cereal endosperm, the most important destinations are the protein storage organelles, i.e. the protein bodies and protein storage vacuoles, since these have developed to facilitate stable protein accumulation [3]. In rice, it has been shown that the two major classes of storage proteins, prolamins and glutelins, accumulate in different storage compartments, and are sorted in distinct ways [62]. Prolamins accumulate in protein bodies inside the rough endoplasmic reticulum (ER), and these eventually bud off to form separate organelles, whereas glutelins accumulate in protein storage vacuoles, which are derived from the smooth ER, and are conveyed to these organelles by transport vesicles budding from the Golgi apparatus [3]. There have been few studies of recombinant protein localization within the endosperm, but in one interesting report a recombinant antibody targeted to the secretory pathway in rice endosperm, and tagged for retrieval to the ER with a KDEL tetrapeptide tag, was found mainly in protein bodies and to some minor extent also in protein storage vacuoles [63]. The lack of a KDEL sequence in endosperm-expressed proteins generally results in secretion to the apoplast, where in most cases the recombinant protein accumulates in the space under the cell wall. A variety of signal peptides has been used to target proteins to the secretory pathway, including endogenous signal peptides from the transgene (e.g. immunoglobulin signal peptides for recombinant antibodies) [4,63] and heterologous plant signal peptides such as the barley α-amylase signal sequence (BAASS) [56].

4.3.3
Production Considerations for Cereals

Most of the useful information concerning the practical and commercial aspects of molecular farming in cereals has come from ProdiGene Inc., which has conducted detailed studies into the economic aspects of molecular farming in maize [64,65]. The development of a product line takes about 30–35 weeks, including 4–6 weeks for vector construction, 5–7 weeks to identify transformants, 4–6 weeks for the regeneration of plants and transfer to greenhouse, and 12–17 weeks for the T_0 plants to reach maturity and set seed. At this point, T_1 seeds are removed for molecular analysis of the integrated transgene, in order to select transgenic events for further characterization. After a further 17 weeks, the T_1 plants are mature and seeds can be removed for protein extraction and scale-up. At this point, about 10 mg of protein can be obtained for preliminary analysis. After another 17 weeks, the T_2 plants are mature and T_3 seeds can be removed for analysis and scale-up. It is routine to obtain about 1 g of pure recombinant protein by this time, but theoretical yields can be up to 1 kg. Also at this time, molecular, genetic and biochemical analysis of samples makes it possible to establish a master seed line for future production. After 12–18 months, the T3 generation is mature, and it should be possible to obtain 100 g of recombinant protein on a routine basis. Production lines can then be optimized and established in the field.

Another practical issue that needs to be addressed is the compliance of cereal-based production with regulatory guidelines established by bodies such as APHIS and the USDA. Again, ProdiGene Inc. has taken a leading role, developing a procedure in which the producer crop is isolated from other crops, and identity preservation is used from planting through to product extraction, to prevent mingling and contamination. For pharmaceutical products, further strict regulations govern quality assurance and quality control [8].

4.4
Examples of Recombinant Proteins Produced in Cereals

Molecular farming is often subdivided into pharmaceutical and industrial components, the former dealing with medically relevant proteins (e.g. human blood products, antibodies, vaccine candidates) and the later with bulk enzymes, technical proteins and biopolymers. However, there is no strict boundary between these two areas, while other products do not fit clearly into either category. What is clear is that a large and increasingly diverse spectrum of recombinant proteins is being produced in cereals, mostly as laboratory or greenhouse experiments, but also in field trials and large-scale commercial enterprises.

4.4.1
ProdiGene and Maize

The first proteins from transgenic plants to reach commercial status were avidin and β-glucuronidase (GUS) both of which are used as diagnostic agents in molecular biology. An important principle demonstrated by these case studies is that molecular farming in cereals can be an economical alternative even when the natural source of a protein is abundant (i.e. egg whites for avidin, and *Escherichia coli* for GUS) and where a market is already established.

Avidin is an abundant glycoprotein that is routinely purified from hens' eggs. It is widely used in molecular biology due to its very high affinity for biotin. Recombinant avidin was expressed in transgenic maize to establish whether the molecular farming approach could compete with the established source [6]. In the best-expressing transgenic line, avidin represented over 2% of the aqueous protein extracted from dry maize seeds, equivalent to approximately 230 mg per kg of transgenic seed. The *Ubi1* promoter was used to drive transgene expression and the protein was targeted to the secretory pathway using the BAASS. Consequently, the mature protein accumulated in the intercellular spaces. The recombinant protein was shown to be nearly identical to native avidin in terms of molecular weight and biotin binding properties. The protein remained stable in seeds stored at 10 °C and was not affected by commercial processing practices (dry milling, fractionation and hexane extraction). Interestingly, avidin expression in transgenic maize plants correlated with partial or complete male sterility. The cost of producing avidin by molecular farming in maize is estimated to be only 10% of the cost of extraction from egg whites. Recombinant avidin produced in maize is now available from Sigma-Aldrich.

The β-glucuronidase enzyme (GUS) is widely used as a screenable marker in transformation and promoter analysis, so unlike avidin, its expression and function has been confirmed in virtually all plant species that have been transformed. The use of plants as a commercial source of the enzyme was first demonstrated by Witcher et al. [7]. The *Ubi1* promoter was used to drive transgene expression and the protein, lacking a targeting signal, accumulated in the cytosol. In the best maize lines, the recombinant enzyme accounted for up to 0.7% of water-soluble protein extracted from dry seed, equivalent to about 80 mg per kg dry seeds. Purified recombinant GUS was similar in molecular mass, physical and kinetic properties to native GUS isolated from *E. coli*. Transgenic seed containing recombinant GUS could be stored at an ambient temperature for up to two weeks and for at least three months at 10 °C without significant loss of enzyme activity, and could be subject to normal maize processing practices as discussed above without loss of GUS activity [66].

Transgenic maize lines have also been used to investigate the large-scale production of a variety of other products, although none has yet reached commercial status. Perhaps the closest to commercial production are the protease trypsin, the protease-inhibitor aprotinin and fungal laccase (which is used to modify lignin). The aprotinin gene, driven by the *Ubi1* promoter, was introduced into maize embryos by particle bombardment [67]. The protein was targeted to the extracellular matrix, and in the best-expressing lines, seeds accumulated recombinant aprotinin at levels up to

0.7% soluble protein. Biochemical analysis of the purified recombinant protein revealed similar properties in terms of molecular weight, N-terminal amino acid sequence, isoelectric point, and trypsin inhibition activity as native aprotinin. The cost of aprotinin production in transgenic plants is comparable to the cost of extracting the protein from its natural source. Trypsin and laccase have been expressed in maize using constitutive and seed-specific promoters [56, 68]. Trypsin expression presents a difficulty because the enzyme is a protease, so it could degrade endogenous proteins and itself (autoproteolysis) thus reducing yields. Therefore, the enzyme was expressed as an inactive precursor, trypsinogen, using the seed-specific promoter, a strategy which is patented by ProdiGene [69]. The laccase 1 isozyme from *Trametes versicolor* reached the highest expression levels (about 1% total soluble protein) when driven by the embryo-preferred globulin promoter, combined with the BAASS to target the protein for secretion to the cell wall matrix [56].

The production of a secretory antibody in maize has also been reported [8]. Immature maize embryos were transformed with *A. tumefaciens* containing five transgenes, encoding the four components of the antibody (heavy chain, light chain, joining chain, secretory component) and a selectable marker. The four antibody transgenes were driven by the *Ubi1* promoter in combination with the BAASS, and the transgenic seeds were analyzed by enzyme-linked immunosorbent assay to detect assembled antibodies, showing that the antibody accumulated to up to 0.3% total soluble protein in T_1 seeds. Based on ProdiGene's success with other products, the selection of high-performance lines and backcrossing should allow this yield to be increased as much as 70-fold over six generations. Maize has also been investigated as a potential source of oral recombinant vaccines, by expressing the Lt-B protein of enterotoxigenic *E. coli* [70, 71].

4.4.2
Recombinant Proteins Expressed in Rice

Rice grains were used only occasionally for the expression of recombinant proteins until this system was adopted as a production platform by Ventria Bioscience. In early studies, rice plants were used for the expression of α-interferon [72] and recombinant antibodies [4], while more recently a cedar pollen allergen has been expressed in grains [73]. The PI' promoter was used to drive α-interferon expression [72], the antibodies were expressed using either the maize *Ubi1* or enhanced CaMV 35S promoters (with the maize promoter showing five times the activity of the viral promoter) [4], while the cedar pollen antigen was expressed using the seed-specific *GluB-1* promoter, leader and signal peptide [73]. This construct was expected to direct the recombinant protein to type II protein bodies, but instead it accumulated in type I protein bodies, suggesting some competition between the signal sequence and the structure of the recombinant protein in determining its final destination.

In 2002, Ventria Bioscience published two papers describing the expression of human proteins in rice grains [74,75]. Lysozyme was expressed under the control of the glutelin 1 promoter, leader and signal peptide, and an expression level of 0.6% dry weight (equivalent to 45% of soluble proteins) was achieved [74]. As was the case for maize production lines, the superior transgenic events were selected and maintained over several generations, and showed no loss of expression. Furthermore, the product itself was similar to the native enzyme under all tests and was shown to be active against a laboratory strain of *E. coli*. Lactoferrin was expressed using the same regulatory elements as above [75] and accumulated to a level of 0.5% dry weight of dehusked grains. As for lysozyme, all biochemcial and functional tests showed that the protein was similar in its properties to the native enzyme, and receptor-binding activity in human Caco-2 cell cultures was conserved.

In a further interesting development, Yang *et al.* [76] describe the expression, in rice grains, of a REB transcriotion factor (rice endosperm bZIP), which acts upon the globulin (*Glb*) promoter. When the *Glb* promoter was placed upstream of the *gus*A gene for β-glucuronidase, GUS activity was 2–2.5 times higher in the presence of REB in a cell-based transient assay. This enhancement was abolished when the upstream promoter motif GCCACGT(A/C)AG was deleted from the *Glb* promoter, and a gain of activity was observed when this sequence was added to the normally unresponsive *Gt1* promoter. In transgenic rice grains expressing lysozyme under the control of the *Glb* promoter, the presence of REB resulted in a 3.7-fold increase in lysozyme accumulation.

4.4.3
Recombinant Proteins Produced in Wheat

As discussed above, wheat has been used only rarely for molecular farming. Thus far, the only example of a pharmaceutical protein produced in wheat is a single chain Fv antibody, which was expressed using the *Ubi1* promoter and achieved a maximum expression level of 1.5 μg g^{-1} dry weight [77]. Transgenic wheat producing *Aspergillus* phytase has also been reported [78].

4.4.4
Recombinant Proteins Produced in Barley

There are few reports of barley used for the production of recombinant proteins but some of those reports show high expression levels which, due to the low producer price of barley, could make this a viable production crop. In early reports, recombinant glucanase and xylanase were expressed at very low levels (0.004%) [79,80] but as stated above, a recombinant diagnostic antibody (SimpliRED, for the detection of HIV) has been expressed at levels exceeding 150 μg g^{-1} [15], and a recombinant cellulase enzyme was expressed at levels exceeding 1.5% total seed protein [16].

4.5
Conclusions

Cereal expression systems are among the most advantageous for field-based recombinant protein production, since they combine intrinsic biosafety features (self-pollination in rice, barley and wheat, seed-specific protein expression) with practical benefits, particularly the tendency for recombinant proteins expressed in seeds to remain stable and active for prolonged periods at ambient temperatures. The commercial success of the first maize-derived recombinant proteins produced by ProdiGene has demonstrated that plants, and cereals in particular, represent an economically viable production system which provides a real alternative to mammalian cells, microbial cultures and indeed other crop systems such as tobacco, oilseed rape and potato. It is likely that rice and barley will be the next crops to emerge as commercial production platforms, particularly through the efforts of Ventria Bioscience.

References

[1] J.K.-C. Ma, P.M.W. Drake, P. Christou, *Nature Rev. Genet.* **2003**, *4* (10), 794–805.

[2] R.M. Twyman, E. Stoger, S. Schillberg et al., *Trends Biotechnol.* **2003**, *21* (12), 570–578.

[3] K. Müntz, *Plant Mol. Biol.* **1998**, 38 (1–2) 77–99.

[4] E. Stoger, M. Sack, Y. Perrin et al., *Mol. Breeding* **2000**, *9* (3), 149–158.

[5] U. Commandeur, R.M. Twyman, R. Fischer, *AgBiotechNet* **2003**, *5* (ABN 110), 1–9.

[6] E.E. Hood, D.R. Witcher, S. Maddock et al., *Mol. Breeding* **1997**, *3* (4), 291–306.

[7] D.R. Witcher, E.E. Hood, D. Peterson et al., *Mol. Breeding* **1998**, *4* (4), 301–312.

[8] E.E. Hood, S.L. Woodard, M.E. Horn et al., *Curr. Opin. Biotechnol.* **2002**, *13* (6), 630–635.

[9] E.E. Hood, *Enzyme & Microbial Technol.* **2002**, *30* (3), 279–283.

[10] E.E. Hood, in: P. Christou. H. Klee (eds) *Handbook of Plant Biotechnology*, Wiley-VCH, NY (in press).

[11] P. Christou, *Rice Biotechnology and Genetic Engineering*, Technomic Publishing Co. Inc., Lancaster, PA

[12] J. Yu, S.N. Hu, J. Wang et al., *Science* **2002**, *296* (5565), 79–92.

[13] S.A. Goff, D. Ricke, T.H. Lan et al., *Science* **2002**, *296* (5565), 92–100.

[14] M. Terashima, Y. Murai, M. Kawamura et al., *Appl. Microbiol. Biotechnol.* **1999**, 52 (4), 516–523.

[15] P.H.D. Schunmann, G. Coia, P.M. Waterhouse, *Mol. Breeding* **2002**, *9* (2), 113–121.

[16] G.P. Xue, M. Patel, J.S. Johnson et al., *Plant Cell Rep.* **2003**, *21* (11), 1088–1094.

[17] H. Lorz, B. Baker, J. Schell, *Mol. Gen. Genet.* **1985**, *199* (2), 178–182.

[18] I. Potrykus, M.W. Saul, J. Petruska et al., *Mol. Gen. Genet.* **1985**, *199* (2), 183–188.

[19] K. Shimamoto, R. Terada, T. Izawa et al., *Nature* **1989**, *338* (6212), 274–276.

[20] S.K. Datta, K. Peterhans, K. Datta et al., *Bio/Technol.* **1990**, *8* (8), 736–740.

[21] S. Omirulleh, M. Abraham, M. Golovkin et al., *Plant Mol. Biol.* **1993**, **21** (3), 415–428.

[22] T.M. Klein, E.D. Wolf, R. Wu et al., *Nature* **1987**, *327* (6117), 70–73.

[23] D. McCabe, P. Christou, *Plant Cell Tiss. Org. Cult.* **1993**, *33* (3), 227–236.

[24] T.M. Klein, M. Fromm, A. Weissinger et al., *Proc. Natl Acad. Sci. USA* **1988**, *85* (12), 4305–4309.

[25] P. Christou, T.L. Ford, M. Kofron, *Bio/Technol.* **1991**, *9* (10), 957–962.

[26] V. Vasil, A.M. Castillo, M.E. Fromm et al., *Bio/Technol.* **1992**, *10* (6), 667–674.

[27] Y.C. Wan, P.G. Lemaux, *Plant Physiol.* **1994**, *104* (1), 37–48.

[28] J.P. Hernalsteens, L. Thiatoong, J. Schell et al., *EMBO J.* **1984**, *3* (13), 3039–3041.

[29] W. Schafer, A. Gorz, G. Kahl, *Nature* **1987**, *327* (6122), 529–532.

[30] D.M. Raineri, P. Bottino, M.P. Gordon et al., *Bio/Technol.* **1990**, *8* (1), 33–38.

[31] Y. Hiei, S. Ohta, T. Komari et al., *Plant J.* **1994**, *6* (2), 271–282.

[32] H. Rashid, S. Yokoi, K. Toriyama et al., *Plant Cell Rep.* **1996**, *15* (10), 727–730.

[33] J.J. Dong, W.M. Teng, W.G. Buchholz et al., *Mol. Breeding* **1996**, *2* (3), 267–276.

[34] Y. Ishida, H. Saito, S. Ohta et al., *Nature Biotechnol.* **1996**, *14* (6), 745–750.

[35] M. Cheng, J.E. Fry, S.Z. Pang et al., *Plant Physiol.* **1997**, *115* (3), 971–980.

[36] S. Tingay, D. McElroy, R. Kalla et al., *Plant J.* **1997**, *11* (6), 1369–1376.

[37] S.B. Gelvin, *Annu. Rev. Plant. Physiol.* **2000**, *51*, 223–256.

[38] T. Komari, Y. Hiei, Y. Saito et al., *Plant J.* **1996**, *10* (1), 165–174.

[39] J. T. O'Dell, F. Nagy, N. H. Chua, *Nature* **1985**, *313* (6005), 810–812.

[40] M.A. Lawton, M.A. Tierney, I. Nakamura et al., *Plant Mol. Biol.* **1987**, *9* (4), 315–324.

[41] R. Kay, A. Chan, M. Daly et al., *Science* **1987**, *236* (4806), 1299–1302.

[42] B. Verdaguer, A. deKochko, R.N. Beachy et al., *Plant Mol. Biol.* **1996**, *31* (6), 1129–1139.

[43] P.M. Schenk, T. Remans, L. Sagi et al., *Plant Mol. Biol.* **2001**, *47* (3), 399–412.

[44] P.M. Schenk, L. Sagi, T. Remans et al., *Plant Mol. Biol.* **1999**, *39* (6), 1221–1230.

[45] P. Vain, K.R. Finer, D.E. Engler et al., *Plant Cell Rep.* **1996**, *15* (7), 489–494.

[46] J. Callis, M. Fromm, V. Walbot, *Genes & Dev.* **1987**, *1* (10), 1183–1200.

[47] A. Tanaka, S. Mita, S. Ohta et al., *Nucleic Acids Res.* **1990**, *18* (23), 6767–6770.

[48] S. Takumi, M. Otani, T. Shimada, *Plant Sci.* **1994**, *103* (2), 161–166.

[49] Y.Z. Li, H.M. Ma, J.L. Zhang et al., *Plant Sci.* **1995**, *108* (2), 181–190.

[50] D.P. Xu, D. Mcelroy, R.W. Thornburg et al., *Plant Mol. Biol.* **1993**, *22* (4), 573–588.

[51] P.H.D. Schunmann, B. Surin, P.M. Waterhouse, *Funct. Plant Biol.* **2003**, *30* (4), 453–460.

[52] A.H. Christensen, P.H. Quail, *Transgenic Res.* **1996**, *5* (3), 213–218.

[53] D.A. Russell, M.E. Fromm, *Transgenic Res.* **1997**, *6* (2), 157–168.

[54] Z.W. Zheng, N. Murai, *Plant Sci.* **1997**, *128* (1), 59–65.

[55] Z.W. Zheng, Y. Kawagoe, S.H. Xiao et al., *Plant J.* **1993**, *4* (2), 357–366.

[56] E.E. Hood, M. R. Bailey, K. Beifuss et al., *Plant Biotechnol. J.* **2003**, *1* (2), 129–140.

[57] Y.-S. Hwang, D. Yang, C. McCullar et al., *Plant Cell Rep.* **2002**, *20* (9), 842–847.

[58] Y.-S. Hwang, S. Nichol, S. Nandi et al., *Plant Cell Rep.* **2001**, *20* (7), 647–654.

[59] P.H.D. Schunmann, G. Coia, P.M. Waterhouse, *Mol. Breeding* **2002**, *9* (2), 113–121.

[60] R. Fischer, N.J. Emans, R.M. Twyman et al., in: M. Fingerman, R. Nagabhushanam (eds) *Recent Advances in Marine Biotechnology, Volume 9 (Biomaterials and Bioprocessing)*, Science Publishers Inc., Enfield NH, pp 279–313, **2003**.

[61] S. Schillberg, N. Emans, R. Fischer, *Phytochem. Rev.* **2002**, *1* (1), 45–54.

[62] T. Okita, S.S. Choi, H. Ito et al., *J. Exp. Bot.* **1998**, *49* (324), 1081–1090.

[63] E. Torres, P. Gonzalez-Melendi, E. Stoger et al. *Plant Physiol.* **2001**, *127* (3), 1212–1223.

[64] R.L. Evangelista, A.R. Kusnadi, J.A. Howard et al., *Biotechnol. Prog.* **1998**, *14* (4), 607–614

[65] A.R. Kusnadi, Z.L. Nikolov, J.A. Howard, *Biotechnol. Bioeng.* **1997**, *56* (5), 473–484.

[66] A.R. Kusnadi, R.L. Evangelista, E.E. Hood et al., *Biotech. Bioeng.* **1998**, *60* (1), 44–52.

[67] G.Y. Zhong, D. Peterson, D.E. Delaney et al., *Mol. Breeding* **1999**, *5* (4), 345–356.

[68] S.L. Woodard, J.M. Mayor, M.R. Bailey et al., *Biotechnol. Appl. Biochem.* **2003**, *38* (2), 123–130.

[69] J.A. Howard, E.E. Hood, US Patent 6,087,558, **1998**.

[70] S.J. Streatfield, J.R. Lane, C.A. Brooks et al., *Vaccine* **2003**, *21* (7–8), 812–815.

[71] B.J. Lamphear, S.J. Streatfield, J.A. Jilka et al., *J. Control. Release* **2002**, *85* (1–3), 169–180.

[72] Z. Zhu, K. Hughes, L. Huang et al., *Plant Cell Tiss. Org. Cult.* **1994**, *36* (2), 197–204.

[73] A. Okada, T. Okada, T. Ide et al., *Mol. Breeding* **2003**, *12* (1), 61–70.

[74] J. Huang, S. Nandi, L. Wu et al., *Mol. Breeding* **2002**, *10* (1–2), 83–94.

[75] S. NANDI, Y.A. SUZUKI, J. HUANG et al., *Plant Sci.* **2002**, *163* (4), 713–722.

[76] D. YANG, L. WU. Y.S. HUANG et al., *Proc. Natl Acad. Sci USA* **2001**, *98* (20), 11438–11443.

[77] E. STOGER, C. VAQUERO, E. TORRES et al., *Plant Mol. Biol.* **2000**, *42* (4), 583–590.

[78] H. BRINCH-PEDERSEN, A. OLESEN, S.K. RASMUSSEN, *Mol. Breeding* **2000**, *6* (2), 195–206.

[79] L.G. JENSEN, O. OLSEN, O. KOPS et al., *Proc. Natl Acad. Sci. USA* **1996**, *93* (8), 3487–3491.

[80] M. PATEL, J.S. JOHNSON, R.I.S. BRETTELL et al., *Mol. Breeding* **2000**, *6* (1), 113–123.

5

The Field Evaluation of Transgenic Crops Engineered to Produce Recombinant Proteins

JIM BRANDLE

5.1
Introduction

It has now been 16 years since the first mammalian protein was expressed in plants [1]. The concept of plant recombinant protein (PRP) production developed from that early experiment. The first reports of PRPs were published in the late 1980s when antibodies were produced in tobacco [2] and human serum albumin was expressed in tobacco and potato [3]. That work has now evolved into a burgeoning industry, which aims to produce a myriad of protein-based industrial and biopharmaceutical products in crop plants, aquatic plants and algae [4]. Crop plants are particularly suitable for the production of recombinant proteins because they offer what amounts to infinite scalability and low up-front production costs. Unlike fermentation systems, growing plants are not constrained by physical facilities. Recombinant proteins that had only been available in microgram quantities can now be produced in plants by the kilogram. However, if the advantages of scale are to be realized, then crops expressing genes for a wide range of recombinant proteins must leave the greenhouse for evaluation in confined field trials.

5.2
Regulation of Field-testing

Field-testing is a critical component of the commercialization and scale-up process for agricultural biotechnology. It provides an opportunity to examine the interaction between the transgene and the production conditions, to examine the environmental impact of the plants, to develop production projections and GMP processes, and to create material for pilot scale manufacturing, pre-clinical and early clinical experimentations. Field-testing played a major role in the introduction of the first generation of input traits from agricultural biotechnology, such as herbicide and insect resistance. More than 10,000 transgenic field trials have been conducted since 1989 and they provided insight into efficacy and agronomic performance for a wide range of transgenic crops [5]. Field-testing has been similarly important in the development of crops engineered for recombinant protein production, although there were few trials until the mid 1990s. As a result of those PRP field trials, two products (β-glucuronidase and avidin), both of which are laboratory reagents, are now commercially available. There is no large-scale commercial production of any PRP crop thus far, but as the many potential products move through the development process, large-scale production will follow.

In most countries, a permit is required to grow a genetically modified crop in the field. Typically, genetically modified means that a recombinant DNA technique has been used in the creation of that crop, but the exact definition does vary by country. In Canada the permitting body for field trials is the Canadian Food Inspection Agency (CFIA). In the United States it is the United States Department of Agriculture's Animal and Plant Health Inspection Service (USDA-APHIS), and similar organizations exist in Europe, Asia and South America [6]. Since 1990 there have been 246 field trial applications worldwide involving the production of industrial or biopharmaceutical proteins, with 76% of those trials conducted in the United States, 20% in Canada, 3% in Europe and 1% in Argentina (Table 5.1). The number of applications to field test PRP crops peaked in North America between 1995 and 1996 and again between 1999 and 2001. Following those peaks, the number of field trial applications in North America again began to decline, resulting in a cycling pattern of trial numbers. In Canada, the latest decline was attributed to a "pause" in field testing during 2001, while the CFIA developed new regulatory protocols for PRP crops. The Canadian moratorium has now been lifted and field test numbers have rebounded. In the United States, the current decline may reflect normal fluctuations in the product development cycle, which have been seen in previous years, or they may result from newly recognized liability issues. That liability is an issue became

Tab. 5.1 Summary of field trial applications and notifications for experiments involving transgenic plants expressing genes for recombinant proteins

Year	USA	Canada	Europe	Argentina
1990	0	0	0	0
1991	1	0	0	0
1992	1	0	0	0
1993	2	0	0	0
1994	4	1	0	0
1995	18	2	0	0
1996	19	11	1	0
1997	7	5	0	0
1998	22	2	1	1
1999	20	8	2	0
2000	32	6	0	1
2001	35	3	3	0
2002	12	6	1	0
2003	2	16	1	0
Total	175	60	9	2

apparent during what has become known as the "Prodigene Affair". In the fall of 2002, a small number of volunteer corn plants appeared in a soybean crop in the year following a field trial involving transgenic corn containing a biopharmaceutical. The trial was conducted by a Texas based crop biotechnology company called Prodigene [7]. The soybeans were harvested and may have contained traces of the dry tissue from the transgenic corn plants that grew in the field. As a part of their field trial permit requirements, Prodigene was required to have removed any volunteer corn plants, but failed to do so and received a substantial fine as a consequence. Although risk to the public was probably minimal, the incident did bring risks associated with PRP production to the forefront for public debate.

The regulation of crops producing pharmaceutical or industrial proteins was governed by generic directives in both the United States and Canada until 2003. The philosophy underlying the North America system is the regulation of the product and not the process, whereas in Europe it is the opposite. In either case the information required for field trials by the various regulatory agencies is similar [5]. For field testing PRP crops, proponents submitted applications that contained information relating to the unmodified plant, the modified plant, the construct used in the transformation, reproductive isolation, the nature of the site and its potential environmental impact, trial protocol, and post-harvest land use and monitoring. Each application was considered on a case-by-case basis and was assessed for environmental safety. Field trials for PRP crops typically required only environment safety assessment, which included such things as potential for weediness or invasiveness, gene flow to wild relatives, and non-target impacts or impacts on biodiversity. If the material produced in the trial was for use in early clinical experiments, then efficacy and safety would be evaluated by other regulatory agencies such as Heath Canada or the Food and Drug Administration (FDA) in the United States [8]. Otherwise these agencies are not involved in the field-testing of PRP crops. In-depth examinations of the principles used to regulate genetically modified crops around the world and the key issues associated with the assessment of ecological risk are provided in two recent reviews [5, 9].

Containment is essential for the field production of PRP crops, as was first identified by Menassa and co-workers who made a strong case for the use of male-sterile tobacco in order to keep PRP products out of the human food chain [10]. Recombinant proteins produced for medical or industrial use can have untoward biological activity in humans and other animals, and the risks of either acute or chronic exposure are often unknown. Therefore, it is critical to keep crops producing such materials out of the food and feed chain, and the environment. Commandeur *et al.* have detailed the risk factors specifically associated with the production of PRPs and made a distinction between the ecological/health risk and those associated with product safety [11]. They divided the risk associated with PRP crops into two categories: the first related to the spread of the transgene beyond the platform crop leading ultimately to human exposure and the second was product safety, which is the risk that the products of molecular farming *per se* were harmful to humans or other animals. They suggested numerous means to reduce the risk of transgene escape including the use of self-pollinated, non-food crops with minimum exposure to wild relatives along with various means to minimize the amount of genetic material introduced into the crop. The issues surrounding PRP crop production were discussed in depth at a recent Pew Initiative Workshop focused on the benefits and risks of plant recombinant pharmaceutical production [12]. The major benefit identified was the capability of plants to produce low-cost recombinant proteins. As a result of favorable costs, plant-based production systems could be used to meet the growing demand for new protein based medicines, particularly those that are too cost-prohibitive when produced in conventional mammalian cell culture systems. Risks associated with the escape of the PRP crop were recognized and concerns were expressed about the ability of industry to contain these crops and prevent their entry into the food chain. USDA regulators pointed to new, more stringent, field-testing regulations for PRP crops and indicated that PRP crops will be the subject of regulatory oversight throughout the research and commercial production cycle, all of which would serve to ensure containment.

From a field testing and field production standpoint, the risk profile for transgenic crops producing biopharmaceuticals and even industrial enzymes is quite different from that of the first generation transgenic crops now in unconfined release. Risk is the product of probability and consequences. Under a standard set of production conditions the probability of an adverse event, like pollen dispersal for example, is the same for PRP transgenics as it is for transgenics carrying first generation traits. However, the consequences of that event may be far worse, given that the recombinant protein may be biologically active in humans, and therefore the risk is higher. Recognizing that the risk level of PRP crop field trials using standard production methods is higher than that associated with conventional transgenic crops, both the US and Canadian regulatory authorities deviated from the use of generic regulations for field tests involving PRP crops, and in 2003 issued amended regulations. In the United States, USDA-APHIS strengthened their permit conditions by: requiring more inspection visits, increasing the amount of information required about harvest and storage procedures, implementing mandatory training programs for staff involved with the permit, increasing buffer zones from 800 to 1600 m for corn-based production, restricting the production of food and feed crops at the test site in the post-harvest growing season, increasing the size of the fallow zone from 7.5 to 15 m and requiring the use of dedicated equipment for harvest [13]. Miller was critical of the changes and felt that they were unnecessarily restrictive [14]. He argued that the new regulations were the result of the Prodigene affair and that the resulting real risk to consumers was "extremely low". Others, like the Union of Concerned Scientists, were calling for more stringent measures such as disallowing food crops as production platforms, indoor production, mandatory male sterility and designated grow zones [15].

In Canada the interim regulatory amendments governing field trials with "plants with novel traits" that produce industrial or pharmaceutical proteins were more conservative than those issued in the United States [16]. For example, the CFIA is recommending that major food or feed species, and crops pollinated by honeybees, are not to be used for PRP production. The use of non-food, fiber, small acreage or new crops is encouraged. They asked that proponents consider the potential for escape of the crop or the transgene into the environment when choosing a crop platform.

Plants used to produce PRPs should be "amenable to confinement". Isolation distances were increased, and the cultivation of food and feed crops following a PRP crop was discouraged. New hazard and exposure data for human and livestock health assessment may also be required from PRP-containing traditional food or feed crops prior to the approval of field trials. Exposure risk concerns the potential for PRPs to be present in human food or animal feed, and where exposure can occur, what mechanisms are used to limit biological activity. Hazards included direct toxicity and allergenicity in humans or animals as well as hazards presented by the co-product streams that result from processing. These latter requirements could place a major burden on proponents to prove their materials are safe prior to even confined field trials.

5.3
Design of Field Trials

There are three considerations in the design of field trials for PRP crops. The first is eliminating harm to human health, the second is mitigation of harm to the environment and the third is experimental design. Human exposure could result from accidental release of the material, which can be eliminated through careful material handling procedures and record keeping. In Canada for example, inspectors must witness the destruction of residual plant material and records of the dispensation of retained material must be kept. Other issues include the exposure of workers involved in the trial and they can be resolved through the use of protective gear or by harvest equipment design [17]. Ensuring that the environment is not harmed requires a clear understanding of the biology of the host crop species. Documents outlining taxonomy, genetics, reproductive biology, potential for inter-specific hybridization, weediness potential and field production systems are available from the regulatory authorities in Canada and the United States [18]. These documents help to determine areas of potential risk during the production cycle and proponents then develop means of limiting that risk. For example, although tobacco is not an outcrossing species, it is a prolific seed producer and trial protocols often include the removal of flowers or harvest prior to flowering to limit the potential for seeds to escape into the environment [10, 19]. Physical isolation is another method of limiting escape. In Canada and the United States, isolation distances for various crops used in the production of recombinant proteins are prescribed [13, 16]. Isolation can also be achieved by conducting trials outside of normal production zones for a given crop, although the feasibility of such a procedure is questionable, since the agricultural environment providing the isolation may not be suitable because it is isolated from the normal production zone. In addition to physical distance, male-sterility has been used in tobacco to re-enforce containment during the field-testing of tobacco plants producing proteins in their leaves [10]. Seed-based production systems such as corn have also used sterility-based containment for field-testing [20]. Many molecular containment methods have been developed but none have as yet been deployed in PRP crop field trials.

From a research perspective, field trials, like other types of experimentation, are designed around a hypothesis. For example, the yield of recombinant protein from a selected transgenic line is theoretically 300 kg ha^{-1}. Therefore, one may hypothesize that the yield will be the same in agricultural production and a procedure is designed to validate the hypothesis. There are four phases in this process, i.e. the selection of: test materials, traits to measure, measurement procedures, and data analysis methods [21]. In our example, the test materials would be any number of transgenic lines selected for high recombinant protein yields. The traits to measure could be the yield of target tissue or the concentration of recombinant protein, which together would provide the yield of recombinant protein per unit area. The remaining two phases represent the experimental design stage and allow experimental error to be estimated and controlled, and the results of the experiment to be interpreted. In field trials, experimental error is estimated through the use of replication and randomization. Multiple plots containing the same treatments, and assignments of plots at random to locations within the trial, allow the real differences between the treatments to be separated from experimental error. In this example, the trial should be repeated in space and time (locations and years) to provide a more accurate representation of recombinant protein yield in a normal range of agricultural conditions. The use of a process called blocking to account for environmental gradients in the field, along with optimum plot management and methods of data analysis help to control experimental error and enhance the researcher's ability to see treatment differences. The various experimental designs employed in field research, and their analysis and interpretation, are discussed by Gomez and Gomez [21].

5.4
Results of Field Trials

The few studies that have been published to date tend to include field trials as a part of large pilot studies aiming to demonstrate the feasibility of producing a particular protein in a given system. All of the trials used some reproductive measure like male-sterility or flower removal to control the spread of pollen. The results of multi-location experiments with corn expressing avidin were among the first to be published [20]. The trials were conducted throughout the United States and made use of off-season nurseries to speed up the trial process. The authors concluded that avidin accumulation in corn was stable across the multiple environments, although generations and locations appeared to be confounded. Gastric lipase was produced in a field test in France, but the focus of the experiments was extraction efficiency and characterization, and the results of field performance were not presented [18]. Tobacco was used to produce human interleukin-10 in a single location field trial conducted in Canada [10]. The authors examined the performance of five male sterile transgenic lines in an effort to estimate IL-10 production levels and to determine if male sterility was affecting crop vigor. Male sterility had no affect on biomass yield and IL-10 was produced at 0.5–1.0 g ha^{-1}. Lee and co-workers used field trials of transgenic white clover expressing a vaccine antigen-GFP fusion protein in order to produce material to test for storage stability [22]. The antigen present in the dried clover was found to be stable for as long as six months.

There has been a large number of field trials of transgenic crops that accumulate recombinant proteins, but the results of only a few have been reported so far. Of particular interest for the future are reports of detailed trial protocols that address confinement concerns; multi-location, multi-year performance and stability data; environmental impact and non-target impact studies; and GLP production. The impact of production conditions on extraction efficiency and product quality are also significant issues that should be addressed through field-testing.

References

[1] D. D. Lefebvre, B. L. Miki, J. F. Laliberte, *Bio/Technology* **1987**, *5* (10), 1053–1056.

[2] A. Hiatt, R. Cafferkey, K. Bowdish, *Nature* **1989**, *342* (6245), 76–78.

[3] P.C. Sijmons, B.M. Dekker, B. Schrammeijer et al., *Bio/Technology* **1990**, *8* (3), 217–221.

[4] G. Giddings, G. Allison, D. Brooks et al., *Nature Biotechnol.* **2000**, *18* (11), 1151–1155.

[5] J. P. Nap, P. L. J. Metz, M. Escaler et al., *Plant J.* **2003**, *33* (1), 1–18.

[6] Lists of field-test notifications in the European Union, the USA, Canada, Australia and Japan respectively. Available from: http://engl.jrc/it; http://www.nbiap.vt.edu/; http://www.inspection.gc.ca/; http://www.health.gov.au/ogtr/index.htm; www.s.affrc.go.jp/docs/sebtan/eguide/edevelp.htm. List of field tests applications for many countries in Asia, South America and Africa are available from: http://binas.unido.org/binas/home/php.

[7] B. Cassidy, D. Powell, Pharmaceuticals from plants: The Prodigene affair. **2002**. Available from: http://www.foodsafetynetwork.ca/gmo/prodigene.htm

[8] United States Department of Health and Human Services, Food and Drug Administration, Guidance for Industry. Drugs, biologics, and medical devices derived from bioengineered plants for use in humans and animals. **2002**.

Available from: http://www.fda.gov/cber/gdlns/bioplant.htm

[9] A. J. Conner, T. R. Glare, J. P. Nap, *Plant J.* **2003**, *33* (1), 19–46.

[10] R. Menassa, V. Nguyen, A. Jevnikar et al., *Mol. Breeding* **2001**, *8* (2), 177–185.

[11] U. Commandeur, R. M. Twyman, R. Fischer, *AgBiotechNet* **2003**, *5*, ABN110.

[12] M. Fernadez, L. Crawford, C. Hefferan, Pharming the field. A look at the benefits and risks of bioengineering plants to produce pharmaceuticals. Proceedings of workshop sponsored by the Pew Initiative on food and biotechnology and the USDA. **2003**. Available from: http://pewagbiotech.org/events/0717/Conference Report.pdf

[13] United States Department of Agriculture. USDA strengthens 2003 permit conditions for field testing genetically engineered plants. Available from: http://www.aphis.usda.gov/lpa/news/2003/03/gepermits_brs.html

[14] H.I. Miller, *Nature Biotechnol.* **2003**, *21* (5), 480–481.

[15] Union of Concerned Scientists, Pharm and industrial crops: The next wave of agricultural biotechnology. **2003**. Available from: http://www.ucsusa.org/food_and_environment/biotechnology/page.cfm?pageID=1033

[16] Canadian Food Inspection Agency. Interim amendment to DIR2000–07 for confined research field trials of PNTs for plant molecular farming. **2003**. Available from: http://www.

inspection.gc.ca/english/plaveg/pbo/
dir/dir0007ie.shtml

[17] L. CROSBY, *BioPharm Int.* **2003**, *16* (4),
60–67.

[18] United Sates Department of Agricul-
ture. Biology of Crop Plants. **2003**.
Available from: http://www.aphis.
usda.gov/ppq/biotech/biology.html;
Canadian Food Inspection Agency Biol-
ogy Documents, Companion Docu-
ments for Regulatory Directive 94–08:
Assessment Criteria for Determining
Environmental Safety of Plants with
Novel Traits. Available from: http://

www.inspection.gc.ca/english/plaveg/
pbo/dir/biodoce.shtml.

[19] V. GRUBER, P. P. BERNA, T. ARNAUD
et al., *Mol. Breeding* **2001**, *7* (4), 329–
340.

[20] E.E. HOOD, D. R WITCHER, S. MADDOCK
et al., *Mol. Breeding* **1997**, *3* (4), 291–
306.

[21] K.A. GOMEZ, A.A. GOMEZ. *Statistical
Procedures for Agricultural Research*,
John Wiley and Sons, USA, **1984**.

[22] R.W.H. LEE, A.N. POOL, A. ZIAUDDIN
et al., *Mol. Breeding* **2003**, *11* (4), 259–
266.

6
Plant Viral Expression Vectors: History and New Developments

Vidadi Yusibov and Shailaja Rabindran

6.1
Introduction

Proteins such as antibodies, enzymes, hormones and vaccine antigens can be used to prevent, diagnose and treat a range of diseases. Such molecules are therefore of paramount importance in health and medicine. Historically, many of these proteins have been isolated from human or animal sources. However, the low quantities present in such source material coupled with safety risks and high purification costs have limited the availability of protein therapeutics and vaccines for many types of disease.

In the mid-1970s, recombinant DNA technology revolutionized research involving the above molecules by making it possible to produce recombinant proteins in bacterial expression systems. Although, until recently, prokaryotic expression systems have been the most common method for recombinant protein production, they have limitations because of the absence of eukaryotic posttranslational modification and the improper folding of many complex human proteins. During the last three decades, many research laboratories have therefore focused on developing alternative platforms for recombinant protein expression, which can overcome the shortcomings of bacteria. The first systems to emerge from these studies were those based on animal and insect cell cultures [1,2]. Several products, including monoclonal antibodies, vaccines and therapeutics, have been produced using these systems [3,4], but the high production costs combined with the requirement for highly sophisticated manufacturing facilities for each target protein has encouraged the development of different production systems.

Transgenic animals [5] have been considered as an alternative because they can produce complex human proteins in large amounts. However, potential disadvantages of transgenic animals include economic and time constraints as well as risks that the products could be contaminated with human pathogens.

In recent years, plants have emerged as a promising new system for the production of recombinant proteins. Plants are widely described as "green factories" because they provide possible solutions to several of the safety and economic concerns raised by animal systems. The advantages of plants include economical large-scale

Molecular Farming. Edited by Rainer Fischer, Stefan Schillberg
Copyright © 2004 WILEY-VCH Verlag GmbH & Co. KGaA, Weinheim
ISBN: 3-527-30786-9

production, the absence of contaminating mammalian pathogens, and the ability to produce complex, biologically-active proteins.

There are several different strategies for the production of recombinant proteins in plants. The traditional approach is to use transgenic plants, in which the gene of interest is incorporated into the plant genome and is passed from generation to generation as a stable trait (reviewed in [6–10]). A number of therapeutic proteins and vaccine antigens have been produced in transgenic plants and have been successfully tested in animal and human trials. Nevertheless, the transgenic approach has some shortcomings, including the length of time required to obtain the transgenic producer lines, low levels of expression, and inherent difficulties in the modification of an existing product. In some expression hosts, scaling up production also takes a long time.

One alternative strategy that has emerged in the last decade is the use of plant virus expression vectors to synthesize recombinant proteins in plants. The foreign gene is inserted into the viral genome so that, upon infection of the host plant cell, the transgene is replicated and expressed along with native viral genes. This method of transient expression adds several further advantages to plant-based expression, including improved time efficiency, higher levels of target protein expression, flexibility and convenience in the modification of existing products (or the development of new ones), ease of scale-up, flexibility in the selection of a production host and the potential for protein manufacture in contained facilities [11]. Furthermore, target proteins (in particular antigenic peptides) can be genetically fused to viral structural proteins, such as coat proteins, so that the plant virus is used not only for expression but also for the delivery of the vaccine antigens. To date, the coat proteins from a number of plant RNA viruses have been successfully used as carriers for antigenic peptides derived from various pathogens (for reviews see [11–13]). This approach is economically attractive, and benefits from higher levels of safety and efficacy in generating pathogen-specific immune responses.

This chapter evaluates some of the approaches used to produce vaccines and heterologous proteins in plants using plant virus-based expression systems, and discusses novel strategies that are being considered for the development of better vectors.

6.2
Plant RNA Viruses as Expression Vectors

The majority of viruses that infect plants have single-stranded, positive-sense RNA genomes. It has therefore been necessary to use infectious cDNA clones for the in vitro manipulation of RNA viruses, allowing them to be developed as effective tools for the commercial production of target proteins in plants. This approach has also been used to study the genetic and metabolic profiles of both viruses and their host plants. Siegel [14] conceptualized the potential use of RNA viruses as expression vectors. Brome mosaic virus (BMV) and Tobacco mosaic virus (TMV) were the first two RNA viruses to be converted into expression vectors. These vectors have since been promoted from elegant laboratory tools (for studying virus movement or replication) to the category of expression vectors for producing biopharmaceuticals in plants. After more than a decade of work by many research groups, the genomes of a number of plant RNA viruses have been engineered to express target sequences (Table 6.1). Several extensive reviews have been published on this subject [10–12,5]. Many of these articles have demonstrated the feasibility of plant viruses as expression tools for the production of foreign proteins in plants. Based on genome structure and identification of virus gene functions, several approaches have been employed for the expression of foreign sequences using plant viruses as expression systems. These include: i) replacing nonessential viral genes with target sequences, ii) inserting target sequences into the viral genome as an additional gene with an additional promoter, iii) fusing target sequences with viral genes encoding structural proteins, iv) fusing the target sequences and viral structural gene with a cleavage site or read-through sequence, iv) functional complementation of defective viral components, and v) trans-complementation of viral genes through transgene expression in the host plant.

Tab. 6.1   Examples of plant viruses used in the development of expression vectors

Virus	Strategies used	Pathogen or protein/ epitope	References
Tobacco mosaic virus	CP replacement, Second subgenomic promoter (sgp), CP fusion, Readthrough fusion with CP, fusion with cleavage site, fusion with CtxB	malaria, rabies virus, MHV, FMDV, HCV, BHV-1, *Ps. aeruginosa*, neuropeptide, TMOF, allergens, MABs, α-trichosanthin, cytokines α-galactosidase-A, scFvs	11, 16, 17, 52, 59–61
Potato virus X	second sgp, CP fusion, CP fusion with FMDV 2A element, IRES elements	GFP, rotavirus, scFv,	25, 26, 62, 63
Zucchini yellow mosaic virus	Fusion with CP with protease cleavage	anti-HIV proteins	37
Plum pox virus	Fusion with CP with protease cleavage	CPV	40
Cowpea mosaic virus	Fusion with CP, fusion to CP with protease cleavage site, complementation	GFP, HIV, MEV, HRV, MV, CPV, *S. aureus*, *Ps. aeruginosa*	11, 12, 13, 31, 43, 58
Alfalfa mosaic virus	Fusion with CP, second sgp	HIV, rabies, measles, RSV	32, 54, 64
Tomato bushy stunt virus	CP fusion	HIV	65, 66

CP: coat protein; CtxB: cholera toxin B subunit; scFv: single chain Fv antibody fragment; TMOF: trypsin modulating oostatic factor; MAB: monoclonal antibody; GFP: green fluorescent protein; CPV: Canine parvovirus; BHV: Bovine herpes virus; FMDV: Foot and mouth disease virus; HCV: Hepatitis C virus; HRV: Human rhino Virus; MEV: Mink enteritis virus; MHV: Murine hepatitis virus; MV: Measles virus; RSV:Respiratory syncytial virus

6.2.1
Tobacco mosaic virus (TMV)

Several groups have demonstrated the feasibility of TMV as an expression vector in plants (for reviews see [11,16,17]). The most widely used TMV-based vectors have an additional heterologous subgenomic promoter that directs expression of the foreign gene [18,9]. One of the first active proteins produced in plants using this vector was α-trichosanthin, an inhibitor of HIV [20]. An improvement on the original TB2 vector resulted in the development of 30B [19], which is widely used both as a laboratory tool to study gene function and for the production of therapeutics and vaccines. Due to size limitations, there is unstable long distance movement of the chimeric virus in tobacco. Using DNA shuffling, Toth *et al.* [21] created improved 30B-based vectors that had better movement characteristics in tobacco. For further manipulations to ensure the production of functional proteins, signal peptide sequences were incorporated to target the protein either to the endoplasmic reticulum (ER) or the apoplast [17].

TMV particles have also been used as an epitope-presentation system. Detailed X-ray crystallographic analysis of the coat protein identified suitable sites for the insertion of foreign peptides. The helical arrangement of 2130 copies of coat protein around the viral RNA allows the presentation of multiple copies of the foreign epitope on the virus surface. At the same time, this can be detrimental to virion stability because the extra peptide can destabilize the virion structure. To circumvent this problem, a more stable system was developed exploiting the readthrough sequence from the replicase gene, so that both wild type and fusion peptide-containing coat proteins were produced [22,23]. Recently, Bendahmane *et al.* [24] showed that by modifying the pI:charge ratio of hybrid coat proteins so that it resembled that of the wild type TMV coat protein, a more stable hybrid virus was produced.

6.2.2
Potato virus X (PVX)

PVX, a member of the potexvirus group, is another single stranded RNA virus that is widely used as an expression vector. Santa Cruz *et al.* [25] created a PVX-based vector in which the target molecule was genetically fused to the amino terminus of the PVX coat protein via the foot and mouth disease virus (FMDV) 2A peptide, which facilitates cotranslational processing of the polyprotein. Although this is an elegant way to produce foreign proteins, the extra C-terminal sequences would be undesirable, particularly for proteins used in therapeutic applications. Recently, Toth *et al.* [26] created another PVX-based vector in which an internal ribosome entry site (IRES) was inserted between the foreign and coat protein genes. On infection, this virus produced a bicistronic mRNA, from which both coat protein and the foreign protein were translated at detectable levels. PVX vectors have been used to produce single-chain Fv antibody fragments (scFvs) in *N. benthamiana* at up to 0.2 µg g^{-1} in leaf tissue [27,28].

6.2.3
Cowpea mosaic virus (CPMV)

Cowpea mosaic virus (CPMV), a bipartite RNA virus, was one of the first viruses used to express foreign peptides as viral coat protein fusions [29]. Since then, CPMV vectors have been improved and chimeric virus particles (CVPs) containing different target peptides inserted at various loop structures within the coat protein have been produced in plants. These CVPs have elicited strong immune responses in model animals and have achieved protection against disease challenge [12,13]. CPMV has also been used to express full-length proteins. In one version of the vector, the foreign genes were inserted between the movement protein and large coat protein genes with artificial proteolytic cleavage sites engineered on either side of the foreign gene to facilitate the release of soluble protein [30]. In another version, the foreign gene was expressed as a fusion to the C-terminus of the CPMV S protein, using the FMDV 2A peptide as a bridge [31].

6.2.4
Alfalfa mosaic virus (AlMV)

Vectors based on Alfalfa mosaic virus (AlMV), a tripartite RNA virus, are particularly useful for producing target molecules genetically fused to the viral coat protein. The AlMV capsid comprises multiple 24-kDa coat protein units, which form particles of different sizes and shapes based on the size of the encapsidated RNA. The N-terminus of the coat protein is located at the surface of the virion and is a useful site for the insertion of peptides without interfering with virion assembly. There have been several studies using AlMV viral particles for the production and delivery of antigenic epitopes [32,22]. More recently, we have used AlMV as a screening tool to map antigenic determinants for subunit vaccine development (Munz et al., unpublished data). Both B cell and T cell epitopes are being identified as potential vaccine candidates from a variety of pathogens to formulate multivalent vaccines. The ability to insert peptides of up to 50 amino acids in length as coat protein fusions, together with the ease of particle recovery, makes this system very amenable for the development of vaccines. The availability of P12 transgenic tobacco plants [34] that contain AlMV RNAs 1 and 2 integrated into the plant genome allows for trans-complementation so that infections can be initiated by the delivery of RNA3 alone. The virus particles thus produced are not infectious to non-transgenic plants, therefore improving the containment of the hybrid virus. AlMV is undoubtedly a very efficient tool for the production and presentation of target sequences in the form of coat protein fusions.

6.3
Biological Activity of Target Molecules

Much progress has been made in the development of plant virus expression vectors over the last decade and they can now be regarded as commercially viable systems sufficient for the production of large quantities of target protein. During this period, a number of key issues in commercial product development have been improved, including the yield, biological activity and immunogenicity of the products, the genetic stability of the expression system, and the efficiency of downstream processing to recover the product from virus-infected plants.

Studies have shown that plants can make biologically active recombinant proteins through both transgenic and transient expression approaches. Although the plant post-translational machinery is similar to that of mammalian cells, there are some notable differences, e.g. differences in glycosylation, particularly the absence of sialation, which may impact the activity of certain proteins. The absence of mammalian enzymes may prevent complex maturation processes that are critical for the biological activity of proteins such as insulin. Fortunately these shortcomings affect the activity of only a limited number of proteins.

Most therapeutic proteins expressed in plants have full biological activity. For example, α-galactosidase A (Gal A), a human lysosomal enzyme deficient in patients with Fabry's disease, was expressed using a TMV vector. Treatment of the disease involves enzyme replacement therapy, which is expensive. Correct disulfide bond formation, glycosylation and dimerization are essential for the activity of this protein. When Gal A protein was targeted to the endoplasmic reticulum, enzyme activity was nominal [16], but when portions of a putative C-terminal propeptide were deleted, and the protein was targeted to the apoplast, high levels of enzyme activity were observed. Targeting to the apoplast also facilitates Gal A purification because leaf cells do not normally secrete proteins. The specific activity of Gal A was similar to that of the native enzyme purified from human tissue, and the predicted glycosylation sites were properly modified.

The pituitary glycoprotein follicle-stimulating hormone (FSH) was produced in plants as a single-chain molecule (sc-bFSH) using a TMV vector [35]. Native bFSH comprises two polypeptides expressed from genes at different loci. The cDNAs encoding the α and β subunits of the protein were cloned so that the subunits were fused in the configuration β-α to produce sc-bFSH, a 30-kDa protein. The sc-bFSH protein accumulated to 3% TSP (total soluble protein) in the intercellular wash fluid (IF) and was biologically active, indicating that it was correctly folded. Deglycosylation patterns of the IF extract with N-glycosidase F, which digests all oligosaccharide species except those containing core α(1,3)-fucose, indicated that the N-linked glycans present in the recombinant protein had two types of cores, with and without α (1,3)-fucose. When immunoaffinity-purified sc-bFSH was analyzed by MALDI-MS, the presence of two complex N-glycan structures was demonstrated. Plant N-glycan structures include β(1,2)-xylose and core α(1,3)-fucose. Because these glycans are not native to mammalian systems, it is expected that they may be allergenic in mammals. It is also expected that mammalian proteins with native glycosylation patterns will be more stable when administered for therapy. Recently, 'mammalianized' plants were developed that expressed mammalian β-1,4-galactosyltransferase [36]. This exciting development provides a possible solution to the production of correctly glycosylated mammalian proteins, including FSH and antibodies, in plants.

Arazi *et al.* [37] used a Zucchini yellow mosaic virus (ZYMV) vector to produce two anti HIV proteins – MAP30 (*Momordica* anti-HIV protein, 30 kDa) and GAP31 (*Gelonium* anti-HIV protein, 31 kDa) that are naturally present in other plant species. These two proteins have antiviral and antitumor activities and are effective against viruses, tumor cells and microbes. Purified preparations from ZYMV-infected squash had specific activities comparable to the native proteins, and demonstrated anti-HIV activity, which was measured by inhibition of syncytial formation and production of the viral core protein p24. Each recombinant protein was also able to inhibit the growth of both Gram-positive and Gram-negative bacteria as well as the yeast *Candida albicans*, and was not toxic to human cells at a wide range of concentrations.

6.4
Efficacy of Plant Virus-produced Antigens

As indicated above, target molecules and antigens can be engineered into plant virus vectors and produced: i) as free soluble proteins such as vaccine antigens, or ii) as fusions with viral coat protein subunits so that they are incorporated into the virus particles. Soluble antigens could consist of pathogen sequences only or fusions of pathogen sequences with molecules that provide adjuvant or other activities, such as the pentameric cholera toxin B subunit (CTB), which stimulates a mucosal immune response. A significant amount of data has been accumulated using both approaches.

6.4.1
Vaccine Antigens

Several soluble antigens that have been produced in plants using plant viruses as expression systems have shown immunogenic and protective properties in test animals [11]. The plant-derived proteins, which include known immunogens from pathogens or allergens from plants [38], have been tested either as crude plant extracts containing recombinant protein or as purified material from the infected plants.

Wigdorovitz *et al.* [39] used a TMV expression vector to produce VP1, the 26-kDa structural protein from FMDV, and tested it in mice. Mice injected intraperitoneally with leaf extracts prepared from infected plants mounted an antibody response against the plant-derived protein. All immunized mice were protected when challenged with virulent FMDV. One-year old calves immunized with plant extracts containing VP1 also developed FMDV-specific antibody responses.

In another study, Fernandez-Fernandez *et al.* [40] developed a vaccine against VP60 of rabbit hemorrhagic disease virus (RHDV) using plum pox virus (PPV) as the expression system. VP60 was produced as a polyprotein that was processed to release the heterologous protein. Rabbits immunized with extracts of *Nicotiana clevelandii* containing VP60 mounted an efficient immune response that protected them against a lethal challenge with RHDV. Nine of ten rabbits that received plant extracts containing VP60 (two 1-ml doses of extract containing ~2 g of fresh leaves) produced VP60-specific antibodies while animals vaccinated with extracts of plants infected with wild type PPV did not produce antibodies against RHDV. When the animals vaccinated with PPV-V60 were challenged, the antibody response increased and no clinical symptoms of the disease were observed, whereas all but one of the control animals died after challenge. No RHDV was detected in the livers of the surviving animals two weeks after challenge. The serological responses of animals vaccinated with plant extracts containing VP60 were almost as high as those of animals immunized with a commercial vaccine.

In exciting developments, McCormick *et al.* [41,42] produced individualized vaccines for the treatment of cancer using a TMV-based expression vector. B-cell tumors (such as non-Hodgkin's lymphoma) express a unique cell surface immunoglobulin (Ig) that acts as a tumor-specific marker. When such Igs are conjugated to carriers such as KLH and administered to patients with an adjuvant, an immune response is triggered in the patient with a favorable clinical outcome. It is difficult to produce Igs as vaccines, so single-chain variable region (scFv) vaccines that have the hypervariable domains from the tumor-specific Ig have been developed as alternatives. These scFv vaccines have the ability to elicit an anti-idiotypic response in animals and can block tumor progression in mouse lymphoma models. Using a TMV expression vector, 38C13 scFv (from a mouse B-cell lymphoma) was recovered from the IF of infected plants. The major fraction of protein produced in the plants was soluble and properly folded. When administered to mice, it induced specific anti-38C13 responses. A strong IgG2a isotype response was observed, which is often correlated with increased tumor protection. A 90% survival rate was reported in the group of mice that received the plant-produced vaccine plus adjuvant and were then challenged with 38C13 tumor cells two weeks after the third vaccination. There was also a 70% survival rate in mice receiving the scFv alone and an 80% survival rate in those receiving the standard vaccine. This approach has successfully been tested in a group of non-Hodgkin's lymphoma patients in an FDA approved phase I human clinical trial (Barry Holtz, Conference on Plant-Made Pharmaceuticals, Quebec City, Canada, 2003).

6.4.2
Particle-based Vaccine Antigen Delivery

Several plant virus coat proteins, including those of TMV, CPMV, AlMV and Tomato bushy stunt virus (TBSV), have been used to produce and deliver antigenic determinants from a variety of viral and bacterial pathogens. These data have been summarized in numerous publications and several reviews [12,13]. The ease of virus purification coupled with enhanced peptide immunogenicity when fused to carrier molecules makes this approach very attractive for vaccine development.

CPMV particles that contained a 17-mer neutralizing epitope, 3L17, from the VP2 capsid of Mink enteritis virus (MEV) fused to the S protein were generated. When mixed with adjuvant, these particles protected all the test animals from clinical disease when challenged with virulent MEV. A modified construct, which presented the peptide on the surface of both L and S coat protein subunits, induced an antibody response that was higher than that of a peptide-KLH conjugate [43]. The predominance of IgG_{2a} indicated early activation of TH1 cells. These results were validated by cell proliferation and IFN-γ release from mice cells exposed to CVPs in vitro. Intranasal immunization resulted in better mucosal responses than serum antibody responses. The significant outcome of these studies was that when peptides are presented on viral particles it is possible to shift the bias towards a TH1 response (which mediates macrophage and cytotoxic T cell activation), and that CVPs can protect against both systemic and mucosal infections.

CPMV particles have also been used to present epitopes from bacterial pathogens such as *Pseudomonas aeruginosa* and *Staphylococcus aureus*. When fused to CPMV, the D2 peptide from the *S. aureus* fibronectin-binding protein (FnBP) induced high titers of FnBP-specific antibodies in mice and rats immunized subcutaneously [44]. The sera inhibited binding of fibronectin to immobilized recombinant FnBP, and rat serum was able to block the adherence of *S. aureus* to fibronectin. The response had a strong TH1 bias probably because CPMV CP elicits TH1/IgG_{2a}-type responses. The isotype of anti-D2 IgG in induced mice was predominantly IgG_{2a} and IgG_{2b}. These studies highlight the potential of plant virus-based vaccines to protect against *S. aureus* infections that include invasive endocarditis, septicaemia, peritonitis and bovine mastitis. When presented on CPMV particles, a linear B-cell epitope from the outer membrane protein F of *P. aeruginosa* induced peptide-specific antibodies in C57BL/6 mice that bound complement and increased phagocytosis of *P. aeruginosa* by human neutrophils in vitro [45]. In a mouse model of chronic pulmonary infection, the particles afforded protection when challenged with two different immunotypes of the pathogen. The levels of protection were similar to those observed when the peptide was coupled to KLH.

In another study, CPMV particles that displayed a peptide derived from the epidermal growth factor receptor variant III (EGFRvIII) elicited specific antibody responses in mice against the peptide and protected mice from tumor challenge [45].

Most infectious diseases involve colonization or invasion through mucosal surfaces by a pathogen. As a first line of defense it is important to develop a strong mucosal response against the pathogen. Such responses can be achieved when the oral or nasal route is used for immunization. Due to the acidic nature of the stomach and the presence of proteolytic enzymes in the gastrointestinal tract, special formulation of vaccine antigens will be required for efficient delivery and the generation of a suitable immune response, if the oral route is employed. The alternate method of intranasal immunization would be more effective because it requires less of the immunogen and conditions are more favorable to maintain stability of the viral particles. It has been found that when recombinant virus particles are administered intranasally, antibodies can be detected at distal sites, such as the bronchial, intestinal and vaginal lavages [43].

Chimeric TMV particles containing the 5B19 epitope from the spike protein of murine hepatitis virus (MHV) fused near the C-terminus of TMV coat protein subunits were used to immunize mice intranasally (three doses per week for ten weeks) [46]. High IgG titers and moderate IgA titers specific to the peptide could be detected. When mice were challenged intranasally with MHV, five of six mice that were immunized intranasally with chimeric TMV particles for ten weeks survived the challenge. Two further mice that received immunization for six weeks and one that was immunized for four weeks survived the challenge, indicating that longer periods of immunization with the chimeric TMV particles resulted in higher antibody titers that protected the animals from disease challenge.

Nemchinov *et al.* [47] used a TMV expression vector to develop a subunit vaccine against hepatitis C virus (HCV). A consensus sequence matching hypervariable region 1 (HVR1) of HCV, encoding a potential neutralizing epitope of 27 amino acids, was fused to the C-terminus of CTB. Mice immunized intranasally with plant extract containing ~0.5–1 µg CTB/HVR1 developed anti-HVR1 antibodies. These experiments were carried out without adjuvant and the amounts of immunogen used per dose were less than 0.1 µg of HVR1 epitope. The same epitope has been engineered as a fusion with other plant viruses such as AlMV and is undergoing testing.

The sera of mice fed with fresh spinach leaves infected with AlMV particles presenting a rabies virus epitope contained IgG and IgA. Mucosal IgA was also detected [48]. Human volunteers (in FDA approved trials) fed with spinach containing recombinant particles generated both IgG and IgA responses specific to the pathogen [49]. The trials also suggested that plant virus particle-based vaccines could be effectively used in prime-boost regimens. In more recent work, recombinant AlMV particles containing an epitope from the G protein of human respiratory syncytial virus (RSV) induced protective immunity in mice [33].

Mice immunized intranasally with chimeric PVX particles expressing a six-amino-acid neutralizing epitope from gp 41 of HIV-1 produced high levels of HIV-1-specific IgG and IgA antibodies [50]. The anti-H66 IgG titers ranged from 2000 to > 30,000. Mice immunized intranasally produced IgA in the serum and in fecal extracts.

Excellent progress has been made in a relatively short period of time in demonstrating the potential of plant virus vectors not only as expression tools but also as elegant and efficient means for the administration of vaccine antigens by different routes. The targets seem to be unlimited.

6.4.3
Other Uses of Plant Virus Particles

In a unique study Khor *et al.* [51] demonstrated that CVPs could also be used as antiviral agents. The cellular receptor for measles virus (MV) has been identified as CD46. Two different peptides from CD46, when presented on CPMV particles, inhibited the infection of HeLa cells by MV in vitro in a dose-dependent manner. The extent of inhibition was 18–180-fold more effective than soluble CD46 peptide, probably due to increased stability. The CVPs also protected mice models of human MV infection when challenged intracranially with MV. The results showed that the chimeric CPMV particles could block MV entry into neurons in the brains of the treated animals. This technology has paved the way for the creation of antiviral agents against several important viruses, particularly those for which the cellular receptors have been characterized. Virus particles that can present multiple peptides on their surfaces could simultaneously evoke humoral responses, cell-mediated responses, and inhibit viral entry, thus acting as a vaccine as well as an antiviral agent.

In yet another application of plant virus peptide presentation systems, Borovsky [52] used TMV to present a peptide, trypsin modulating oostatic factor (TMOF), that terminates trypsin biosynthesis in the mosquito gut and causes larval mortality. This unique study uses plant virus particles for the biological control of insect pests.

6.5
Plant Viruses as Gene Function Discovery Tools

Plant virus vectors have been used extensively as laboratory tools to identify the functions of unknown genes, to localize viral proteins by fusing them to green fluorescent protein (GFP), to manipulate biosynthetic pathways in plants, to screen genomic libraries rapidly, and to study gene silencing. Gene function can be elucidated using plant viral vectors either by overexpressing the gene or by silencing the gene. This has been an important feature in functional genomics strategies to identify plant genes, and plant viral vectors have proven to be invaluable tools for this purpose. The genomes of several viruses have been engineered to study post-transcriptional gene silencing, termed virus-induced gene silencing (VIGS).

Kumagai *et al.* [53] used a TMV expression vector to overexpress tomato phytoene synthase (PSY). When *Nicotiana benthamiana* plants were inoculated with in vitro transcripts of these constructs, infected leaves developed a bright orange phenotype, and accumulated high levels of phytoene. Plants that were inoculated with a construct that expressed a partial segment of phytoene desaturase cDNA in the antisense orientation developed a white phenotype, reflecting the inhibition of carotenoid biosynthesis due to *pds* gene silencing. TMV, PVX, TRV and AlMV are some of the plant viruses being used as expression vectors to identify and manipulate plant gene function (reviewed in [11]).

6.6
New Approaches to the Development of Viral Vectors

During the last decade there has been significant progress in the development of plant virus expression systems for the production of biopharmaceuticals. These studies have revealed the tremendous potential of plant viruses but have also raised some obvious and inherent shortcomings of the technology. A major limitation is the size of the foreign protein that can be produced. Often, large genes are not stably maintained in the plant viral genome. New approaches are being developed in an attempt to overcome some of these shortcomings, e.g. vector stability, host range and the efficiency of target molecule expression.

The design of chimeric viruses to create a functional vector system combining components of different viruses and thus expanding the virus host range was one of the early approaches employed by Yusibov *et al.* [32]. Antigenic determinants from rabies virus and HIV-1 engineered as fusions with the AlMV coat protein were expressed using a TMV vector. Thus the system had two coat proteins, one serving as a carrier molecule for target peptides and other for long distance movement. This extra coat protein system allows the presentation of larger peptides on the surface of AlMV particles but also provides greater genetic stability. A chimera of TMV that expressed the AlMV coat protein instead of the native coat protein was also created [54]. This hybrid virus had a broader host range and infected plant species such as spinach and soybean.

The development of two-component functional complementation-based expression vectors using either homologous or heterologous virus components is an ongoing attempt to circumvent the size limitations of a single component vectors and also to facilitate the production of multi-subunit proteins [55–57]. Liu and Lomonossoff [58] described the agrodelivery of two subgenomic components of CPMV from a mixed suspension of bacteria, each harboring different subgenomic complements.

Further improvements to virus expression systems include trans-complementation of some of the virus functions from transgenic host plants (P12 plants for AlMV). By integrating parts of the viral vector into the plant chromosome, this system has the potential for multiple technical solutions that could overcome limitations of classical viral vectors [33]. Viral vectors can be used as molecular switches for tightly controlled, high-level transgene expression (Hull *et al.* unpublished data).

A novel transient expression technology has been developed at Icon Genetics AG, Munich, Germany (Prof. Gleba, Icon Genetics, personal communication). This technology uses T-DNAs that encode parts of a plant virus genome transiently co-delivered by two or more *Agrobacterium* recombinants. Because more than one bacterium can infect a cell, the viral vector is assembled in planta. This technology is also versatile because multiple genes can be co-expressed and functional proteins assembled in the plants. In another version, two viral sequences encoded on separate T-DNAs, one containing the coat protein gene but lacking the origin of assembly sequence (OAS) and the other containing the foreign gene and OAS but lacking the coat protein gene, are co-infiltrated into plants. Functional complementation between the two viruses results in long distance movement and expression of the foreign gene in upper leaves. In yet another version of the vector, a functional amplicon is produced only after transient delivery of a T-DNA encoding viral integrase by agroinfiltration. This approach has been utilized to express several proteins and has clearly produced greater quantities of the target protein. For example, up to 5 mg g^{-1} fresh weight of GFP could be obtained in leaves.

6.7
Conclusion

There is no doubt that plants represent one of the most productive and yet inexpensive sources of biomass. The absence of contaminating animal pathogens, the eukaryotic translational machinery and the ease of plant virus manipulation make plants and plant viruses very useful for the economical production of large amounts of commercial products. A whole new area in agriculture called molecular farming (pharming), i.e. the use of plants to produce pharmaceuticals, could soon revolutionize the pharmaceutical industry. Plant virus-based vectors are powerful tools for the production of biopharmaceuticals, and some products are now close to commercialization. It is clear that as the industry moves forward, the focus of research should turn towards more practical problems such as the development of efficient extraction and purification systems. The glycosylation patterns of plant-derived proteins, allergy testing, and safety and containment issues need to be addressed before the plant virus approach becomes commercially viable. This would require the collaborative efforts of many researchers, including plant virologists, immunologists, chemical engineers and protein chemists.

References

[1] T.A. KOST, J.P. CONDREAY, *Curr. Opin. Biotechnol.* **1999**, *10* (5), 428–433.

[2] F. HESSE, R. WAGNER, *Trends Biotechnol.* **2000**, *18* (4), 173–180.

[3] K. KOTHS, *Curr. Opin. Biotechnol.* **1995**, *6* (6), 681–687.

[4] B.D. KELLEY, *Curr. Opin. Biotechnol.* **2001**, *12* (2), 173–174.

[5] M.A. DALRYMPLE, I. GARNER, *Biotechnol. Genet. Eng. Rev.* **1998**, *15*, 33–49.

[6] C.J. ARNTZEN, *Nature Biotechnol.* **1997**, *5* (3), 221–222.

[7] A.M. WALMSLEY, C.J. ARNTZEN, *Curr. Opin. Biotechnol.* **2000**, *11* (2), 126–129.

[8] A.M. WALMSLEY, C.J. ARNTZEN, *Curr. Opin. Biotechnol.* **2003**, *14* (2), 145–150.

[9] H.S. MASON, H. WARZECHA, T. MOR et al., *Trends Mol. Med.* **2002**, *8* (7), 324–329.

[10] P. AWRAM, R.C. GARDNER, R.L. FORSTER et al., *Adv. Virus Res.* **2002**, *58*, 81–124.

[11] G.P. POGUE, J.A. LINDBO, S.J. GARGER et al., *Annu. Rev. Phytopathol.* **2002**, *40*, 45–74.

[12] C. PORTA, G.P. LOMONOSSOFF, *Rev. Med. Virol.* **1998**, *8* (1), 25–41.

[13] G.P LOMONOSSOFF, W.D.O. HAMILTON, in: J. Hammond, P. McGarvey, V. Yusibov (eds) *Plant Biotechnology*, Springer Verlag, Berlin, pp. 177–189, **1999**.

[14] A. SIEGEL, *Phytopathology* **1983**, *73* (5), 775–775.

[15] H.B. SCHOLTHOF, K-B.G. SCHOLTHOF, A.O JACKSON, *Annu. Rev. Phytopathol.* **1996**, *34*, 299–323.

[16] G.P. POGUE, J.A. LINDBO, W.O. DAWSON et al., in: *Plant Molecular Biology Manual*, Kluwer Academic Publishers, Dordrecht, The Netherlands, Section L4, pp 1–27, **1998**.

[17] V. YUSIBOV, S. SHIVPRASAD, T.H. TURPEN et al., in: J. Hammond, P. McGarvey, V. Yusibov (eds) *Plant Biotechnology*, Springer Verlag, Berlin, pp. 81–94, **1999**.

[18] J. DONSON, C.M. KEARNEY, M.E. HILF et al., *Proc. Natl Acad. Sci. USA* **1991**, *88* (16), 7204–7208.

[19] S. SHIVPRASAD, G.P. POGUE, D.J. LE-WANDOWSKI et al., *Virology* **1999**, *255* (2), 312–323.

[20] M.H. KUMAGAI, T.H. TURPEN, N. WEIN-ZETTL et al., *Proc. Natl Acad. Sci. USA* **1993**, *90* (2), 427–430.

[21] R.L. TOTH, G.P. POGUE, S. CHAPMAN, *Plant J.* **2002**, *30* (5), 593–600.

[22] H. HAMAMOTO, Y. SUGIYAMA, N. NAKA-GAWA et al., *Bio/Technology* **1993**, *11* (8), 930–932.

[23] T.H. TURPEN, S.J. REINL, Y. CHAROENVIT et al., *Bio/Technology* **1995**, *13* (1), 53–57.

[24] M. BENDAHMANE, M. KOO, E. KARRER et al., *J. Mol. Biol.* **1999**, *290* (1), 9–20.

[25] S. SANTA CRUZ, S. CHAPMAN, A.G. RO-BERTS et al., *Proc. Natl Acad. Sci. USA* **1996**, *93* (13), 6286–6290.

[26] R.L. TOTH, S. CHAPMAN, F. CARR et al., *FEBS Lett.* **2001**, *489* (2–3), 215–219.

[27] S. HENDY, Z.C. CHEN, H. BARKER et al., *J. Immunol. Methods* **1999**, *231* (1–2), 137–146.

[28] P. ROGGERO, M. CIUFFO, E. BENVENUTO et al., *Prot. Express. Purif.* **2001**, *22* (1), 70–74.

[29] R. USHA, J.B. ROHL, V.E. SPALL et al., *Virology* **1993**, *197* (1), 366–374.

[30] J., VERVER, J., WELLINK, J., VAN LENT et al., *Virology* **1998**, *242* (1), 22–27.

[31] K. GOPINATH, J. WELLINK, J. PORTA et al., *Virology* **2000**, *267* (2), 159–173.

[32] V. YUSIBOV, A. MODELSKA, K. STEPLEWSKI et al., *Proc. Natl Acad. Sci. USA* **1997**, *94* (11), 5784–5788.

[33] H. BELANGER, N. FLEYSH, S. COX et al., *FASEB J.* **2000**, *14* (14), 2323–2328.

[34] J. F. BOL, *J. Gen. Virol.* **1999**, *80* (5), 1089–1102.

[35] D. DIRNBERGER, H. STEINKELLNER, L. ABDENNEBI et al., *Eur. J. Biochem.* **2001**, *268* (16), 4570–4579.

[36] M. BARDOR, L. FAYE, P. LEROUGE, *Trends Plant Sci.* **1999**, *4* (9), 376–380.

[37] T. ARAZI, P. LEE-HUANG, P.L., HUANG et al., *Biochem. Biophys. Res. Commun.* **2002**, *292* (2), 441–448.

[38] M. KREBITZ, U. WIEDERMANN, D. ESSL et al., *FASEB J.* **2000**, *14* (10), 1279–1288.

[39] A. WIGDOROVITZ, D.M. PEREZ FILGUEIRA, N. ROBERTSON et al., *Virology* **1999**, *264* (1), 85–91.

[40] M. R. FERNANDEZ-FERNANDEZ, M. MOURINO, J. RIVERA et al., *Virology* **2001**, *280* (2), 283–291.

[41] A.A. MCCORMICK, M.H. KUMAGAI, K. HANLEY et al., *Proc. Natl Acad. Sci. USA* **1999**, *96* (2), 703–708.

[42] A.A. MCCORMICK, S.J. REINL, T.I. CAMERON et al., *J. Immunol. Methods* **2003**, *278* (1–2), 95–104.

[43] B.L. NICHOLAS, F.R. BRENNAN, J.L. MARTINEZ-TORRECUADRADA et al., *Vaccine* **2002**, 20 (21–22), 2727–2734.

[44] F.R. BRENNAN, L.B. GILLELAND, J. STACZEK et al., *Microbiology* **1999**, *145* (8), 2061–2067.

[45] F.R. BRENNAN, T.D. JONES, W.D. HAMILTON, *Mol. Biotechnol.* **2001**, *17* (1), 15–26.

[46] M. KOO, M. BENDAHMANE, G.A. LETTIERI et al., *Proc. Natl Acad. Sci. USA* **1999**, *96* (14), 7774–7779.

[47] L.G. NEMCHINOV, T.J. LIANG, M.M. RI-

FAAT, *Arch. Virol.* **2000**, *145* (12), 2557–2573.

[48] A. MODELSKA, B. DIETZSCHOLD, N. FLEYSH et al., *Proc. Natl Acad. Sci. USA* **1998**, *95* (5), 2481–2485.

[49] V. YUSIBOV, D.C. HOOPER, S.V. SPITSIN et al., *Vaccine* **2002**, *20* (25–26), 3155–3164.

[50] C. MARUSIC, P. RIZZA, L. LATTANZI et al., *J. Virol.* **2001**, *75* (18), 8434–8439.

[51] I.W. KHOR, T. LIN, J.P. LANGEDIJK et al., *J. Virol.* **2002**, *76* (9), 4412–4419.

[52] D. BOROVSKY, *J. Exp. Biol.* **2003**, *206* (21), 3869–3875.

[53] M.H. KUMAGAI, J. DONSON, G. DELLA-CIOPPA, *Proc. Natl Acad. Sci. USA* **1995**, *92* (5), 1679–1683.

[54] S. SPITSIN, K. STEPLEWSKI, N. FLEYSH et al., *Proc. Natl Acad. Sci. USA* **1999**, *96* (5), 2549–2553.

[55] D.J. LEWANDOWSKI, W.O. DAWSON, *Virology* **1998**, *251* (2), 427–437.

[56] D.J. LEWANDOWSKI, W.O. DAWSON, *Virology* **2000**, *271* (1), 90–98.

[57] V. YUSIBOV, N. FLEYSH, S. SPITSIN et al., in: J.T. Romer, J.A. Saunders, B.F. Matthews (eds), *Regulation of Phytochemicals by Molecular Techniques*, Elsevier Science Ltd, London, pp. 59–78, **2001**.

[58] L. LIU, G.P. LOMONOSSOFF, *J. Virol. Methods* **2002**, *105* (2), 343–348.

[59] A.A. LIM, S. TACHIBANA, Y. WATANABE et al., *Gene* **2002**, *289* (1–2), 69–79.

[60] F. PEREZ FILGUEIRA, D. ZAMORANO, P. DOMINGUEZ, et al., *Vaccine* **2003**, *21*(27–30), 4201–4209.

[61] L. WU, L. JIANG, Z. ZHOU et al., *Vaccine* **2003**, *21* (27–30), 4390–4398.

[62] S. CHAPMAN, T. KAVANAGH, D. BAULCOMBE, *Plant. J.* **1992**, *2* (4), 549–557.

[63] L. SMOLENSKA, I.M. ROBERTS, D. LEARMONTH et al., *FEBS Lett.* **1998**, *441* (3), 379–382.

[64] J. SANCHEZ-NAVARRO, R. MIGLINO, A. RAGOZZINO et al., *Arch. Virol.* **2001**, *146* (5), 923–939.

[65] T. JOELSON, L. AKERBLOM, P. OXELFELT et al., *J. Gen. Virol.* **1997**, *78* (6), 1213–1217.

[66] G. ZHANG, C. LEUNG, L. MURDIN et al., *Mol. Biotechnol.* **2000**, *14* (2), 99–107.

7
Production of Pharmaceutical Proteins in Plants and Plant Cell Suspension Cultures

Andreas Schiermeyer, Simone Dorfmüller and Helga Schinkel

7.1
Introduction

Molecular farming or *pharming*™ is the production of pharmaceutically important and commercially valuable proteins in plants [1]. Since the first production of a pharmaceutical protein in transgenic tobacco [2] a broad range of proteins with potential medical applications has been expressed in plants. These range from small peptides such as enkephalins [3] to complex, multisubunit molecules such as secreted antibodies of the IgA type [4]. In this chapter we will address the decisions that must be made during the production and marketing of such proteins. This begins with the choice of production host, which can range from cellular green algae [5] to a variety of crop species [6]. For pharmaceutical applications, it is necessary to work under current good manufacturing practice (cGMP) conditions and the use of plant suspension cultures must therefore be considered as an alternative to whole plants. Another crucial consideration is the design of the expression cassette, which must be optimized to achieve high-level accumulation of functional recombinant protein. Intrinsic factors such as how the protein is processed in the plant need to be taken into account, as well as the ability of the host species to carry out specific forms of post-translational modification.

When sufficiently high levels of expression and protein accumulation are achieved, efficient downstream processing protocols must be developed to insure product quality and the economic feasibility of production. As the demand for safe, recombinant pharmaceutical proteins continues to expand, the market potential of plant-produced recombinant proteins is considerable. Molecular farming can produce recombinant proteins at a lower cost than traditional expression systems based on microbial or animal cell culture, and without the risk of contamination with human pathogens.

Molecular Farming. Edited by Rainer Fischer, Stefan Schillberg
Copyright © 2004 WILEY-VCH Verlag GmbH & Co. KGaA, Weinheim
ISBN: 3-527-30786-9

7.2
Plant Species Used for Molecular Farming

Many species can be exploited for molecular farming (Table 7.1), with tobacco being the most widely used host plant to date. The advantages of tobacco plants include the high biomass yield (which can reach 50–100 tonnes ha^{-1} in a high-density population depending on the cultivar [7]), the ease of stable transformation either by co-cultivation with *Agrobacterium tumefaciens* [8], or transiently by infiltration with transgenic agrobacteria [9] or transfection with viral vectors [10, 11]. Another benefit is that tobacco is not used as a food or feed crop, ensuring that a transformed line expressing a highly potent drug will not contaminate food resources by outcrossing or during the processing steps. A drawback of this species is its content of nicotine and other alkaloids. Although there are cultivars available with reduced alkaloid content [12, 13] it is necessary to remove all traces of these toxic compounds during downstream processing, especially if the recombinant protein is intended for clinical applications. An additional disadvantage is the limited shelf life of transgenic leaf material after harvest [14], a problem that can be solved in other plant species by targeting the desired protein to natural storage organs like seeds or tubers. For tobacco leaves, downstream processing must commence immediately after harvest to ensure the stability of the recombinant protein, although Fiedler *et al.* [15] reported no losses of scFv (single chain fragment variable) antibody specificity or antigen binding capacity in dried leaves three weeks after harvest.

In the quest to find other plants that are suitable as bioreactors, various monocotyledonous and dicotyledonous species have been tested. These include corn [16], rice and wheat [17], alfalfa [18], potato [19, 20], oilseed rape [21], pea [22], tomato [23] and soybean [24]. The major advantage of cereal crops is that recombinant proteins can be directed to accumulate in seeds, which are evolutionar specialized for storage and thus protect proteins from proteolytic degradation. Recombinant proteins are reported to remain stable in seeds for up to five months at room temperature [17] and for at least three years at refrigerator temperature without significant loss of activity [25]. In addition, the seed proteome is less complex than the leaf proteome, which makes purification quicker and more economical [26].

It has been shown recently that the yields of recombinant protein in transgenic plants vary according to intrinsic factors such as the developmental stage and extrinsic factors such as the climate [27]. Such factors also affect the precise nature of posttranslational modification, particularly glycosylation. Elbers and colleagues [28] reported that the glycosylation profile, expressed as the ratio of complex to high mannose type glycans, increased with the age of tobacco leaves. Therefore, it is necessary to consider measures that ensure protein homogeneity, e.g. restricting expression to defined plant organs or specific compartments. Alternatively, it is possible to use an inducible expression system to restrict protein expression to a defined time point or period. Examples of such production methods include the postharvest system developed by Croptech [29] and the amylase-based system for controlled expression in suspension cells [30].

Tab. 7.1 Pharmaceutical proteins produced in transgenic plants

Protein	Regulatory elements	Additional elements to the construct	Species/Tissue	Accumulation level	Literature
α creatine kinase, MAK33 mAb (Fab)	CaMV 35S promoter/different termination sequences (ocs and rbcS)	LP of 2S2 storage protein of *Arabidopsis thaliana* — Fd and κ chain with 3'ocs; Fd chain with 3'ocs, κ chain with 3'rbcS; Fd chain with 3'ocs, κ chain-KDEL with 3'rbcS	*A. thaliana* (leaves); *A. thaliana* (leaves); *A. thaliana* (leaves)	3.6% of TSP; 6.5% of TSP; 5.9% of TSP	31
α human creatine kinase (Fab, IgG)	CaMV 35S promoter/ ocs terminator	LP of the 2S2 storage protein of *A. thaliana* — Fab; IgG; Fab; IgG	*Nicotiana tabacum* (callus); *N. tabacum* (callus); *A. thaliana* (callus); *A. thaliana* (callus)	0.128% of TSP; 0.477% of TSP; 0.188% of TSP; 0.217% of TSP	32
α substance P (single-domain Ab)	CaMV 35S promoter/ nos terminator	*pelB* LP	*Nicotiana benthamiana*	1% of TSP	33
Ig of 38C13 mouse B cell lymphoma (scFv)	Viral vector TTO1A, a hybrid fusion of TMV and tomato mosaic virus, N-terminal: rice α-amylase LP		*N. benthamiana* (leaves)	12.3–30.2 µg g^{-1} FW	11
α *Streptococcus mutans* adhesin (IgA, IgG)	CaMV 35S promoter/ nos terminator	Native mouse immunoglobulin LP; The γ chain constant region was replaced by an α chain constant region	*N. tabacum*	IgG: 7.7 µg mL^{-1} plant extract; IgG1/A: 1.5 µg mL^{-1} plant extract; IgG2/A: 2.1 µg mL^{-1} plant extract	34

Tab. 7.1 (continued)

Protein	Regulatory elements	Additional elements to the construct	Species/Tissue	Accumulation level	Literature
α *Streptococcus mutans* adhesin (sIgA-G)	CaMV 35S promoter/nos terminator	Native mouse immunoglobulin LP, four transgenic lines expressing either the κ, the hybrid IgA-G heavy chain, the murine J chain or the secretory component were crossed to obtain a sIgA expressing line	*N. tabacum*	200–500 µg g^{-1} FW	35
α human IgG (IgG)	CaMV 35S promoter/T7 terminator	5'murine LP of HC and LC LC and HC separately transformed	*Medicago sativa*	1% of TSP (full-size IgG)	18
α HSV-2 glycoprotein B (IgG)	CaMV 35S promoter/nos terminator	LP of tobacco extensin	*Glycine max*	Not mentioned	36
Human hemoglobin (α + β)	CaMV enhanced 35S promoter/CaMV 35S terminator	Transit peptide of small subunit of RubisCO of *Pisum sativum*	*N. tabacum*	0.05% of seed protein	37
EGF (synthetic gene)	CaMV 35S promoter/nos terminator	Codon-optimized for high level peptide expression in *Escherichia coli*	*N. tabacum* (leaves)	0.001% of TSP	38
Human glutamic acid decarboxylase (hGAD)	CaMV enhanced 35S promoter/TEV 5'UTR/nos terminator		Low alkaloid tobacco *Solanum tuberosum*	0.4% of TSP of leaves or tuber	12
Isoform of hGAD (hGAD65)	CaMV 35S promoter/nos terminator		*N. tabacum* (leaves)	0.01–0.04% of TSP	39
			Daucus carota (tap root)	0.01% of TSP	
			N. tabacum (leaves)	0.04% of TSP	40
		GAD67$_{1-87}$(rat)/hGAD65$_{88-585}$ Fusion	*N. benthamiana* (leaves)	0.19% of TSP 2.2% of TSP	

Expression mediated by potato virus X infection

Tab. 7.1 (continued)

Protein	Regulatory elements	Additional elements to the construct	Species/Tissue	Accumulation level	Literature
Human collagen Pro α (I) chain	CaMV 35S promoter/ CaMV 35S terminator	Pathogenesis-related protein (PR) LP or native LP	tobacco	>100 μg g⁻¹ tissue	41
Hepatitis B surface antigen	CaMV enhanced 35S promoter/ TEV 5'UTR/nos terminator Patatin promoter		*N. tabacum* (leaves) *S. tuberosum* (tuber)	66 μg g⁻¹ TSP 1.1 μg g⁻¹ FW 0.33 μg g⁻¹ FW	42 43
	CaMV 35S promoter/TEV 5'UTR/nos terminator	Soybean vegetative storage protein LP (vspαS) vspαS plus putative vacuolar targeting signal (vspαL)		0.8 μg g⁻¹ FW 2.4 μg g⁻¹ FW	
		Transit peptide of small subunit of RubisCO		n.d.	
		KDEL		1.25 μg g⁻¹ FW 0.33 μg g⁻¹ FW	
	CaMV 35S promoter/Ω TMV 5'UTR/nos terminator				
	CaMV 35S promoter/TEV 5'UTR/vspB terminator			6.5 μg g⁻¹ FW	
	CaMV 35S promoter/TEV 5'UTR/pin II gene terminator			16.0 μg g⁻¹ FW	
Binding subunit of heat-labile enterotoxin B	CaMV 35S promoter/TEV 5'UTR/vspB terminator	With and without KDEL	*N. tabacum* (leaves) *S. tuberosum*	5 μg g⁻¹ TSP (without KDEL) 14 μg g⁻¹ TSP (with KDEL) 30 μg g⁻¹ TSP (without KDEL) 110 μg g⁻¹ TSP (with KDEL)	44

Tab. 7.1 (continued)

Protein	Regulatory elements	Additional elements to the construct	Species/Tissue	Accumulation level	Literature
Cholera toxin B subunit	CaMV 35S promoter/nos terminator	Kozak sequence, KDEL	*Lycopersicon esculentum*	Leaves: 0.02% of TSP; Fruits: 0.04% of TSP	45
IL-10	CaMV 35S promoter/nos terminator	Native LP, with and without KDEL	*N. tabacum*	800 ng g^{-1} TSP (without KDEL); 55 µg g^{-1} TSP (with KDEL)	13
IL-2	Patatin promoter	Native LP	*S. tuberosum*	115 units g^{-1} FW	46
Enkephalins	Promoter and terminator of 2S1 albumin gene of *A. thaliana*	Integrated in 2S1 albumin of *A. thaliana* and flanked by tryptic cleavage sites	*A. thaliana* (seed); *Brassica napus* (seed)	2.9% of TSP, (200 nmol g^{-1}); 50 nmol g^{-1}	3
T84.66 α CEA (scFv)	Legumin A seed specific promoter/Ω sequence TMV 5′UTR	Light chain LP of mAb24, KDEL	*P. sativum* (seed)	9 µg g^{-1} FW	22
T84.66 α CEA (diabody)	CaMV enhanced 35S promoter/CHS 5′UTR/Pw 3′UTR of TMV/CaMV 35S terminator	Light chain LP of mAb24, with and without KDEL	*N. tabacum* (leaves)	1–5 µg g^{-1} (apoplast/transient); 4–12 µg g^{-1} (ER/transient); 0.44–0.93 µg g^{-1} (apoplast/stable); 3–9.3 µg g^{-1} (ER/stable)	47
Aprotinin	Maize ubiquitin promoter/pinII terminator	Maize ubiquitin 5′intron, barley α-amylase LP	Extracellular matrix of embryo (maize seeds)	In T5: 0.35% of extractable seed proteins	16

Tab. 7.1 (continued)

Protein	Regulatory elements	Additional elements to the construct	Species/Tissue	Accumulation level	Literature
Human serum albumin	CaMV enhanced 35S promoter/AlMV RNA synthetic leader sequence/nos terminator	Native or PR-S leader peptide	*S. tuberosum* (leaves)	0.02% of TSP	48
Hirudin	Oleosin promoter/nos terminator	Fused to native oleosin	*B. napus*	1% of seed protein	21
β-Interferon	CaMV 35S promoter/CaMV 35S terminator	Native LP	*N. tabacum*	170 ng g^{-1} FW	49
α human Rhesus D (IgG)	CaMV 35S promoter (LC, HC and tandem construct)/E9 polyA terminator of *P. sativum*	Murine immunoglobulin LP	*A. thaliana*	Cross of LC + HC: 0.6% of TSP, tandem: 0.3% of TSP	50
	mas1'2' dual promoter (bidirectional)/nos and E9 terminator			Bidirectional: 0.12% of TSP	
Dog gastric lipase	CaMV enhanced 35S promoter/CaMV 35S terminator	LP of rabbit gastric lipase precursor	*N. tabacum* (leaves; secretion)	7% of acid extractable proteins	51
		LP of sweet potato sporamin precursor	*N. tabacum* (leaves; vacuole)	5% of acid extractable proteins	
α Hepatitis B surface antigen (scFv)	CaMV 35S promoter/TMV Ω 5' UTR/nos terminator		*N. tabacum* (cytosol)	n.d.	52
		Sporamin LP	*N. tabacum* (apoplast)	0.031% of TSP	
		sporamin pre-pro-peptide LP	*N. tabacum* (vacuole)	0.032% of TSP	
		Sporamin LP, KDEL	*N. tabacum* (ER)	0.22% of TSP	

Tab. 7.1 (continued)

Protein	Regulatory elements	Additional elements to the construct	Species/Tissue	Accumulation level	Literature
α Hepatitis B surface antigen (scFv)	CaMV 35S promoter/TMV Ω 5'UTR/nos terminator	KDEL N-terminal ER LP from calreticulin	*N. tabacum* (root exudates)	630–760 ng g^{-1} dry weight root/day	53
Human interferon (hIF-α2b and hIF-α8)	CaMV 35S promoter/ nos terminator	Native LP of hIF-α2b and hIF-α8 resp.	*S. tuberosum*	560 IU g^{-1} tissue	54

Abbreviations used in table 1: AlMV: alfalfa mosaic virus; CaMV: cauliflower mosaic virus; CEA: carcinoembryogenic antigen; CHS: chalcone synthase; Fab: antigen binding fragment; Fd: variable domain and first constant domain of the heavy chain; FW: freshweight; HC: heavy chain; hIF: human interferon; HSV: herpes simplex virus; IL: interleukin; LC: light chain; LP: leader peptide; mAb: monoclonal antibody; mas: mannopine synthase; n. d.: not detectable; nos: nopaline synthase; ocs: octopine synthase; *pel*B: pectate lyase; pin II: potato proteinase inhibitor II; PR-S: pathogenesis related protein S; Pw: pseudoknot wildtype of TMV; rbcS: ribulose-1,5-bisphosphate carboxylase small subunit; RubisCO: Ribulose-1,5-bisphosphate Carboxylase Oxigenase; scFv: single chain Fragment variable; TEV: tobacco etch virus; TMV: tobacco mosaic virus; TSP: total soluble protein; UTR: untranslated region; vspB: vegetative storage protein B.

7.3
Cell Culture as an Alternative Expression System to Whole Plants

Plant cell cultures have not been used as frequently as intact plants for molecular farming but there are several examples that demonstrate their suitability as hosts for the production of recombinant proteins. Free cell suspension is regarded as the most suitable cultivation system for large-scale biotechnology applications, but tentative experiments have also been performed with hairy roots and shooty teratomas [55]. Examples of foreign proteins successfully expressed in suspension cells include monoclonal antibodies and their derivatives, cytokines, hormones and enzymes [56–60]. The tobacco suspension cell line BY-2 appears to be a particularly suitable host cell line because of its high growth rate [61]. Like plant tissues, BY-2 suspension cells can be transformed easily by cocultivation with *A. tumefaciens* [62]. In addition to this tobacco cell line, soybean cell culture has been used for the production of hepatitis B virus surface antigens [24] and rice suspension cells have been employed for the expression of human α-1-antitrypsin [30, 63, 64] and human lysozyme [65]. In contrast to the tobacco transformation process, gene transfer to rice is often performed by particle bombardment. Table 7.2 provides an overview of pharmaceutical proteins that have been produced in plant cell cultures so far and the expression cassette elements used in each case. For additional informations see also chapter 2.

Possible contamination by chemical or biological substances is one of the most important concerns when producing pharmaceutical proteins. Plant cell cultures ensure the production of the desired protein in a controlled, sterile and sealed environment and can be adapted to cGMP conditions. Therefore, the risk of contamination is minimized and the production conditions can be modified more easily in a contained reactor than in the field. Another advantage is the ability to freeze plant suspension cells in liquid nitrogen [66, 67] so that master and working cell banks can be established, a prerequisite for cGMP procedures [68].

Although plant cell culture is not as cost effective as plant cultivation in the open field, it will become an economical process if higher protein yields can be achieved [58]. The cultivation medium of plants is chemically defined, consisting of a carbon source, minerals, vitamins and phytohormones [69]. Furthermore, it is protein-free and relatively inexpensive. In contrast, animal cells often require complex supplements such as fetal calf serum and/or expensive growth factors, although serum-free cultivation is possible in case of Chinese hamster ovary (CHO) cells [70].

When trying to increase the yields of recombinant protein in plant suspension cultures, one should consider optimizing the nutrient supply and including product-stabilizing agents. Sharp and Doran [55] showed that antibody accumulation in hairy root cultures was improved by increasing the dissolved oxygen tension to 150% air saturation and that loss of the antibody could be minimized by inhibiting protein transport in the secretory pathway with the antibiotic brefeldin A. The beneficial effects of stabilizing solutes were demonstrated among others by Magnuson *et al.* [71] who achieved a 35-fold increase in scFv yield by adding polyvinylpyrrolidone (PVP) to the culture medium. Other proteins were stabilized to a lesser degree by supplementing the medium with the protease inhibitor bacitracin [72], bovine serum albu-

Tab. 7.2 Pharmaceutical proteins produced in plant suspension cultures

Protein	Regulatory elements	Additional elements of the construct	Species	Accumulation level	Literature
Human α_1 antitrypsin	Promoter of rice α amylase gene *RAmy3D*/ terminator of *RAmy3D*	LP of rice α amylase gene *RAmy3D*	*Oryza sativa*	4.6–5.7 mg g^{-1} dry cell weight	63
		Additional first intron of *RAmy3D*		No significant increase	
		LP of rice α amylase gene *RAmy3D*		18.2–24.2 mg g^{-1} dry cell weight	80
Human erythropoietin	CaMV 35S promoter/CaMV 35S terminator	Native LP	*N. tabacum* cv BY-2	26 ng g^{-1} TSP	81
α hepatitis B surface antigen-(scFv)	CaMV 35S promoter/ nos terminator	Sporamin LP	*N. tabacum* cv Havana SR	1 mg L^{-1} in culture supernatant and 5 mg kg^{-1} wet weight from cell extracts	146
hIL-12	CaMV enhanced 35S promoter	Native LP	*N. tabacum* cv Havana SR	175 μg L^{-1}	59
hIL-2, hIL-4	CaMV 35S promoter/ T7 terminator	Native LP	*N. tabacum* NT-1	0.1 mg L^{-1} (hIL-2) and 0.18 mg L^{-1} (hIL4)	60
Human serum albumin	CaMV enhanced 35S promoter/ AlMV RNA synthetic leader sequence/nos terminator	Native or PR-S LP	*N. tabacum*	250 μg g^{-1} TSP	48
Bryodin 1	CaMV 35S promoter/ nos terminator	Extensin LP	*N. tabacum* NT-1	30 mg L^{-1}	82

Tab. 7.2 (continued)

Protein	Regulatory elements	Additional elements of the construct	Species	Accumulation level	Literature
hG-CSF	CaMV enhanced 35S promoter/ TMV Ω 5′ UTR/nos terminator		N. tabacum	$105~\mu g~L^{-1}$ extracellular	57
GM-CSF	CaMV 35S promoter/TEV 5′ UTR/T7 terminator		N. tabacum cv BY-2	$150~\mu g~L^{-1}$ intracellular and $250~\mu g~L^{-1}$ extracellular	73
	CaMV enhanced 35S promoter/ TMV Ω 5′ UTR/nos terminator	Native GM-CSF LP	N. tabacum cv Havana SR	$180~\mu g~L^{-1}$ extracellular, after adding of gelatin increase to $738~\mu g~L^{-1}$	74

Abbreviations used in table 2: AlMV: alfalfa mosaic virus; BY-2: *Nicotiana tabacum* cv Bright Yellow 2; CaMV: cauliflower mosaic virus; FW: freshweight; GM-CSF: granulocyte macrophage colony stimulating factor; hG-CSF: human granulocyte colony stimulating factor; hIL: human interleukin; LP: leader peptide; nos: nopaline synthase; PR-S: pathogenesis related protein S; *RAmy 3D*: rice amylase gene promoter 3D; scFv: single chain Fragment variable; TEV: tobacco etch virus; TMV: tobacco mosaic virus; TSP: total soluble protein; UTR: untranslated region.

min (BSA) [73], gelatin [74] or dimethylsulfoxide (DMSO) [75]. Fischer *et al.* [76] reported a transient threefold increase in recombinant antibody levels after supplementing cell cultures with amino acids prior to harvest. All these studies show, however, that the benefits obtained by certain expression cassette elements or defined culture conditions cannot be generalized for all types of pharmaceutical proteins. These factors have to be investigated and optimized individually for every product.

In order to facilitate purification of a recombinant protein from plant cell culture, the recombinant protein can be directed to the extracellular space (apoplast and culture medium) by attaching an appropriate signal sequence or using a native signal sequence if one is present [2, 48, 77]. Only about 100–1000 endogenous proteins are thought to be secreted from cultured plant cells [78], so the purification of a recombinant protein from the culture medium is much easier than purification from whole cell extracts. The plant cell wall is a natural barrier to protein secretion. Proteins >20 kDa tend to be retained [79] although higher-molecular-weight proteins such as monoclonal antibodies are secreted [71]. Secretion can be further enhanced by the addition of DMSO [75].

Although cell cultures produce a much lower amount of biomass than plants cultivated in the open field, tobacco suspension cells have been cultivated at volumes of up to 20 m^3 (see [61] and references therein).

7.4
From Gene to Functional Protein: Processing Steps in Plants

In order to make molecular farming commercially profitable, recombinant proteins must be produced at a sufficiently high yield and in an active form. It has become clear that, for high-level protein accumulation, the stability of transgene expression can be as important as the expression level itself. The quantity of protein is determined by the rate of protein synthesis, assembly as well as proteolytic degradation [83].

Processes that influence the production of a stable and functional end product in plants and plant cell suspension cultures include transgene integration, transcriptional and translational activity, posttranslational modification [84–88]. The first hurdle is stable transgene integration and expression in the genome of the transgenic plants or transformed cells, since transgenes are often silenced either immediately after integration or in subsequent generations [89]. A novel and very interesting method for dealing with this problem in transient gene expression is the use of tomato bushy stunt virus p19 protein [90]. This is a known suppressor of posttranscriptional gene silencing (PTGS) and transient expression of a range of proteins was increased 50-fold in the presence of p19. Another approach is the inclusion of scaffold attachment regions (SARs) or matrix attachment regions (MARs) within the expression construct. Such elements, which attach to the nuclear matrix and may help divide interphase chromatin into functional domains, have been shown to enhance transformation efficiency [91], to achieve stable gene expression [92] and to increase transgene expression levels up to 650-fold [93], although the level of enhancement appears to be dependent on the promoter [94].

Increased transcription levels are assumed to result in increased protein synthesis. One approach to reach this goal is to raise the transgene copy number by the use of amplification-promoting sequences derived from a spacer sequence of tobacco ribosomal DNA [95]. Posttranscriptional processes such as capping, splicing and polyadenylation are important for high protein yields, and it is also important to maximize mRNA stability [84].

The synthesis of a protein may be hampered by suboptimal translation initiation or termination sites. Lukaszewicz *et al.* [96] studied the context sequence of the initiation codon and Sawant *et al.* [97] were able to increase the expression level of a reporter gene by inserting a sequence commonly found in highly expressed plant proteins. Other problems can include the presence of cryptic splice or polyadenylation sites, and unfavorable codon bias (i.e. clusters of rare codons). If problems of this type are encountered, the construction of a synthetic gene can be considered [63, 65]. The artificial synthesis of a whole gene is expensive, but codon optimization at the 5' end of the open reading frame can be sufficient to enhance gene expression. Batard *et al.* [98] demonstrated an increase in wheat P450 enzyme activity in transgenic tobacco after optimization of the first 111 bp of the wheat open reading frame. This corresponds to 7% of the entire open reading frame.

An useful alternative to nuclear transformation is the direct transformation of plastids, since plastid genomes are not affected by epigenetic silencing and they have a codon bias different to that of nuclear genes [99]. By creating transplastomic plants, very high levels of recombinant proteins have been obtained. In the case of pharmaceutical proteins, 7% total soluble protein (TSP) has been achieved for human somatotropin [100] and 10% TSP has been achieved for human serum albumin (HSA) [101]. This exceeds the levels obtained through nuclear transformation, although the production of homoplastic plants is unfortunately time consuming and has only been achieved for a few species [102]. For further discussion of the chloroplast transgenic system, see chapter 8.

Posttranslational modifications can have dramatic effects on the accumulation of a recombinant protein. If proteins are not properly cleaved, pro-proteins will accumulate instead of mature proteins. For example, it has been shown that one of the causes for improper processing of soybean glycinins in potato tubers is the absence of the necessary proteases in the host plant [103]. Some proteins need specific modifications to become active, but these are not carried out in plants. These include the addition of certain sugar molecules [104] or the hydroxylation of proline for the production of native-like collagen in tobacco plants [105]. Differences in glycan structures between plants and mammals (e.g. the presence of xylose and fucose residues in plants) have raised concerns regarding the potential immunogenicity of recombinant proteins [106], especially those used as human therapeutics [86, 107]. In order to obtain an antibody with a glycan profile similar to the mammalian type, Bakker *et al.* [104] expressed human β-1,4 galactosyltransferase in tobacco plants and crossed them with plants expressing a murine antibody. The result was a plantibody with partially galactosylated N-glycans.

The folding of polypeptide chains and the assembly of multiple subunits are critical requirements when complex and multimeric proteins such as full size antibodies are expressed in plants. Correct folding of the protein may be favored by retention in the endoplasmic reticulum (ER), which can be achieved by adding a C-terminal K/HDEL ER-retention signal [15]. This prolongs the exposure of the protein to endogenous chaperones like BiP (binding protein) [108]. Bouquin *et al.* [50] expressed the heavy and light chains of an antibody individually in *Arabidopsis thaliana* plants. Strongly expressing parental lines were selected and crossed to obtain a hybrid line producing the full-size antibody. The multimeric antibody accumulated to higher levels than either of the individual chains in the parental lines, indicating that assembly of the mature protein had a stabilizing effect in comparison to the individual subunits.

Expressing recombinant proteins as N-terminal fusions with ubiquitin [109] is another strategy that can help to achieve proper folding. Additionally, this is an elegant way to obtain proteins that do not start with a methionine residue, since the ubiquitin moiety is cleaved off by endogenous ubiquitin-specific proteases [100].

If the desired protein is prone to proteolytic degradation, the use of appropriate targeting signals to direct the protein to different cellular compartments (e.g. the ER, chloroplast, vacuole or apoplast) or the use of tissue specific promoters (e.g. the hordein promoter for protein expression in barley grains) could be considered [14, 110]. Targeting is especially important if the recombinant protein is toxic to the production host. For example, targeting the non-pharmaceutical protein avidin to the cytosol in transgenic tobacco plants was toxic, but plants were regenerated successfully when this molecule was targeted to the vacuole [111]. The expression of a secreted version of avidin in maize (*Zea mays*) resulted in partial or complete male sterility [112].

7.5
Case Studies of Improved Protein Yields

We will focus on recombinant antibodies (rAbs) and their derivatives (e.g scFvs, diabodies) in this section because these classic plant-produced pharmaceuticals have been expressed in many diverse production systems using a variety of regulatory elements [2, 113]. However, even in the case of this extensively studied class of proteins, direct comparisons between systems are difficult to make, since different genes, regulatory elements and production hosts have been used. Some investigators have compared the expression of a particular rAb in different plant species, or have targeted the antibody to different tissues or subcellular compartments in one particular host. Generally, such comparisons have revealed significant variation in expression and accumulation levels (Table 7.1). Stoger *et al.* [23] expressed the scFv antibody T 84.66, which recognizes the carcinoembryogenic antigen (CEA), and systematically compared several plant systems, regulatory elements and subcellular targeting signals. They concluded that the major factor determining overall accumulation levels was the stability of the antibody, which could be controlled in part by subcellular targeting. In agreement with others, the highest levels of active antibody were obtained by targeting the protein to the secretory pathway using its native signal peptide, and then retaining the antibody in the ER lumen using a KDEL tag. Presumably because of the favorable biochemical environment, the lack of proteases and the presence of chaperones, rAbs accumulate to higher levels in the ER than in any other compartment [15, 108]. Furthermore, this strategy can limit the extent to which core glycans are replaced by plant-specific carbohydrates like $\alpha(1,3)$-fucose and $\beta(1,2)$-xylose [114]. Without ER retention, antibodies are secreted to the apoplast where accumulation levels are typically 10–100-fold lower than in the ER. The cytosol appears to be an inappropriate compartment for antibodies, resulting in very low accumulation levels [115, 116]. An exception is described by De Jaeger *et al.* [117] who targeted a phage display-derived scFv to the cytosol and achieved a level of 1% TSP.

Stoger *et al.* [23] also observed that between species, the amounts of scFv per unit fresh weight were in the same range, and did not correlate with the total protein content in the plant. For example, even though pea is a much more proteinaceous crop than rice, the amounts of antibody measured as a percentage of TSP were considerably lower.

Within each species, individual promoters resulted in distinct, tissue-dependent accumulation patterns. The cauliflower mosaic virus (CaMV) 35S promoter, for example, led to high-level accumulation in callus and leaves whereas the maize ubiquitin-1 promoter was the best choice for producing recombinant proteins in cereal seeds even though it is not in itself seed-specific [23]. The lack of such comparative studies for proteins other than rAbs makes it difficult to generalize an optimal expression strategy for all proteins. Tables 7.1 and 7.2 list recombinant proteins expressed in plants and provide details of the production system, promoters and other regulatory elements used in each case.

7.6
Downstream Processing

When choosing a plant expression system, downstream processing should be taken into account since this will contribute a substantial proportion of the overall costs [118]. A cost distribution for the production of β-glucuronidase in maize has been made by Evangelista *et al.* [26]. This is not a pharmaceutical protein, but the principles of cost analysis should hold true anyway. It was stated that protein extraction and purification would account for 40% and 48%, respectively, of the total annual operating costs, assuming a product purity of 83%. When higher purity is needed, as for pharmaceutical proteins, downstream processing could account for an even larger proportion of the total costs.

Various strategies have been proposed to simplify downstream processing. We have already mentioned the use of tissue-specific promoters [110, 119] or targeting signals that allow protein secretion or accumulation in a particular organelle. A useful example is protein targeting to oil bodies [120] which is used in combination with the oleosin-partitioning technique for protein isolation from oilseed crops. For new products that have yet to undergo clinical testing, the addition of an affinity tag can be considered to simplify protein isolation [121]. Such tags have unique properties, which allow fusion proteins to be isolated from cellular extracts or media by affinity to particular ligands. Alternatively, the tag may confer a novel physical property on the protein. An example is the synthetic elastin-like polypeptide (ELP) tag. When a temperature transition occurs, the tags fold into spirals that aggregate. These aggregates can easily be isolated by centrifugation [122]. However, if such tags are used, they must not adversely affect the function of the protein. Importantly, the presence of affinity tags on pharmaceutical proteins may result in non-compliance with regulatory issues [68].

A minimalist extraction buffer is often recommended for protein purification [25], since most additives provide only a marginal improvement in yields, but will increase costs significantly. The removal of phenolics, for example by tangential-flow ultrafiltration/diafiltration [25], is an important step that should be carried out as early as possible in the purification procedure, since these molecules can become covalently linked to amino acid side chains and can oxidize certain residual groups [123].

When using plant cell cultures for the production of biopharmaceuticals, targeting the protein to the apoplast (and from there to the culture medium) is beneficial for later purification [124]. Plant cell culture media are protein-free, which makes the purification of the product relatively easy. Unfortunately, secreted recombinant proteins are often unstable in the cell culture medium [71]. One strategy to overcome this problem was presented by James *et al.* [125] in form of an affinity chromatography bioreactor. By initiating protein purification while the cells are still growing, the exposure of the protein to destabilizing agents in the culture medium is minimized.

7.7
Market Potential of Plant-derived Pharmaceuticals

The production of pharmaceutical proteins in plants has clear advantages over traditional systems in terms of cost-efficiency and product safety, since there is no risk of contamination with human pathogens. Furthermore plants are much less likely than mammalian cells to be affected by the expression of certain human proteins, such as growth factors and cell cycle inhibitors [29]. Therefore, plants provide a strategic complement to existing microbial and animal production systems.

Antibodies account for more than 20% of all biotechnology-derived molecules, and the market is expected to reach $5 billion by 2005 [25]. More than 250 companies are working on over 700 therapeutic antibodies, of which 220 are in clinical trials [126]. These numbers demonstrate the ongoing commercial expansion of pharmaceutical protein production and indicate a growing demand for production capacity. It has been estimated that plant-derived antibodies can be produced in a 250 m^2 greenhouse at about one tenth of the cost of hybridoma-derived antibodies [18, 87]. Planet Biotechnology Inc. presents an even rosier picture, estimating that the cost of plant-derived antibodies will be 5% of the cost of antibodies produced in animal cells. This calculation, however, was based on yields of 500 μg g^{-1} leaf fresh weight [87], a rather ambitious level for antibody production in plants (see Table 7.1). Once again, the greatest proportion of the cost reflects downstream processing and purification. When purification is needed, it is estimated that a recombinant protein must be expressed at greater than 0.01% of fresh weight [127] or 1% of TSP [87] for molecular farming to be commercially feasible.

To date, several antibodies with therapeutic potential have been produced in plants [87]. Among them are the Guy's 13 antibody that recognizes *Streptococcus mutans* (the causative agent of dental caries), humanized antibodies that recognize viral antigens (herpes simplex and respiratory syncitial virus), an antibody against CEA and a tumor-specific vaccine for the treatment of lymphoma (for detailed overview see Ma *et al.* [147]). Although none of these antibodies is produced commercially at the current time, the lymphoma scFv has completed phase I clinical trials [147] and the Guy's 13 antibody has reached phase II clinical trials in the USA (*www.planetbiotech-nology.com/products.html*). Regarding other pharmaceutical proteins, enterotoxin B and aprotinin have both reached phase I trials, while gastric lipase is undergoing clinical phase IIa trials in France and in Germany (*www.meristem-therapeutics.com/GB/intro.htm*). Although no plant-derived pharmaceutical protein is yet available on the market, two technical proteins (avidin and β-glucuronidase) are successfully produced in corn for commercial purposes [128].

7.8
Containment Strategies for Molecular Farming

The issue that concerns people the most, with respect to the safety of plant biotechnology, is the possibility that transgenes could spread in the environment by outcrossing. Methods to contain foreign genes within transgenic plants are reviewed by Daniell [129] and include maternal inheritance, male or seed sterility and induced transgene excision before flowering. Chloroplast genetic engineering takes advantage of the fact that the chloroplasts of most crops are maternally inherited [130]. Male sterility can be achieved by crossing plants that produce the pharmaceutical protein of interest with plants that are male sterile [13] or through the use of genetic methods to prevent the production of functional pollen [131]. Seed sterility can be achieved using Monsanto's Terminator Technology™ or the recoverable block of function system [132]. The Cre-loxP system has been used for transgene excision [133].

Another strategy is the cultivation of transgenic plants in greenhouses. Although this increases production costs considerably, it could still be the cheapest solution overall. Khoudi *et al.* [18] calculated the cost of 1 g of purified antibody produced in greenhouses, and found that it was only 10% of the cost of hybridoma technology. A further advantage of molecular farming in greenhouses is that the product itself would be contained, a desirable outcome in the case of biopharmaceuticals that are potentially harmful, capable of persisting in the environment or of accumulating in non-target organisms. Such safety issues came to the fore after the ProdiGene fiasco in late 2002 [134], prompting the US Department of Agriculture in March 2003 to tighten its permit conditions relating to the field-testing of plants engineered to produce pharmaceutical and industrial compounds (*www.aphis.usda.gov/ppd/rad/webrepor.html*). Another means to improve safety standards is the use of inducible promoters, which could be used to delay protein expression until the crop has reached the factory [135]. This would also eliminate protein degradation problems caused by the delay between harvesting and processing.

Selectable markers are an essential part of the transformation construct, since they allow the selective propagation of transformed cells. Antibiotic resistance is the most widely used selectable phenotype, but there are concerns that resistance genes could be transferred into bacteria and might spread in the environment. For example Mercer *et al.* [136] reported that oral bacteria can be transformed with naked DNA in human saliva. The persistence of DNA in the field was reviewed recently by Dale *et al.* [137], who stated that there is no evidence for the incorporation of functional plant (trans)genes into the meat products of animals fed with genetically-modified feed. Nevertheless, the use of such markers has been legally challenged in Europe, and the commercial use of antibiotic resistance markers will be prohibited from 2005 (EU Directive 2001/18/EC released 21/03/2001). Other types of marker gene may be used, such as metabolic markers that allow the use of an alternative carbon source [138, 139] or growth-regulation markers that express enzymes involved in the bio-synthesis of plant hormones [140, 141]. Another approach is the use of removable markers [133, 142]. For a review on these selection procedures see [143]. Also, a re-cent publication [144] proves the feasibility of marker-free transformation, which must be regarded as the best possible technique to prevent an unintended spread of antibiotic resistance genes.

7.9
Concluding Remarks

Numerous issues must be taken into account in order to make molecular farming a commercial success. These issues include the yield and quality of the target protein, downstream processing methods, storage and purification, transgene containment and the cost of production. Currently, there are few examples of plant-derived phar-maceutical proteins that have been produced in a manner that addresses such con-cerns. We are facing a growing demand for therapeutics and diagnostics, but almost the entire capacity for cell-based production is already in use, leaving little room for the estimated 50–60 new protein pharmaceuticals that will be approved over the next 6–7 years [145]. A shift to plant-bioreactors could therefore become a necessity. Although molecular farming still faces a number of challenges, one can clearly see the benefits and possibilities of the technology. Further research is needed to meet these challenges and facilitate the cost-effective and safe production of pharmaceuti-cals in plants.

References

[1] E. FRANKEN, U. TEUSCHEL, R. HAIN, *Curr. Opin. Biotechnol.* **1997**, *8* (4), 411–416.

[2] A. HIATT, R. CAFFERKEY, K. BOWDISH, *Nature* **1989**, *342* (6245), 76–78.

[3] J. VANDERKERCKHOVE, J. VAN DAMME, M. VAN LIJSEBETTENS et al., *Bio/Technology* **1989**, *7* (9), 929–932.

[4] J. K. MA, B. Y. HIKMAT, K. WYCOFF et al., *Nature Med.* **1998**, *4* (5), 601–606.

[5] S. P. MAYFIELD, S. E. FRANKLIN, R. A. LERNER, *Proc. Natl Acad. Sci. USA* **2003**, *100* (2), 438–442.

[6] R. FISCHER, S. SCHILLBERG, N. EMANS, *Outlook Agric.* **2001**, *30* (1), 31–36.

[7] S. J. SHEEN, *Beitr. Tabakforsch.* **1983**, *12* (1), 35–42.

[8] R. B. HORSCH, J. E. FRY, N. L. HOFFMANN et al., *Science* **1985**, *227* (4691), 1229–1231.

[9] J. KAPILA, R. DE RYCKE, M. VAN MONTAGU et al., *Plant Sci.* **1997**, *122* (1), 101–108.

[10] S. HENDY, Z. C. CHEN, H. BARKER et al., *J. Immunol. Meth.* **1999**, *231* (1–2), 137–146.

[11] A. A. MCCORMICK, M. H. KUMAGAI, K. HANLEY et al., *Proc. Natl Acad. Sci. USA* **1999**, *96* (2), 703–708.

[12] S. W. MA, D. L. ZHAO, Z. Q. YIN et al., *Nature Med.* **1997**, *3* (7), 793–796.

[13] R. MENASSA, V. NGUYEN, A. JEVNIKAR et al., *Mol. Breeding* **2001**, *8* (2), 177–185.

[14] U. CONRAD, U. FIEDLER, *Plant Mol. Biol.* **1998**, *38* (1–2), 101–109.

[15] U. FIEDLER, J. PHILLIPS, O. ARTSAENKO et al., *Immunotechnology* **1997**, *3* (3), 205–216.

[16] G. Y. ZHONG, D. PETERSON, D. E. DELANEY et al., *Mol. Breeding* **1999**, *5* (4), 345–356.

[17] E. STOGER, C. VAQUERO, E. TORRES et al., *Plant Mol. Biol.* **2000**, *42* (4), 583–590.

[18] H. KHOUDI, S. LABERGE, J. M. FERULLO et al., *Biotechnol. Bioeng.* **1999**, *64* (2), 135–143.

[19] O. ARTSAENKO, B. KETTIG, U. FIEDLER et al., *Mol. Breeding* **1998**, *4* (4), 313–319.

[20] C. DE WILDE, K. PEETERS, A. JACOBS et al., *Mol. Breeding* **2002**, *9* (4), 271–282.

[21] D. L. PARMENTER, J. G. BOOTHE, G. J. H. VAN ROOIJEN et al., *Plant Mol. Biol.* **1995**, *29* (6), 1167–1180.

[22] Y. PERRIN, C. VAQUERO, I. GERRARD et al., *Mol. Breeding* **2000**, *6* (4), 345–352.

[23] E. STOGER, M. SACK, Y. PERRIN et al., *Mol. Breeding* **2002**, *9* (3), 149–158.

[24] M. L. SMITH, M. E. KEEGAN, H. S. MASON et al., *Biotechnol. Prog.* **2002**, *18* (3), 538–550.

[25] J. W. LARRICK, D. W. THOMAS, *Curr. Opin. Biotechnol.* **2001**, *12* (4), 411–418.

[26] R. L. EVANGELISTA, A. R. KUSNADI, J. A. HOWARD et al., *Biotechnol. Prog.* **1998**, *14* (4), 607–614.

[27] L. H. STEVENS, G. M. STOOPEN, I. J. W. ELBERS et al., *Plant Physiol.* **2000**, *124* (1), 173–182.

[28] I. J. W. ELBERS, G. M. STOOPEN, H. BAKKER et al., *Plant Physiol.* **2001**, *126* (3), 1314–1322.

[29] C. L. CRAMER, J. G. BOOTHE, K. K. OISHI, *Curr. Top. Microbiol. Immunol.* **1999**, *240*, 95–118.

[30] M. TERASHIMA, Y. EJIRI, N. HASHIKAWA et al., *Biochem. Eng. J.* **1999**, *4* (1), 31–36.

[31] K. PEETERS, C. DE WILDE, A. DEPICKER, *Eur. J. Biochem.* **2001**, *268* (5), 4251–4260.

[32] M. DE NEVE, M. DE LOOSE, A. JACOBS et al., *Transgenic Res.* **1993**, *2* (4), 227–237.

[33] E. BENVENUTO, R. J. ORDAS, R. TAVAZZA et al., *Plant Mol. Biol.* **1991**, *17* (4), 865–874.

[34] J. K. MA, T. LEHNER, P. STABILA et al., *Eur. J. Immunol.* **1994**, *24* (1), 131–138.

[35] J. K. MA, A. HIATT, M. HEIN et al., *Science* **1995**, *268* (5211), 716–719.

[36] L. ZEITLIN, S. S. OLMSTED, T. R. MOENCH et al., *Nature Biotechnol.* **1998**, *16* (13), 1361–1364.

[37] W. DIERYCK, J. PAGNIER, C. POYART et al., *Nature* **1997**, *386* (6620), 29–30.

[38] K. HIGO, Y. SAITO, H. HIGO, *Biosci. Biotechnol. Biochem.* **1993**, *57* (9), 1477–1481.

[39] A. PORCEDDU, A. FALORNI, N. FERRADINI et al., *Mol. Breeding* **1999**, *5* (6), 553–560.

[40] L. Avesani, A. Falorni, G. B. Tor-
nielli et al., *Transgenic Res.* **2003**, *12* (2),
203–212.

[41] F. Ruggiero, J. Y. Exposito, P. Bournat et al., *FEBS Lett.* **2000**, *469* (1),
132–136.

[42] H. S. Mason, D. M. Lam, C. J. Arntzen, *Proc. Natl Acad. Sci. USA* **1992**, *89*
(24), 11745–11749.

[43] L. J. Richter, Y. Thanavala, C. J. Arntzen, et al., *Nature Biotechnol.* **2000**, *18*
(11), 1167–1171.

[44] T. A. Haq, H. S. Mason, J. D. Clements
et al., *Science* **1995**, *268* (5211), 714–716.

[45] D. Jani, L. S. Meena, Q. M. Rizwan-ul-
Haq et al., *Transgenic Res.* **2002**, *11* (5),
447–454.

[46] Y. Park, H. Cheong, *Protein Expr.
Purif.* **2002**, *25* (1), 160–165.

[47] C. Vaquero, M. Sack, F. Schuster
et al., *FASEB J.* **2002**, *16* (3), 408–410.

[48] P. C. Sijmons, B. M. Dekker,
B. Schrammeijer et al., *Bio/Technology*
1990, *8* (3), 217–221.

[49] O. Edelbaum, D. Stein, N. Holland
et al., *J. Interferon Res.* **1992**, *12* (6),
449–453.

[50] T. Bouquin, M. Thomsen, L. K. Nielsen et al., *Transgenic Res.* **2002**, *11* (2),
115–122.

[51] V. Gruber, P. P. Berna, T. Arnaud
et al., *Mol. Breeding* **2001**, *7* (4), 329–
340.

[52] N. Ramirez, M. Ayala, D. Lorenzo
et al., *Transgenic Res.* **2002**, *11* (1), 61–
64.

[53] X. J. Hu, Z. C. Zhang, Y. M. Bao et al.,
Biotechnol. Lett. **2002**, *24* (18), 1531–
1534.

[54] K. Ohya, T. Matsumura, K. Ohashi
et al., *J. Interferon & Cytokine Res.* **2001**,
21 (8), 595–602.

[55] J. M. Sharp, P. M. Doran, *Biotechnol.
Prog.* **2001**, *17* (6), 979–992.

[56] R. Fischer, N. Emans, F. Schuster
et al., *Biotechnol. Appl. Biochem.* **1999**,
30 (2), 109–112.

[57] S. Y. Hong, T. H. Kwon, J. H. Lee et al.,
Enzyme & Microbial Tech. **2002**, *30* (6),
763–767.

[58] E. James, J. M. Lee, *Adv. Biochem. Eng.
Biotechnol.* **2001**, *72*, 127–156.

[59] T. H. Kwon, J. E. Seo, J. Kim et al., *Biotechnol. Bioeng.* **2003**, *81* (7), 870–875.

[60] N. S. Magnuson, P. M. Linzmaier,
R. Reeves et al., *Protein Expr. Purif.*
1998, *13* (1), 45–52.

[61] T. Nagata, Y. Nemoto, S. Hasezawa,
Int. Rev. Cytol. **1992**, *132*, 1–30.

[62] G. An, *Plant Physiol.* **1985**, *79* (2), 568–
570.

[63] J. Huang, T. D. Sutliff, L. Wu et al.,
Biotechnol. Prog. **2001**, *17* (1), 126–133.

[64] M. M. Trexler, K. A. McDonald,
A. P. Jackman, *Biotechnol. Prog.* **2002**,
18 (3), 501–508.

[65] J. Huang, L. Wu, D. Yalda et al., *Transgenic Res.* **2002**, *11* (3), 229–239.

[66] I. Furutani, H. Yasuhara, S. Matsu-
naka, *Shokubutsu Soshiki Baiyo* **1996**,
13 (1), 331–336.

[67] F. Van Iren, E. W. Schrijnemakers,
P. J. Reinhoud et al., *Biotechnologia*
1996, *1* (32), 60–68.

[68] L. Miele, *Trends Biotechnol.* **1997**, *15* (2),
45–50.

[69] T. Murashige, F. Skoog, *Physiol. Plant.*
1962, *15*, 473–497.

[70] S. C. O. Pak, S. M. N. Hunt,
M. W. Bridges et al., *Cytotechnology*
1996, *22* (1–3), 139–146.

[71] N. S. Magnuson, P. M. Linzmaier,
J. W. Gao et al., *Protein Expr. Purif.*
1996, *7* (2), 220–228.

[72] H. Xu, F. U. Montoya, Z. P. Wang
et al., *Protein Expr. Purif.* **2002**, *24* (3),
384–394.

[73] E. A. James, C. Wang, Z. Wang et al.,
Protein Expr. Purif. **2000**, *19* (1), 131–138.

[74] J. H. Lee, N. S. Kim, T. H. Kwon et al.,
J. Biotechnol. **2002**, *96* (3), 205–211.

[75] M. F. Wahl, G. H. An, J. M. Lee, *Biotechnol. Lett.* **1995**, *17* (5), 463–468.

[76] R. Fischer, Y. C. Liao, J. Drossard, *J.
Immunol. Methods* **1999**, *226* (1–2), 1–
10.

[77] M. B. Hein, Y. Tang, D. A. McLeod
et al., *Biotechnol. Prog.* **1991**, *7* (5), 455–
461.

[78] Y. Okushima, N. Koizumi, T. Kusano
et al., *Plant Mol. Biol.* **2000**, *42* (3), 479–
488.

[79] N. Carpita, D. Sabularse, D. Montezinos et al., *Science* **1979**, *205* (4411),
1144–1147.

[80] M. Terashima, Y. Murai, M. Kawamura et al., *Appl. Microbiol. Biotechnol.*
1999, *52* (4), 516–523.

[81] S. Matsumoto, K. Ikura, M. Ueda et al., *Plant Mol. Biol.* 1995, *27* (6), 1163–1172.

[82] J. A. Francisco, S. L. Gawlak, M. Miller et al., *Bioconj. Chem.* 1997, *8* (5), 708–713.

[83] A. Rozkov, T. Schweder, A. Veide et al., *Enzyme & Microbial Tech.* 2000, *27* (10), 743–748.

[84] A. R. Kusnadi, Z. L. Nikolov, J. A. Howard, *Biotechnol. Bioeng.* 1997, *56* (5), 473–484.

[85] R. Fischer, N. Emans, *Transgenic Res.* 2000, *9* (4–5), 279–299.

[86] G. Giddings, *Curr. Opin. Biotechnol.* 2001, *12* (5), 450–454.

[87] H. Daniell, S. J. Streatfield, K. Wycoff, *Trends Plant Sci.* 2001, *6* (5), 219–226.

[88] S. Schillberg, N. Emans, R. Fischer, *Phytochem. Rev.* 2002, *1* (1), 45–54.

[89] H. Vaucheret, C. Beclin, T. Elmayan et al., *Plant J.* 1998, *16* (6), 651–659.

[90] O. Voinnet, S. Rivas, P. Mestre et al., *Plant J.* 2003, *33* (5), 949–956.

[91] K. Petersen, R. Leah, S. Knudsen et al., *Plant Mol. Biol.* 2002, *49* (1), 45–58.

[92] G. C. Allen, S. Spiker, W. F. Thompson, *Plant Mol. Biol.* 2000, *43* (2–3), 361–376.

[93] Z. Q. Cheng, J. Targolli, R. Wu, *Mol. Breeding* 2001, *7* (4), 317–327.

[94] S. L. Mankin, G. C. Allen, T. Phelan et al., *Transgenic Res.* 2003, *12* (1), 3–12.

[95] N. Borisjuk, L. Borisjuk, S. Komarnytsky et al., *Nature Biotechnol.* 2000, *18* (12), 1303–1306.

[96] M. Lukaszewicz, M. Feuermann, B. Jerouville et al., *Plant Sci.* 2000, *154* (1), 89–98.

[97] S. V. Sawant, K. Kiran, P. K. Singh et al., *Plant Physiol.* 2001, *126* (4), 1630–1636.

[98] Y. Batard, A. Hehn, S. Nedelkina et al., *Arch. Biochem. Biophys.* 2000, *379* (1), 161–169.

[99] J. L. Oliver, A. Marin, J. M. Martinez-Zapater, *Nucleic Acids Res.* 1990, *18* (1), 65–73.

[100] J. M. Staub, B. Garcia, J. Graves et al., *Nature Biotechnol.* 2000, *18* (3), 333–338.

[101] A. Fernández-San Millán, A. Mingo-Castel, M. Miller et al., *Plant Biotechnol. J.* 2003, *1* (1), 71–79.

[102] P. Maliga, *Trends Biotechnol.* 2003, *21* (1), 20–28.

[103] S. Utsumi, S. Kitagawa, T. Katsube et al., *Plant Sci.* 1994, *102* (2), 181–188.

[104] H. Bakker, M. Bardor, J. W. Molthoff et al., *Proc. Natl Acad. Sci. USA* 2001, *98* (5), 2899–2904.

[105] C. Merle, S. Perret, T. Lacour et al., *FEBS Lett.* 2002, *515* (1–3), 114–118.

[106] M. Bardor, C. Faveeuw, A. C. Fitchette et al., *Glycobiology* 2003, *13* (6), 427–434.

[107] M. Cabanes-Macheteau, A. C. Fitchette-Laine, C. Loutelier-Bourhis et al., *Glycobiology* 1999, *9* (4), 365–372.

[108] J. Nuttall, N. Vine, J. L. Hadlington et al., *Eur. J. Biochem.* 2002, *269* (24), 6042–6051.

[109] D. Hondred, J. M. Walker, D. E. Mathews et al., *Plant Physiol.* 1999, *119* (2), 713–724.

[110] H. Horvath, J. Huang, O. Wong et al., *Proc. Natl Acad. Sci. USA* 2000, *97* (4), 1914–1919.

[111] C. Murray, P. W. Sutherland, M. M. Phung et al., *Transgenic Res.* 2002, *11* (2), 199–214.

[112] E. E. Hood, D. R. Witcher, S. Maddock et al., *Mol. Breeding* 1997, *3* (4), 291–306.

[113] K. During, S. Hippe, F. Kreuzaler et al., *Plant Mol. Biol.* 1990, *15* (2), 281–293.

[114] K. Ko, Y. Tekoah, P. M. Rudd et al., *Proc. Natl Acad. Sci. USA* 2003, *100* (13), 8013–8018.

[115] A. Schouten, J. Roosien, F. A. van Engelen et al., *Plant Mol. Biol.* 1996, *30* (4), 781–793.

[116] L. F. Fecker, R. Koenig, C. Obermeier, *Arch. Virol.* 1997, *142* (9), 1857–1863.

[117] G. De Jaeger, E. Buys, D. Eeckhout et al., *Eur. J. Biochem.* 1999, *259* (1–2), 426–434.

[118] T. J. Menkhaus, S. U. Eriksson, P. B. Whitson et al., *Biotechnol. Bioeng.* 2002, *77* (2), 148–154.

[119] J. Phillips, O. Artsaenko, U. Fiedler et al., *EMBO J.* 1997, *16* (15), 4489–4496.

[120] G. J. VAN ROOIJEN, M. M. MOLONEY, *Bio/Technology* **1995**, *13* (1), 72–77.

[121] K. TERPE, *Appl. Microbiol. Biotechnol.* **2003**, *60* (5), 523–533.

[122] D. E. MEYER, K. TRABBIC-CARLSON, A. CHILKOTI et al., *Biotechnol. Prog.* **2001**, *17* (4), 720–728.

[123] C. F. VAN SUMERE, J. ALBRECHT, A. DE-DONDER et al., The chemistry and biochemistry of plant proteins, *Proceedings of the Phytochemical Society Symposium*, Academic Press, **1975**.

[124] P. M. DORAN, *Curr. Opin. Biotechnol.* **2000**, *11* (2), 199–204.

[125] E. JAMES, D. R. MILLS, J. M. LEE, *Biochem. Eng. J.* **2002**, *12* (3), 205–213.

[126] J. V. GAVILONDO, J. W. LARRICK, *Biotechniques* **2000**, *29* (1), 128–145.

[127] I. FARRAN, J. J. SANCHEZ-SERRANO, J. F. MEDINA et al., *Transgenic Res.* **2002**, *11* (4), 337–346.

[128] E. E. HOOD, A. KUSNADI, Z. NIKOLOV et al., *Adv. Exp. Med. Biol.* **1999**, *464*, 127–147.

[129] H. DANIELL, *Nature Biotechnol.* **2002**, *20* (6), 581–586.

[130] P. MALIGA, *Curr. Opin. Plant Biol.* **2002**, *5* (2), 164–172.

[131] C. MARIANI, M. DEBEUCKELEER, J. TRUETTNER et al., *Nature* **1990**, *347* (6295), 737–741.

[132] V. KUVSHINOV, K. KOIVU, A. KANERVA et al., *Plant Sci.* **2001**, *160* (3), 517–522.

[133] E. C. DALE, D. W. OW, *Proc. Natl Acad. Sci. USA* **1991**, *88* (23), 10558–10562.

[134] J. L. FOX, *Nature Biotechnol.* **2003**, *21* (1), 3–4.

[135] C. L. CRAMER, D. L. WEISSENBORN, US Patent Application 5,670,349, **1997**.

[136] D. K. MERCER, K. P. SCOTT, W. A. BRUCE-JOHNSON et al., *Appl. Environ. Microbiol.* **1999**, *65* (1), 6–10.

[137] P. J. DALE, B. CLARKE, E. M. FONTES, *Nature Biotechnol.* **2002**, *20* (6), 567–574.

[138] A. HALDRUP, S. G. PETERSEN, F. T. OKKELS, *Plant Mol. Biol.* **1998**, *37* (2), 287–296.

[139] P. ZHANG, I. POTRYKUS, J. PUONTI-KAERLAS, *Transgenic Res.* **2000**, *9* (6), 405–415.

[140] H. EBINUMA, K. SUGITA, E. MATSUNAGA et al., *Proc. Natl Acad. Sci. USA* **1997**, *94* (6), 2117–2121.

[141] T. KUNKEL, Q. W. NIU, Y. S. CHAN et al., *Nature Biotechnol.* **1999**, *17* (9), 916–919.

[142] J. I. YODER, A. P. GOLDSBROUGH, *Bio/Technology* **1994**, *12* (3), 263–267.

[143] P. A. LESSARD, H. KULAVEERASINGAM, G. M. YORK et al., *Metabolic Eng.* **2002**, *4* (1), 67–79.

[144] N. DE VETTEN, A. M. WOLTERS, K. RAEMAKERS et al., *Nature Biotechnol.* **2003**, *21* (4), 439–442.

[145] K. GARBER, *Nature Biotechnol.* **2001**, *19* (3), 184–185.

[146] N. RAMIREZ, D. LORENZO, D. PALENZUELA et al., *Biotechnol. Letters* **2000**, *22* (15), 1233–1236.

[147] J. K. MA, P. M. DRAKE, P. CHRISTOU, *Nat. Rev. Genet.* **2003**, *4* (10), 794–805.

8
Chloroplast Derived Antibodies, Biopharmaceuticals and Edible Vaccines

Henry Daniell, Olga Carmona-Sanchez and Brittany E. Burns

8.1
Introduction

In 2003, the UN Human Poverty Index showed that 1.2 billion of the developing world's 4.8 billion people were living on less than $1 per day, while a further 2.8 billion were living on less than $2 per day. The global poverty level, and the incidence of disease, is increasing dramatically every year. The World Health Organization estimates that approximately 170 million people worldwide are infected with hepatitis C virus (HCV), with 3–4 million new cases each year, and that more than one third of the world's population is infected with hepatitis B virus (HBV). In Asia, the prevalence of chronic hepatitis B and C is very high (about 150 million and 110 million people infected with HBV and HCV, respectively). A large majority of HCV-infected patients have severe liver cirrhosis and currently there is no vaccine available for this disease. In addition, the rising cost of treatment for severe illnesses calls for the more economical production of therapeutic proteins. For example, the annual demand for insulin-like growth factor 1 (IGF-1) per cirrhotic patient is 600 mg (1.5–2 mg per day) and the cost of IGF-1 per mg is $30,000. Furthermore, the current annual cost of interferon therapy for viral hepatitis is $26,000 (Cowley & Geoffrey, Newsweek, April 22, 2002). It is evident that agricultural scale production of therapeutic proteins and vaccines is necessary to meet this large demand at an affordable cost.

Plants have been used in medicine for many centuries but it is now possible to exploit plants as bioreactors for the production of human therapeutic proteins. The use of plants to produce biopharmaceuticals has several advantages, including the ability of plants to supply large quantities of therapeutic proteins to patients in a cost-efficient manner. In addition, plant-derived products are less likely than those derived from animal cells to be contaminated with human pathogenic microorganisms, since plants do not act as hosts for human infectious agents [1]. Plants offer a number of advantages over other transgenic systems. For instance, the use of a cell fermenter or bioreactor is unnecessary and the need for post-translational modification and purification can be eliminated, thus lowering the production costs. The need for fermentation facilities has seriously limited the introduction of new thera-

Molecular Farming. Edited by Rainer Fischer, Stefan Schillberg
Copyright © 2004 WILEY-VCH Verlag GmbH & Co. KGaA, Weinheim
ISBN: 3-527-30786-9

peutic proteins in the past. However, it has been estimated that one tobacco plant is able to produce more recombinant protein than a 300-liter fermenter of *E. coli* (Crop-Tech, VA) reflecting tobacco's biomass production of 40 tons of leaf fresh weight per acre (based on multiple mowings per season) and the production of up to one million seeds per plant [2]. Consequently, the quantity of recombinant protein that can be harvested is only limited by the number of hectares that can be planted with transgenic crops [3]. Furthermore, human proteins can be expressed without codon optimization, and fully assembled proteins can be produced since post-translational modifications such as glycosylation and the formation of disulfide bonds are possible in plant cells, especially within particular cellular compartments.

8.2
Expression of Therapeutic and Human Proteins in Plants

The first human protein produced in plants – growth hormone – was expressed as a fusion with the *Agrobacterium tumefaciens* nopaline synthase gene product [3,4]. It was not until 1989 that the expression of antibodies in tobacco demonstrated the ability of plants to produce therapeutic proteins for human use [5]. Transgenic tobacco, maize, soybean and alfalfa plants are expected to yield over 10 kg of therapeutic protein per acre, and this should reduce production costs by approximately 90% compared to other systems [6]. However, with the exception of enzymes like phytase in tobacco (14% total soluble protein (TSP) [7]), and eubacterial glucanase in *Arabidopsis thaliana* (26% TSP [8]), expression levels are generally less than 1% TSP in nuclear transgenic plants, especially for human therapeutic proteins [9]. Such examples include the B subunit of *Escherichia coli* heat-labile enterotoxin (LTB) (0.01% TSP [10]), hepatitis B virus envelope surface protein (0.01% TSP [11,12]), human cytomegalovirus glycoprotein B (0.02% TSP [13]), transmissible gastroenteritis coronavirus glycoprotein S (0.06% TSP [14]), human serum albumin (0.02% TSP), human protein C (0.001% TSP), human epidermal growth factor (0.0001% TSP), erythropoietin (0.026% TSP), and human interferon-β (0.000017% of fresh weight) [2, 9, 15, 16]. Therefore, strategies are required to increase expression levels, allowing commercial exploitation of the therapeutic and human proteins mentioned above. Such strategies have been the focus of transgenic plant research over the past several years.

8.3
The Transgenic Chloroplast System

The world of plant genetic engineering research was revolutionized when it was reported that *Bacillus thuringiensis* (Bt) Cry2Aa2 protein could accumulate to a level of 46.1% TSP in transgenic tobacco chloroplasts [17]. This was not only the highest level of recombinant protein ever achieved in plants, but a complete bacterial operon was expressed successfully for the first time resulting in the formation of stable Cry2Aa2 crystals. These results were possible because native bacterial genes were expressed at levels several hundred-fold higher in the chloroplast compartment than in nuclear transgenic plants [17, 18, 19]. Chloroplast transformation not only results in higher protein expression levels compared to nuclear genetic engineering, but also provides several other advantages. Even though transgenic chloroplasts may be present in pollen, foreign genes do not escape to other crops because chloroplast DNA in not passed into the egg cell. In most crop plants, plastid genes are inherited uniparentally in a strictly maternal fashion. Although pollen from such plants contains metabolically active plastids, the plastid DNA itself is lost during the process of pollen maturation and hence is not transmitted to the next generation [20]. Also, the chloroplast can be a good place to store proteins, or their biosynthetic products, that might be harmful to the host plant and adversely affect its physiology if allowed to accumulate in the cytoplasm [21]. This has been demonstrated in the case of the cholera toxin B subunit (CTB), a candidate oral subunit vaccine for cholera, because it accumulated in large quantities within transgenic plastids but was non-toxic to the plant [22]; in contrast, even very small quantities of LTB were toxic when expressed in the cytosol [86]. Similarly, trehalose, which is used in the pharamaceutical industry as a preservative, was very toxic when it accumulated in the cytosol but was non-toxic when it was compartmentalized within plastids [23].

Transgenes are integrated into the spacer regions of the chloroplast genome by homologous recombination using chloroplast DNA flanking sequences. This allows site-specific integration and thus eliminates the position effects that are frequently observed in nuclear transgenic plants. As a result, it is not necessary to screen numerous putative transgenic lines to select those with high-level transgene expression. All chloroplast transgenic lines express the same level of foreign protein, within a range of physiological variations [22]. Yet another advantage is the lack of transgene silencing in chloroplast transgenic plants, which is a serious concern in nuclear transformation. It has been shown that there is no transcriptional gene silencing in chloroplast transgenic lines despite the accumulation of transcripts at a level 169-fold higher than in nuclear transgenic plants [23]. Similarly, there is no transgene silencing at the translational level, despite the accumulation of foreign protein at levels up to 47% of the total plant protein in chloroplast transgenic lines [17].

Purification costs can account for much of the expense involved in biopharmaceutical production. In the case of insulin, for example, chromatography alone accounts for 30% of the production costs and 70% of the set-up costs [24]. It is estimated that the oral delivery of properly folded and fully functional biopharmaceuticals in plant tissues could potentially reduce production costs by 90%. For such oral delivery to be successful, antibiotic selection should be avoided and the biopharmaceutical protein must be expressed in edible parts of the plant that require no cooking or processing. Antibiotic-free selection using a gene of plant origin has been developed recently [25]. Alternatively, antibiotic resistance genes can be eliminated using direct repeats or the Cre-*lox*P system [26, 27].

Another requirement for oral delivery is the ability to express foreign proteins in plastids that are present in non-green tissues. One such example is the expression of an antibiotic resistance gene (*aadA*) in tomato chromoplasts [28]. More recently, stable and highly efficient plastid transformation has been achieved in the non-green tissues of carrot plants [29]. Using carrots as the source of an edible vaccine or therapeutic protein provides many advantages, such as the uniform source of cell culture (which is one of the essential requirements for producing therapeutic proteins, i.e. a homogeneous single source of origin), rapid division, the large biomass using bioreactors, and direct delivery of a precise dose of vaccine antigens or biopharmaceuticals. Furthermore, when carrots are used as the delivery vehicle, cooking is unnecessary and this preserves the structural integrity of therapeutic proteins during consumption. Most importantly, carrot plants do not flower in the first year. Therefore, there are no flowers or reproductive structures during the harvest and this provides complete gene containment, preventing outcross via pollen and gene flow via seeds.

One of the most important features of the transgenic chloroplast system is that chloroplasts are able to carry out the processing of eukaryotic proteins, including the correct folding of subunits and the formation of disulfide bridges [15,22]. Functional assays showed that chloroplast-derived cholera toxin B subunit (CTB) binds to the intestinal membrane G_{M1}-ganglioside receptor, confirming correct folding and disulfide bond formation in the plant-derived CTB pentamers [15, 22]. Formation of disulfide bonds within a single polypeptide, resulting in a fully functional human therapeutic protein, has also been demonstrated for human somatotropin (growth hormone), interferon-α and interferon-γ [30–32]. Human therapeutic proteins as small as 20 amino acids or as large as the anthrax protective antigen (83 kDa) have been expressed in transgenic chloroplasts [33, 34]. The ability to express operons or multigene cassettes [17, 35–37] allows the engineering of foreign pathways and the expression of multi-subunit therapeutic proteins. The successful assembly of a monoclonal antibody in transgenic chloroplasts has demonstrated the ability of this system to produce complex multi-subunit proteins and has also indicated the presence of chaperones required for the assembly of such proteins within plastids [38]. These observations usher in a new era for the production of therapeutic proteins via chloroplast genetic engineering (Table 8.1). The rest of this chapter provides examples of therapeutic proteins expressed in transgenic chloroplasts.

8.3.1
Chloroplast-derived Human Antibodies

The most widely studied therapeutic proteins produced in plants include monoclonal antibodies for passive immunotherapy and antigens for use as oral vaccines [40]. Antibodies against dental caries, rheumatoid arthritis, cholera, *E. coli* diarrhea, malaria, certain cancers, Norwalk virus, HIV, rhinovirus, influenza, hepatitis B virus and herpes simplex virus have been produced in transgenic plants. However, the anti-*Streptococcus mutans* secretory antibody for the prevention of dental caries is the only plant-derived antibody currently in Phase II clinical trials [40]. Until recently, most antibodies were expressed in tobacco, potato, alfalfa, soybean, rice and wheat [9]. It has been estimated that for every 170 tons of harvested tobacco, 100 tons represents harvested leaves. A single hectare could thus yield 50 kg of secretory IgA [3, 41]. Furthermore, it has been estimated that the cost of antibody production in plants is half that in transgenic animals and 20 times lower than in mammalian cell cul-

Tab. 8.1 The production of therapeutic proteins for human or animal health in transgenic chloroplasts (updated from reference 36).

Biopharmaceutical proteins/Vaccines	Gene	Site of Integration	Promoter	5'/3' regulatory elements	% tsp expression	Laboratory
Elastin derived polymer	EG121	trnI/trnA	Prrn	T7gene10/TpsbA	ND	Daniell
Human somatotropin	HST	trnV/rps12/7	Prrn[a], PpsbA[b]	T7gene10[a] or psbA[b]/Trps16	7.0% [a] and 1.0% [b]	Monsanto
Cholera toxin	CtxB	trnI/trnA	Prrn	ggagg/TpsbA	4%	Daniell
Antimicrobial peptide	MSI-99	trnI/trnA	Prrn	ggagg/TpsbA	21.5–47%	Daniell
Insulin like growth factor	IGF-1	trnI/trnA	Prrn	PpsbA/TpsbA	33%	Daniell
Interferon alpha 5	INFα5	trnI/trnA	Prrn	PpsbA/TpsbA	ND	Daniell
Interferon alpha 2b	INFα2B	trnI/trnA	Prrn	PpsbA/TpsbA	19%	Daniell
Human Serum Albumin	Has	trnI/trnA	Prrn[a], PpsbA[b]	ggagg[a], psbA[b]/TpsbA	0.02% [a], 11.1% [b]	Daniell
Interferon gamma	IFN-g	rbcL/accD	PpsbA	PpsbA/TpsbA	6%	Reddy
Monoclonal antibodies	Guy's 13	trnI/trnA	Prrn	ggagg/TpsbA	ND	Daniell
Anthrax protective antigen	Pag	trnI/trnA	Prrn	PpsbA/TpsbA	4–5%	Daniell
Plague F1~V fusion antigen	CaF1~LcrV	trnI/trnA	Prrn	PpsbA/TpsbA	14.8%	Daniell
CPV VP2 protein	CTB-2L21[a], GFP-2L21[b]	trnI/trnA	Prrn	PpsbA/TpsbA	31.1%[a], 22.6%[b]	Daniell/ Veramendi

tures [9]. Based on the chloroplast's ability to produce bioactive and fully assembled proteins, a codon-optimized and humanized gene encoding a chimeric monoclonal antibody (IgA/G, Guy's 13) under the control of a specific 5'-untranslated region, was used successfully to synthesize and assemble monoclonal antibodies in transgenic tobacco chloroplasts, including the formation of disulfide bridges. Guy's 13 was developed to prevent dental caries, which is caused by *Streptococcus mutans* [15, 42, 43]. Daniell and Wycoff [42] confirmed integration of the chimeric antibody gene into the chloroplast genome using PCR and Southern blot hybridization analysis. Protein expression was also confirmed by western blot analysis. Expression of the individual heavy and light chains as well as the fully assembled antibody was shown (Fig. 8.1), indicating the presence of chaperones for proper protein folding and enzymes for formation of disulfide bonds within transgenic chloroplasts [15].

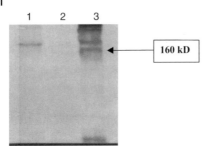

Fig. 8.1 Western blot analysis of transgenic lines showing the expression of an assembled monoclonal antibody in transgenic chloroplasts. Lane 1: Extract from a chloroplast transgenic line, Lane 2: Extract from an untransformed plant. Lane 3: Positive control (human IgA). The gel was run under non-reducing conditions. The antibody was detected with an AP-conjugated goat anti-human kappa antibody.

8.3.2
Chloroplast-derived Biopharmaceuticals

8.3.2.1 Human Serum Albumin

Human serum albumin (HSA) is currently obtained by the fractionation of blood serum. It accounts for 60% of the total protein in blood serum and it is the most widely used intravenous protein [44]. Because of the low expression of HSA in nuclear transgenic plants (approximately 0.02% TSP [45]) it would be desirable to produce this biopharmaceutical protein in transgenic chloroplasts. To achieve higher expression levels, the HSA gene was placed under the translational control of three different regulatory elements: a Shine-Dalgarno sequence (SD), the 5′ *psbA* region, and the *cry2Aa2* UTR [44]. In seedlings transformed with the SD-HSA construct, HSA accumulated to a level of only 0.8% TSP, while 1.6% TSP was achieved using the 5′ *psbA* control region, and 5.9% TSP was achieved using the *cry2Aa2* UTR. On the other hand, HSA reached a maximum of 0.02%, 0.8% and 7.2% TSP in transgenic potted plants when the transgene was regulated by the SD sequence, *cry2Aa2* UTR and 5′ *psbA* sequence, respectively. This demonstrated that excessive proteolytic degradation could reduce overall yields unless compensated by enhanced translation [15]. Furthermore, purification of HSA was possible by centrifugation, due to the formation of inclusion bodies within transgenic chloroplasts (Fig. 8.2, A–C). Inclusion

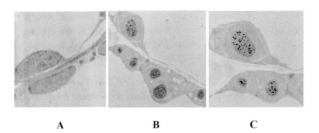

A **B** **C**

Fig. 8.2 HSA accumulation in transgenic chloroplasts. (A–C) Electron micrographs of immunogold-labeled tissues from untransformed leaves (A) and mature leaves transformed with the chloroplast vector pLDApsbAHSA (B-C). Magnifications: A x 10000; B x 5000; C x 6300.

bodies were precipitated by centrifugation and easily separated from the majority of cellular proteins present in the soluble fraction by a single centrifugation step, which may eliminate the need for expensive affinity columns or chromatographic techniques [46].

Because the 5' *psbA* region is light dependent, the accumulation of HSA protein should be increased after continuous illumination. Fernandez-San Millan *et al.* [44] monitored HSA accumulation by ELISA. A maximum of 50 h of continuous light produced an HSA yield of 11.1% TSP in mature leaves. This is the highest expression of HSA demonstrated so far and is 500-fold higher than previous reports of HSA expression in the leaves of nuclear transgenic plants [15]. Due to the high local concentration of the protein, inclusion bodies were formed and this protected it from proteolytic cleavage [44]. Another positive result from this study was the knowledge gained about the regulatory elements (SD, 5' *psbA* region and the *cry2Aa2* UTR). These serve as a model system for enhancing the expression of foreign proteins that are highly susceptible to proteolytic degradation and provide advantages during purification through the use of current purification techniques for inclusion bodies [15].

8.3.2.2 Human Insulin-like Growth Factor-1

Insulin-like growth factor-I (IGF-I) is a naturally occurring, single chain polypeptide with three disulfide bonds. It is produced in the liver and has a molecular mass of 7649 Da [47]. IGF-I is not only critical for the growth of muscle and other tissues [48], but its therapeutic potential currently is being evaluated in diabetes, IGF-I induced neuroprotection, and in the promotion of bone healing [47]. This therapeutic protein is commercially valuable because cirrhotic patients require 600 mg of IGF-1 per year at a cost of $30,000 per mg [49]. In the past, IGF-1 has been expressed in *E. coli* but the mature protein cannot be produced in this system because disulfide bonds do not form in the *E. coli* cytoplasm. Since IGF-I has a high eukaryotic codon content, chloroplast codon optimization was performed by Ruiz [50] and a synthetic *IGF-1* gene was created to further increase expression levels. Integration of the *IGF-1* gene into the tobacco chloroplast genome was achieved through homologous recombination and was confirmed using PCR and Southern blot analysis. Western blotting and chemiluminescence techniques were used to verify the high levels of IGF-I expression in transgenic chloroplasts. ELISA tests were used to quantify IGF-1 in transgenic chloroplasts containing the native and synthetic *IGF-1* genes, revealing levels as high as 32% TSP (Fig. 8.3). Most importantly, these observations showed that the chloroplast translation machinery is quite flexible, unlike the bacterial translation machinery, which could only translate the synthetic chloroplast-codon optimized *IGF-1* gene [15, 50].

8.3.2.3 Human Interferon (IFNα2b)

Malignant carcinoid tumors, a symptom of carcinoid syndrome, presented a therapeutic challenge for many years until the introduction of interferon treatment in 1982 by Oberg and Eriksson [51]. They reported that 47 out of 111 patients (42%) treated with interferon-α (median dose of 6 mega-units (MU) of interferon-α, five

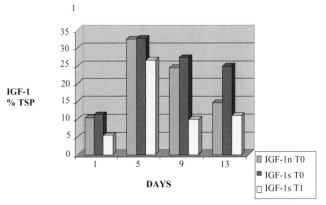

Fig. 8.3 Expression of IGF-1 in transgenic chloroplasts after continuous light exposure for 13 days. IGF-1 expression is shown as a percentage of the total soluble protein. IGF-1n is the native gene and IGF-1s is the chloroplast codon-optimized gene.

times weekly) demonstrated a significant biochemical response and 15% demonstrated more than 50% reduction in tumor size. Survival analysis demonstrated a median survival period of only 8 months in the group of patients treated with chemotherapy, compared to 80+ months ($p < 0.001$) in the groups treated with interferon-α [51]. However, alpha interferons have other therapeutic uses, such as the inhibition of viral replication and cell proliferation, enhancement of the immune response, and most recently, the treatment of patients suffering from West Nile virus. The Food and Drug Administration first approved a specific subtype of interferon-α (IFNα2b) in 1986 for the treatment of hairy cell leukemia in the United States and the recombinant IFNα2b now on the market is produced using an *E. coli* expression system. In plants, nuclear transformation has resulted in very low expression levels (0.000017%) TSP in transgenic tobacco [52] and rice [53]. Due to necessary *in vitro* processing and purification, the average cost of treatment is $26,000 per year. The injection of IFNα2b causes side effects such as fatigue, weight loss and anemia due to diarrhea, as well as flushing and bronchoconstriction [51]. Treatment with alpha interferons became less popular when it was found that up to 20% of patients produce anti-IFNα antibodies, which reduce the effectiveness of the treatment because IFNα2b aggregates with human serum albumin in the blood. However, oral administration could be beneficial in the treatment of various infectious diseases [15].

In the Daniell laboratory, recombinant IFNα2b containing a polyhistidine purification tag and a thrombin cleavage site was expressed in transgenic tobacco chloroplasts, for use as an oral therapeutic (Fig. 8.4, ref 31). For comparison, the gene cassette was integrated into the chloroplast genome of cv. Petit Havana and into a low-nicotine variety of tobacco, LAMD-609 [15, 31, 55]. Western blots performed using a monoclonal antibody to detected monomers and multimers of IFNα2b in both tobacco varieties, and disulfide bond formation was confirmed. Integration of the transgene into both chloroplast genomes was confirmed by Southern blot analysis.

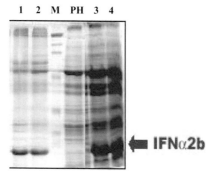

1 2 M PH 3 4

IFNα2b

Fig. 8.4 Coomassie-stained SDS-polyacrylamide gel showing chloroplast transgenic lines expressing IFNα2b. Lanes 1 and 2: Total soluble protein (TSP); Lanes PH, 3 and 4: Total protein (TP).

In the Petit Havana transgenic lines, chloroplast genome homoplasmy occurred in the first generation, and this corresponded to the highest level of IFNα2b expression. ELISAs revealed IFNα2b levels of up to 18.8% TSP in Petit Havana and up to 12.5% TSP in LAMD-609.

IFNα2b activity was confirmed by the ability of the recombinant protein to protect HeLa cells against the cytopathic effects of encephalomyocarditis virus (EMC) and through the identification of interferon-induced transcripts (Fig. 8.5). Chloroplast derived IFNα2b was found to have the same activity as commercially produced Intron A. The mRNA levels of two genes induced by IFNα2b (2′-5′ oligoadenylate synthase and *STAT-2*) were tested by RT-PCR using primers specific for each gene. Chloroplast-derived IFNα2b induced the expression of both genes in a manner similar

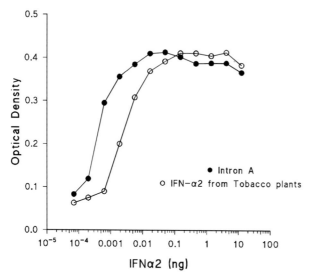

• Intron A
○ IFN–α2 from Tobacco plants

Fig. 8.5 Demonstration of IFNα2b functionality by the ability of IF-Nα2b to protect HeLa cells against the cytopathic effect of encephalomyocarditis virus (EMC). Chloroplast derived IFNα2b was as active as commercially produced Intron A.

to commercial IFNα2b. These levels of expression and functionality are ideal for IFNα2b purification and for the further use of oral IFNα2b delivery in pre-clinical studies.

8.3.2.4 Anti-Microbial Peptides (AMPs): MSI-99

Due to the increasing number of drug-resistant bacteria arising from the misuse of antibiotics and anti-microbial agents, new strategies to combat or prevent different human pathogens need to be developed. Research on defense peptides secreted from the skin of the African clawed frog (*Xenopus laevis*) has resulted in the discovery of magainin, an amphipathic α-helix-containing peptide, which has been investigated as a broad-spectrum topical agent, a systemic antibiotic, a wound healing agent, and an anticancer drug [33, 56, 57]. Magainin has an affinity for negatively charged phospholipids in the outer leaflet of the prokaryotic membrane [33, 58–60]. A sequence encoding the magainin analog MSI-99, a synthetic lytic peptide, was integrated into the chloroplast genome of tobacco variety Petit Havana and expressed at high levels (21.5–47% TSP) [33, 43].

The minimum inhibitory concentration of MSI-99 was investigated. Based on total inhibition of bacterial and fungal cells, MSI-99 was most effective against *P. syringae*, requiring only 1 µg per 1000 bacteria [33]. The amount of antimicrobial peptide required to kill bacteria was used to estimate the level of expression in transgenic plants, since the lytic activity of antimicrobial peptides is concentration dependent. The effectiveness of the chloroplast-derived lytic peptide was tested *in vitro*, using a multi-drug resistant Gram-negative bacterium, *Pseudomonas aeruginosa*, which is an opportunistic pathogen of plants, animals and humans [15]. Cell extracts prepared from T_1-generation plants resulted in 96% inhibition in growth of this pathogen (Fig. 8.6). These results may provide an alternative method for combating drug-resistant human pathogenic bacteria. In addition, the lytic peptide may also be useful as a treatment for

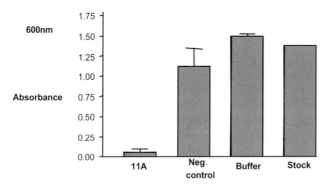

Fig. 8.6 In vitro bioassay of T_1 generation chloroplast transgenic line against *P. aeruginosa*. Bacterial cells from an overnight culture were diluted to A_{600} 0.1–0.3 and incubated for 2 hours at 25 °C with 100 µg of total protein extract. One ml of LB was added to each sample and incubated overnight at 26 °C. Absorbance was recorded at 600 nm. Data was analyzed using a GraphPad Prism.

patients suffering from cystic fibrosis, who are extremely susceptible to *P. aeruginosa* infections.

As stated above, AMPs such as magainin have been studied as possible anticancer agents because of their preference for negatively charged phospholipids. It has been reported that the outer leaflet of melanoma and colon carcinoma cells contain 3–7 fold higher levels of phosphatidylserine than normal cells. Previous studies have shown that analogs of magainin were effective against hematopoietic cancer, melanoma, sarcoma and ovarian teratoma lines [15]. The transgenic chloroplast system could thus provide sufficient expression levels for the commercially feasible production of AMPs as anticancer drugs and as antimicrobial agents.

8.3.3
Chloroplast-derived Vaccine Antigens

Edible vaccines are currently being developed for a number of human and animal diseases, including measles, cholera, foot and mouth disease, hepatitis B and C [1, 61]. Many of these diseases are likely to require booster vaccinations or multiple antigens to induce and maintain protective immunity. This problem could be solved, given the ability of plants to express more than one transgene, allowing the delivery of multiple antigens for repeated inoculations [17, 35, 61, 62]. Other advantages of plant-based vaccines include the reduced need for medical personnel and sterile injection conditions, heat stability, antigen protection through bioencapsulation, the generation of systemic and mucosal immunity, and improved safety via the use of a subunit vaccine. Most importantly, as explained by Webster et al. [61], it is unlikely that an edible vaccine would lead to oral tolerance because it is achieved by dose-specific oral antigen delivery. To explain the concept of oral therapeutic protein delivery, hepatitis B surface antigen (HBsAg) was used for oral immunization. No primary immune response was detected after two 300-µg doses of yeast-derived HBsAg. However, a primary response began after two servings of transgenic potatoes containing 85–300 µg of HBsAg [54]. The main advantage of expressing biopharmaceuticals in plants is bioencapsulation. Since the proteins are presented inside plant material, degradation occurs only slowly and a larger quantity of the protein can survive proteolyisis. For this reason, less of the recombinant protein is needed to achieve the same therapeutic effect as a non-orally delivered protein.

8.3.3.1 Cholera Toxin B Subunit (CTB)
The first step in developing a highly expressed edible vaccine containing an adjuvant is the successful expression of CTB in transgenic plants. Since CTB has previously been expressed in nuclear transgenic plants at levels of 0.01% TSP (leaves) and 0.3% TSP (tubers), expression of this adjuvant should increase if the corresponding gene is integrated into the chloroplast genome [86]. Integration of an unmodified CTB-coding sequence into the chloroplast genome, confirmed by PCR and Southern blot analysis, resulted in the accumulation of CTB as functional oligomers, at levels of up to 4.1% TSP in transgenic chloroplasts [22]. G_{M1}-ganglioside binding assays confirmed that chloroplast-derived CTB binds to the intestinal membrane receptor of cholera toxin

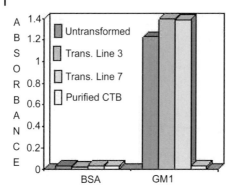

Fig. 8.7 CTB-G_{M1}-ganglioside binding ELISA assay. Plates, coated first with G_{M1}-ganglioside and bovine serum albumin (BSA), respectively, were irrigated with total soluble plant protein from chloroplast transgenic lines (3 and 7) and 300 ng of purified bacterial CTB. The absorbance of the G_{M1}-ganglioside-CTB-antibody complex in each case was measured at 405 nm. Total soluble protein from untransformed plants was used as the negative control.

(Fig. 8.7), establishing unequivocally that chloroplasts are capable of forming disulfide bridges during the assembly of foreign proteins [15]. These results and the stable transformation of plastids in edible plant parts confirm the feasibility of producing in large quantities edible vaccines, or adjuvant-antigen fusion proteins, capable of eliciting an immune response in humans [28, 29]. It has been often shown that an adjuvant increases the immune response and in turn raises the antibody titer.

8.3.3.2 *Bacillus anthracis* Protective Antigen

The anthrax bioterrorist attacks that followed the events of September 11th 2001 resulted in a renewed interest *Bacillus anthracis*, the causative agent of this disease. Research has focused on the development of better vaccines than the one currently available. It has been estimated that the aerosolized release of 100 kg of anthrax spores upwind of Washington DC would cause mortalities of 130,000–3,000,000 [63]. Nonetheless, wild-type *Bacillus anthracis* is susceptible to conventional antibiotics, including penicillin, oxyfloxacin and ciprofloxacin. The problem lies not with the bacterial infection itself, but with three proteins released by the bacteria – protective antigen (PA, 83 kDa), lethal factor (LF, 90 kDa) and edema factor (EF, 89 kDa) – known as anthrax toxins [63].

The Centers for Disease Control list *Bacillus anthracis* as a category A biological agent and estimate the cost of an anthrax attack to exceed $26 billion per 100,000 exposed individuals [34]. Concerns regarding vaccine purity, the need for multiple injections, and the limited supply of PA, underscore the urgent need for an improved vaccine. Therefore, in the Daniell laboratory, the PA gene (*pag*) was inserted into a chloroplast vector along with the *psb*A regulatory signals for enhanced translation [34]. As in other investigations, integration of transgenes into the chloroplast genome was confirmed by PCR and Southern blot analysis. Crude plant extracts contained up to 2.5 mg of full length PA per gram of fresh leaf tissue and showed exceptional stability (several months in stored leaves or crude extracts). Maximum levels of expression were observed in mature leaves under continuous illumination. Co-expression of the ORF2 chaperonin from *Bacillus thuringiensis* did not increase PA accumulation or fold it into cuboidal crystals in transgenic chloroplasts. Trypsin, chymotrypsin and furin proteolytic cleavage sites present in PA were protected in transgenic chloroplasts because only PA 83 was detected. Both CHAPS and SDS detergents extracted PA with equal efficiency and PA was detected in the soluble fraction. Chloroplast-derived PA bound to anthrax toxin receptor, heptamerized, and bound to lethal factor, resulting in macrophage lysis. Up to 25 µg of functional PA per ml of crude extract was observed (Fig. 8.8). With an average yield of 172 mg of PA per plant, 400 million doses of vaccine (free of contaminants) could be produced per acre of transgenic tobacco, using a low yielding experimental cultivar in a greenhouse, which could be further enhanced 20-fold in commercial cultivars cultivated in the field (with multiple harvests).

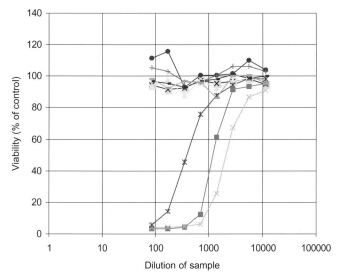

Fig. 8.8 Macrophage cytotoxic assays of plant extracts. Supernatant samples were tested from T_1-generation of chloroplast transgenic line pLD-JW1 (proteins were extracted in buffer containing no detergent and MTT was added after 5 hours). ▬▬ pLD-JW1 (extract stored for 2 days); ╌×╌ pLD-JW1 (extract stored for 7 days); ─✳─ PA 5 µg ml^{-1}; ─×─ Control wild type (extract stored for 2 days); ▴ Control wild type (extract stored for 7 days; ─●─ Control wild type no lethal factor (LF) (extract stored for 2 days); ─■─ Control wild type no LF (extract stored for 7 days); ─┼─ Control pLD-JW1 no LF (extract stored for 2 days); ╌■╌ Control pLD-JW1 no LF (extract stored for 7 days).

8.3.3.3 *Yersinia pestis* F1~V Fusion Antigen

Yersinia pestis, the Gram-negative bacterium responsible for plague, is also listed as a category A biological agent by the Centers for Disease Control. *Y. pestis* is naturally transmitted from one animal host to another either directly or via a flea vector. This pathogen has been used for centuries as a biological warfare agent and it raises concerns today as one of the microorganisms with potential for use against civilian or military populations. If used as a warfare agent, the pneumonic form of plague would be the most probable outcome of infection and is devastating due to the rapid onset of the disease, the high mortality rate, and the rapid spread of the disease. If a person with pneumonic plague is not treated with antibiotics within 18 hours of the onset of symptoms, he or she will die. Therefore, immunization against aerosolized plague has been an area of interest in current research.

The current, killed whole unit vaccine is only slightly effective against the bubonic form of plague and not immunoprotective against pneumonic and septicemic plague [64]. Several subunit vaccines have been evaluated for immunogenicity and protective efficacy against *Y. pestis*. CaF1 and LcrV are the two most favorable choices. F1 is a capsular protein located on the surface of the bacterium with anti-phagocytic properties. The V antigen is a component of the *Y. pestis* Type III secretion system and it may form part of an injectosome. Expression of these antigens as fusion proteins in *Salmonella typhimurium* has been shown to be immunogenic in mice, but has not proven to be safe [65]. However, when these antigens are produced in *E. coli*, they have been shown to be safe and to provide protection against subcutaneous and aerosol challenges of *Y. pestis* in mice [66]. Expressing subunit vaccines in chloroplasts is very promising for three reasons: (i) subunit vaccines, which do not express active toxins, are safe and do not multiply; (ii) bacterial and many viral genes have a high AT content allowing for high-level expression in the chloroplast; and (iii) oral delivery of vaccines yields high mucosal IgA titers along with high systemic IgG titers, enabling the immune system to fight off infectious agents at important entry points (e.g. the lungs and vagina).

Investigations in the Daniell laboratory demonstrated expression of the F1 and V antigens in transgenic chloroplasts as a recombinant F1~V fusion protein, consisting of the F1 protein fused at its carboxyl terminus to the amino terminus of the entire V antigen (67). The gene encoding the fusion proteins was obtained from the U.S. Army Medical Research Institute of Infectious Diseases (USAMRIID) and integrated into the chloroplast genome of Petit Havana and LAMD-609 varieties using a universal transformation vector. Expression levels of F1~V in the two varieties were compared by western blotting and ELISA, after continuous illumination for up to five days. For the continuous illumination analysis, leaf material was sampled on days 0, 1, 3 and 5, and samples were taken from young, mature and old leaves (Fig. 8.9). As stated above, the *psbA* 5' UTR, which was used to control the fusion gene, is light regulated, and continuous illumination therefore enhanced translation. This led to a buildup of F1~V protein until the plant became stressed by the unnatural light period. Mature leaves showed the highest yields, with the largest amount of F1~V accumulating on the third day of continuous illumination, producing an average yield of 14.8% TSP. Functional assays are currently underway.

Average % TSP

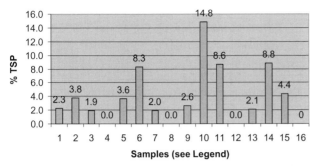

Fig. 8.9 Enzyme linked immunoassays, with protein yields expressed as %TSP and µg of F1V per gram of fresh leaf material. For the continuous illumination experiment, leaf material was sampled on days 0, 1, 3 and 5. Young, mature, and old samples were taken for each experiment. The figure shows average levels of F1V antigen for: Samples 1–4 (Day 0: (1) young, (2) mature, (3) old, and (4) wild type leaf samples); Samples 5–8 (Day 1: (5) young, (6) mature, (7) old and (8) wild type leaf samples]); Samples 9–12 (Day 3: (9) young, (10) mature, (11) old and (12) wild type leaf samples); Samples 13–16 (Day 5: (13) young, (14) mature, (15) old & (16) wild type leaf samples).

Since plague remains endemic in some regions of the world, and because of the increased threat that *Y. pestis* could be used as a biological warfare agent, the development of improved vaccines against plague is a high priority. The ideal vaccine should be deliverable in one or two doses and should have the ability to produce high-titer and long-lasting antibodies quickly. Moreover, such a vaccine should protect against aerosolized transmission of *Y. pestis*.

8.3.3.4 Canine Parvovirus (CPV) VP2 Protein

Canine parvovirus (CPV) infects dogs and other Canidae such as wolves, coyotes, South American dogs and Asiatic raccoon dogs, producing haemorrhagic gastroenteritis and myocarditis. CPV-1 was the first strain of canine parvovirus to be discovered, and was initially described in 1967. It did not pose much of a threat except to newborn puppies. However, the CPV-2 strain, which appeared in the US in 1978, appeared to be a mutated form of the feline parvovirus (more commonly known as feline distemper virus). Infected animals shed CPV-2, which is hardy in the environment, in large numbers, and this led to rapid worldwide dissemination. Attempts to shield puppies from exposure were completely futile. A second mutant, CPV-2a, was recognized in 1979 and was found to be more aggressive. The vaccine was in short supply and many veterinarians had to use feline distemper vaccine as a substitute, since it was the closest one available. The most common form of the virus today is CPV-2b and young animals are commonly vaccinated with an attenuated whole virus vaccine.

Advances in molecular biology have lead to the identification of CPV antigens that are capable of eliciting a protective immune response. A linear antigenic peptide (2L21) from the VP2 capsid protein (amino acids 1–23) of CPV was selected and its DNA sequence was successfully introduced into tobacco chloroplasts [68]. The 2L21 synthetic peptide, chemically coupled to a KLH carrier protein, has been extensively studied and has been shown to protect dogs and minks against parvovirus infection very effectively [69, 70]. This peptide was expressed in nuclear transgenic plants as an N-terminal fusion protein with β-glucuronidase (GUS) [71] but expression levels were inadequate.

Therefore, the 2L21 peptide, fused either to the cholera toxin B subunit (CTB) or the green fluorescent protein (GFP) was expressed in chloroplast transgenic plants, and accumulated to levels up to ten-fold higher than those previously reported in nuclear transformation [71]. The expression levels were dependent on plant age. Both young and senescent plants accumulated lower amounts of proteins than mature plants [68], showing that the time of harvest is important when scaling up the process of protein production. The maximum level of CTB-2L21 was 7.49 mg g^{-1} fresh weight (equivalent to 31.1% TSP) and that of GFP-2L21 was 5.96 mg g^{-1} fresh weight (equivalent to 22.6% TSP). The inserted epitope was detected with a CPV-neutralizing monoclonal antibody, indicating that the epitope is correctly positioned at the C-terminus of the fusion proteins. The resulting chimeric CTB-2L21 protein retained an ability to form pentamers, possessed the G_{M1}-ganglioside binding char-

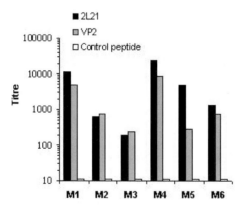

Fig. 8.10 Titers of antibodies at day 50 induced by plant-derived CTB-2L21 recombinant protein. Balb/c mice were intraperitoneally immunized with leaf extract from CTB-2L21 transgenic plants. Animals were boosted at days 21 and 35. Each mouse received 20 μg of CTB-2L21 recombinant protein. Individual samples of mouse serum were titrated against 2L21 synthetic peptide, VP2 protein and a control peptide (amino acids 122–135 of hepatitis B virus surface antigen). Titers were expressed as the highest serum dilution to yield twice the absorbance mean of preimmune sera. M1-M6: mice 1 to 6; 2L21: epitope from the VP2 protein of the canine parvovirus; CTB: cholera toxin B; VP2: protein of the canine parvovirus that includes the 2L21 epitope.

acteristics of the native CTB, and induced antibodies that were able to recognize the VP2 protein of CPV following the intraperitoneal immunization of mice (Fig. 8.10). These results show that plant derived CTB-2L21 recombinant protein is immunogenic via intraperitoneal administration, as shown by its ability to induce a humoral response that cross-reacts with the native VP2 protein. Additional experiments are underway to check the ability of this fusion protein to induce specific immune responses after mucosal delivery. This is the first report of an animal vaccine epitope expressed in transgenic chloroplasts.

8.4
Advances in Purification Strategies for Biopharmaceuticals

The purification of therapeutic proteins expressed in transgenic systems is a major challenge. For example, insulin produced in E. coli is purified by chromatography, which accounts for 30% of operating expenses and 70% of set up costs [24]. Therefore, new strategies for biopharmaceutical purification are currently being investigated. To overcome the high cost of purification, protein-based polymers have been produced in E. coli through genetic engineering [72, 73]. For example, a synthetic polymer gene encoding the polypeptide $(GVGVP)_{121}$ was overexpressed in E. coli to the extent that polymer inclusion bodies occupied 80–90% of the cell volume [72, 73]. The inverse temperature transition property of this polymer makes it an ideal fusion protein for the purification of therapeutic proteins. The protein is soluble in water at temperatures below 25 °C, but the polymer aggregates into a more-ordered, viscoelastic state called a coacevate at 37 °C [74, 75]. Indeed, it is known that polypeptides, proteins, protein-based polymers and any polymers in which the hydrophobic and polar residues are in the correct balance, will fold and assemble (i.e. become more ordered) as the temperature increases and will disassemble and unfold (i.e. become less ordered) as the temperature decreases [76]. This inverse temperature transition property makes it easier and cheaper to harvest polymers in aqueous solutions simply by increasing the temperature and this avoids cumbersome purification procedures and the use of enzymes and organic solvents which may alter the quality of the polymer or any biopharmaceutical protein [77]. The idea of producing biopolymers in plants instead of bacteria seems very appealing since a further decrease in production costs could be achieved. Genetic engineering of plants should allow the production of polymers at much lower costs and in higher volumes than are possible with microbial fermentation [78, 79]. Because of these advantages, a large protein based polymer $(GVGVP)_{121}$ was expressed in transgenic tobacco chloroplasts [80].

Research in the Daniell laboratory [81] confirms the possibility of producing recombinant insulin using this inexpensive purification approach and offers the prospect of low-cost treatment for diabetic patients. The proinsulin gene was fused to a gene encoding the synthetic biopolymer $(GVGVP)_{40}$, expressed in E. coli and the fusion protein was purified by raising the temperature from 4 °C to 42 °C, utilizing the inverse temperature transition property. At 4 °C, the biopolymer exists as an extended molecule, but when incubated at 42 °C it folds into dynamic structures called β-spirals that further aggregate by hydrophobic association to form twisted filaments [82]. Raising the temperature to 42 °C does not affect the tertiary structure of the proinsulin polypeptide. This has been demonstrated previously by the production of human insulin in bacteria using a temperature-responsive promoter. Production of the recombinant protein was induced by temperature shift from 30 °C to 42 °C without any adverse effects on protein stability [83].

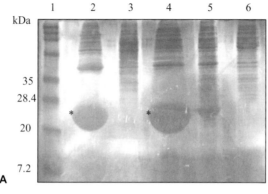

*Fusion protein (~25 kDa)

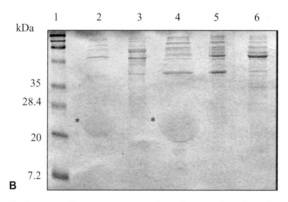

Fig. 8.11 (A and B) Expression and purification of insulin-polymer fusion protein detected in copper (A) and Coomassie (B) stained gels. The same gel was first stained with copper, destained and re-stained with Coomassie R-250. Lane 1: Prestained marker; Lane 2: Purified extract of polymer-insulin fusion protein from the chloroplast vector pSBL-OC-40Pris; Lane 3: Reverse orientation of fusion protein from pSBL-OC-40Pris; Lane 4: Purified extract of polymer-insulin fusion protein from the chloroplast vector pLD-OC-40Pris; Lane 5: Reverse orientation of pLD-OC-40Pris; Lane 6: Purified extract of *E. coli* strain XL-1 Blue containing no plasmid. Two rounds of thermally reversible phase transition purification were performed.

The fusion protein was purified twice using the inverse temperature transition property of the biopolymer, and the protein was analyzed by SDS-PAGE, western blotting and immunodetection using mouse anti-human proinsulin primary antibody. Negative staining of the SDS-polyacrylamide gels with 0.3 M $CuCl_2$ confirmed the presence of the proinsulin-polymer fusion protein (Fig. 8.11, A). Copper-stained gels with a black background appear to have dark bands against a light, semiopaque background when illuminated obliquely from above [84]. The same gel was then stained with Commassie R-250, which would stain only the proinsulin but not the polymer (Fig. 8.11, B). Commassie R-250 does not stain polymer proteins without a fusion, since the sulphonic acid groups on the dye only form ion pairs with lysine and arginine, which are not present in $(GVGVP)_{40}$ [85]. Since Commassie R-250 staining was successful, this confirmed the presence of the fusion protein. Thus, the polymer-proinsulin fusion protein has been confirmed by western blot, and two different staining methods.

8.5
Conclusion

Chloroplast transformation facilitates the high-yield production of therapeutic proteins such as biopharmaceuticals and vaccine antigens. High-level expression in transgenic chloroplasts promises unlimited quantities of therapeutic proteins to people around the world at lower costs. However, purification methods such as chromatography are still very expensive. In order to address this concern, two major approached have been developed. The oral delivery of therapeutic proteins should completely eliminate the need for expensive purification. Alternatively, less expensive purification steps should be developed. This chapter gives examples of both such approaches. The advancements augur well for the production of therapeutic proteins in transgenic chloroplasts.

Acknowledgements

The results reported from investigations in the Daniell laboratory were supported by NIH R 01 GM 63879 and USDA 3611-21000-017-00D grants to HD.

References

[1] G. GIDDINGS, G. ALLISON, D. BROOKS et al., *Nature Biotechnol.* **2000**, *18* (11), 1151–1155.

[2] C. CRAMER, *Curr. Top. Microbiol. Immunol.* **1999**, *240*, 95–118.

[3] S. SCHILLBERG, R. FISCHER, N. EMANS, *Natuwissenschaften* **2003**, *90* (4), 145–155.

[4] A. BARTA, K. SOMMERGRUBER, D. THOMPSON et al., *Plant Mol. Biol.* **1986**, *6* (5), 347–357.

[5] A. HIATT, R. CAFFERKEY, K. BOWDISH, *Nature* **1989**, *342* (6245), 76–78.

[6] J. LARRICK, D. W. THOMAS, *Curr. Opin. Biotechnol.* **2001**, *12* (4), 411–418.

[7] T. C. VERWOERD, P. A. VANPARIDON,

A. J. J. VANOOYEN VANLENT et al.,
Plant Physiol. **1995**, *109* (4), 1199–1205.

[8] M. T. ZIEGLER, S. R. THOMAS,
K. J. DANNA, *Mol. Breeding* **2000**, *6* (1),
37–46.

[9] H. DANIELL, S. STREATFIELD, K. WYCOFF,
Trends Plant Sci. **2001**, *6* (5), 219–226.

[10] T. A. HAQ, H. S. MASON, J. D. CLEMENTS
et al., *Science* **1995**, *268* (5211), 714–716.

[11] H. S. MASON, D. LAM, C. J. ARNTZEN,
Proc. Natl Acad. Sci. USA **1992**, *89* (24),
11745–11749.

[12] Y. THANAVALA, Y. YANG, P. LYONS et al.,
Proc. Natl Acad. Sci. USA **1995**, *92* (8),
3358–3361.

[13] E. TACKABERRY, A. DUDANI, F. PRIOR
et al., *Vaccine* **1999**, *17* (23–24), 3020–
3029.

[14] N. GOMEZ, C. CARRILLO, J. SALINAS
et al., *Virology* **1998**, *249* (2), 352–358.

[15] H. DANIELL, in The *Encyclopedia of
Plant and Crop Sciences*, R.M. Goodman, ed. Marcel Dekker, New York
2004, pp 705–710.

[16] A. KUSNADI, G. NIKOLOV. J. HOWARD,
Biotechnol. Bioeng. **1997**, *56* (5), 473–
484.

[17] B. DE COSA, W. MOAR, S. B. LEE et al.,
Nature Biotechnol. **2001**, *19* (1), 71–74.

[18] K. E. MCBRIDE, Z. SVAB, D J. SCHAAF
et al., *Biotechnol. Technol.* **1995**, *13* (4),
362–365.

[19] M. KOTA, H. DANIELL, S. VARMA et al.,
Proc. Natl Acad. Sci. USA **1999**, *96* (5),
1840–1845.

[20] N. NAGATA, C. SAITO, A. SAKAI et al.,
Planta **1999**, *209* (1), 53–65.

[21] L. BOGORAD, *Trends Biotechnol.* **2000**, *18*
(6), 257–263.

[22] H. DANIELL, S. B. LEE, T. PANCHAL
et al., *J. Mol. Biol.* **2001**, *311* (5), 1001–
1009.

[23] S. B. LEE, H. B. KWON, S. J. KWON et al.,
Mol. Breeding **2003**, *11* (1), 1–13.

[24] D. PETRIDIS, E. SAPIDOU, J. CALANDRA-
NIS, *Biotechnol. Bioeng.* **1995**, *48*, 529–
541.

[25] H. DANIELL, B. MUTHUKUMAR, S.B. LEE,
CURR. GENET. **2001**, 39: 109–116.

[26] S. IAMTHAM, A. DAY, *Nature Biotechnol.*
2000, *18* (11), 1172–76.

[27] S. CORNEILLE, K. LUTZ, Z. SVAB et al.,
Plant J. **2001**, *27* (2), 171–178.

[28] S. RUF, M. HERMANN, U. BERGER et al.,

Nature Biotechnology **2001**, *19* (9), 870–
875.

[29] S. KUMAR, A. DHINGRA, H. DANIELL,
manuscript in review.

[30] J. M. STAUB, B. GARCIA, J. GRAVES et al.,
Nature Biotechnol. **2000**, *18* (3), 333–
338.

[31] R. FALCONER, MS thesis, University of
Central Florida, USA, **2002**.

[32] S. LEELAVATHI, V. S. REDDY, *Mol. Breeding* **2003**, *11* (1), 49–58.

[33] G. DEGRAY, K. RAJASEKARAN, F. SMITH
et al., *Plant Physiol.* **2001**, *127* (3), 852–
862.

[34] H. DANIELL, J. WATSON, V. KOYA et al.,
Vaccine, **2004**, in press.

[35] H. DANIELL, A. DHINGRA, *Curr. Opin.
Biotechnol.* **2002**, *13*, 136–141.

[36] O.N. RUIZ, H. HUSSEIN, N. TERRY,
H. DANIELL, Plant Physiology, **2003**,
132: 1344–1352.

[37] A. LOSSL, C. EIBL, H. J. HARLOFF et al.,
Plant Cell Re. **2003**, *21* (9), 891–899.

[38] H. DANIELL, *Plant Biotechnology: 2002
and Beyond.* Kluwer Academic Publishers, The Netherlands, **2003**, pp 371–
376.

[39] M. TORRES, MS thesis, University of
Central Florida, USA, **2002**.

[40] J. LARRICK, D. W. THOMAS, *Curr. Opin.
Biotechnol.* **2001**, *12* (4), 411–418.

[41] J. K.-C. MA, A. HIATT, M. HEIN
et al., *Science* **1995**, *268* (5211), 716–
719.

[42] H. DANIELL, K. WYCOFF, WO Patent 01-
64929, **2001**.

[43] H. DANIELL, A. DHINGRA, A. FERNAN-
DEZ-SAN MILLAN, in *12th International
Congress on Photosynthesis* Vol. S40–04,
CSIRO Publishing, Brisbane, Australia,
2001, pp 1–6.

[44] A. FERNANDEZ-SAN MILLAN, M. MIN-
GEO-CASTEL, MILLER et al., *Plant Biotechnol. J.* **2003**, *1* (2), 71–79.

[45] P. C. SIJMONS, B. M. DEKKER,
B. SCHRAMMEIJER et al., *Bio/Technology*
1990 *8*, 217–221.

[46] S. KUMAR, H. DANIELL, *Methods Mol.
Biol.* **2003**, 267: 365–385

[47] J. TORRADO, C. CARRASCOSA, *Curr.
Pharm. Biotechnol.* **2003**, *4* (2), 123–140.

[48] J. R. FLORINI, D. Z. EWTON,
S. A. COOLICAN, *Endocrine Rev.* **1996**, *17*
(5), 481–517.

[49] B. Nilsson, G. Forsberg, M. Hartmanis, *Methods Enzymol.* 1991, *198*, 3–16.

[50] G. Ruiz, MS thesis, University of Central Florida, USA, 2002.

[51] K. Oberg, B. Eriksson, *Br. J. Haematol.* 1991, *79* (Suppl. 1), 74–77.

[52] O. Eldelbaum, D. Stein, N. Holland et al., *J. Interferon Res.* 1992, *12* (6), 449–453.

[53] Z. Zhu, K. W. Hughes, L. Huang et al., *Plant Cell Tiss. Organ Cul.* 1994, *36* (2), 197–204.

[54] Q. Kong, L. Richter, Y. F. Yang et al., *Proc. Natl Acad. Sci. USA* 2001, *98* (20), 11539–11544.

[55] G.B. Collins, P.D. Legg, M.J. Kasperbauer, *Crop Sci.* 1974, *14*, 77–80.

[56] M. Zasloff, *Proc. Natl Acad. Sci. USA* 1987, *84* (15), 5449–5953.

[57] L. Jacob, M. Zasloff, *CIBA Found. Symp.* 1994, *186*, 197–223.

[58] P. C. Biggin, M. S. P. Sansom, *Biophys. Chem.* 1999, *76* (3), 161–183.

[59] H. W. Huang, *Biochemistry* 2000, *39* (29), 8347–8352.

[60] A. Tossi, L. Sandri, A. Giangaspero, *Biopolymers* 2000, *55* (1), 4–30.

[61] D. E. Webster, M. C. Thomas, R. A. Strugnell et al., *Med. J. Austr.* 2002, *176* (9), 434–437.

[62] U. Conrad, U. Fiedler, *Plant Mol. Biol.* 1994, *26* (4), 1023–1030.

[63] M. T. Stubbs, *Trends Pharmacol. Sci.* 2002, *23* (12), 539–541.

[64] R. W. Titball, E. D Williamson, *Vaccine* 2001, *19* (30), 4175–4184.

[65] S. E. C. Leary, K. F. Griffin, H. S. Garmory et al., *Microb. Pathogenesis* 1997, 23 (3), 167–179.

[66] E. D. Williamson, S. M. Eley, A. J. Stagg et al., *Vaccine* 1997, *15* (10), 1079–1084.

[67] M. L. Singleton, MS thesis, University of Central Florida, USA, 2003.

[68] A. Molina, S. Herva-Stubbs, H. Daniell et al., *Plant Biotechnol. J.* 2003, in press.

[69] J. P. Langeveld, J. I. Casal, A. D. Osterhaus et al., *J. Virol.* 1994, *68* (7), 4506–4513.

[70] J. P. Langeveld, S. Kamstrup, A. Uttenthal et al., *Vaccine* 1995, *13* (11), 1033–1037.

[71] F. Gil, A. Brun, A. Wigdorovitz et al., *FEBS Lett.* 2001, *488* (1–2), 13–17.

[72] C. Guda, X. Zhang, D. T. McPherson et al., *Biotechnol Lett.* 1995, *17* (7), 745–750.

[73] H. Daniell, C. Guda, D. T. McPherson et al., *Methods Mol. Biol.* 1997, *63*, 359–371.

[74] D. T. McPherson, C. Morrow, D. S. Minehan et al., *Biotechnol. Prog.* 1992, *8* (4), 347–352.

[75] D. T. McPherson, J. Xu, D. W. Urry, *Protein Expr. Purif.* 1996, *7* (1), 51–57.

[76] D. W. Urry, *Prog. Biophys. Mol. Biol.* 1992, *57* (1), 233–257.

[77] X. Zhang, D. W. Urry, H. Daniell, *Plant Cell Rep.* 1996, *16* (3–4), 174–179.

[78] H. Daniell, *Inform* 1995, *6* (12), 1365–1370.

[79] H. Daniell, C. Guda, *Chem. Industry* 1997, *14*, 555–560.

[80] C. Guda, S. B. Lee, H. Daniell, *Plant Cell Rep.* 2000, *19* (3), 257–262.

[81] O. Carmona-Sanchez, MS thesis, University of Central Florida, USA, 2001.

[82] D.W. Urry, D.T. McPherson, J. Xu et al., in *The Polymeric Materials Encyclopedia: Synthesis, Properties and Applications*, Solomone J.C. ed., CRC press, Florida, 1996, pp 2645–2699.

[83] M. Schmidt, K. R. Babu, N. Khanna et al., *J. Biotechnol.* 1999, *68* (1), 71–83.

[84] C. Lee, A. Levin, D. Branton, *Anal. Biochem.* 1987, *166* (2), 308–312.

[85] C. V. Sapan, R. L. Lundblad, N. C. Price, *Biotechnol. Appl. Biochem.* 1999, *29*, 99–108.

[86] S. H. Mason, A. T. Haq, D. J. Clements et al., *Vaccine* 1998, 16 (13), 1336–1343.

9
Plant-derived vaccines: progress and constraints

Guruatma Khalsa, Hugh S. Mason, Charles J. Arntzen

9.1
Introduction

Vaccines against infectious diseases are needed to increase global immunization compliance and to decrease the costs of delivery, while expanding participation by vaccine manufacturers. If implemented well, vaccines can provide great economic benefits, with cost-to-benefit ratios of up to 1:10 [1]. The technological revolution in vaccine development over the last half of the 20th century has centered on the use of mammalian cell cultures. This technology, now used throughout the developed and developing world, enables the production of more uniform and potent vaccines than is possible with other methods. However, subunit vaccines have penetrated the developing world to a much lesser degree than the developed world. This is for two reasons: IP challenges and technology entry barriers, whose costs increase along with the sophistication of molecular biology techniques. Thus, the development and implementation of new vaccines often involves substantial economic and logistic barriers that are difficult to overcome in many poor countries. Transgenic plant-derived proteins represent a promising strategy for vaccine development that combines innovations in medical science and plant biology for the creation of affordable vaccines. A growing number of laboratories are investing in the development of plant-derived protein pharmaceuticals, expanding on the seminal works that first put forward this idea [2,3]. The subject of plant-derived vaccines has been reviewed several times in recent years (e.g. [4–8]), and a comprehensive list of the plant-derived vaccine antigens that have been reported is provided in Tables 9.1 and 9.2. While much research is still needed to optimize the production of vaccines in plants and to validate them in large-scale clinical trials, the results to date are very promising and suggest that the technology justifies commercial development. The oral delivery of transgenic plant tissue expressing vaccine antigens, typically mucosally targeted subunits, can promote specific mucosal secretory IgA (sIgA) and serum IgG antibody responses via the gut lymphoid system. Although the protective efficacy of a plant-derived vaccine has yet to be determined in humans, some challenge studies in animals have shown promising results.

Molecular Farming. Edited by Rainer Fischer, Stefan Schillberg
Copyright © 2004 WILEY-VCH Verlag GmbH & Co. KGaA, Weinheim
ISBN: 3-527-30786-9

Tab. 9.1 Vaccine antigens expressed in plant virus epitope display systems.

Antigen	Expression system	Immunogenicity	Protective challenge	References
Canine parvovirus (CPV) VP2 epitope	Epitope display on cowpea mosaic virus in cowpea leaf	Immunogenic in mice when delivered parenterally or nasally. Dogs developed 3L17- and VP2-specific IgG; sera neutralized CPV *in vitro*.	Protective against lethal challenge in dogs immunized parenterally.	76, 77
Canine parvovirus VP2 epitope	Epitope display on plum pox potyvirus in tobacco leaf	Mice developed CPV-specific antibodies that showed neutralizing activity. Immunogenic in mice and rabbits when delivered parenterally	–	78
Foot and mouth disease virus VP1 epitope	Cowpea mosaic virus in cowpea leaf	No immunogenicity assays performed.	–	79
Hepatitis C virus (HCV) HVR1 epitope of E2 envelope protein	Epitope display on cucumber mosaic virus in tobacco leaf	Cross reactive with a wide range of human anti-HVR1 antibodies.	–	80
Human immuno-deficiency virus (HIV) type 1 ELDKWA epitope	Potato virus X in tobacco leaf	Sera from normal and hu-PBL-SCID mice showed anti-HIV-1 neutralizing activity. Immunogenic in mice when delivered parenterally or nasally.	–	18
Human rhinovirus type 14 VP1 epitope	Epitope display on cowpea mosaic virus in cowpea leaf	Immunogenic in rabbits when delivered parenterally.	–	81
Mink enteritis virus VP2 epitope	Epitope display on cowpea mosaic virus in cowpea leaf	Immunogenic in mink when delivered parenterally.	Protective against challenge with virulent MEV.	82
Pseudomonas aeruginosa membrane protein F	Epitope display on cowpea mosaic virus in cowpea leaf	Elicited specific antibodies. Immunogenic in mice when delivered parenterally.	Mice protected when challenged with model chronic pulmonary infection with *P. aeruginosa*.	19, 83
Pseudomonas aeruginosa membrane protein F	Epitope display on tobacco mosaic virus in tobacco leaf	Elicited specific antibodies against 7 immunotype strains. Immunogenic in mice when delivered parenterally.	Mice protected when challenged with model chronic pulmonary infection with *P. aeruginosa*.	20

Tab. 9.1 (continued)

Antigen	Expression system	Immunogenicity	Protective challenge	References
Rabbit hemorrhagic disease virus VP60 epitope	Epitope display on plum pox potyvirus in tobacco leaf	Rabbits developed specific antibodies that showed neutralizing activity. Immunogenic when delivered to mice parenterally; immunogenic in rabbits when delivered parenterally.	Rabbits immunized parenterally survived lethal challenge.	78, 84
Rabies virus glyco-protein (G) and nucleo-protein (N)	Alfala mosaic virus and tobacco mosaic virus in tobacco and spinach leaf	Elicited specific virus-neutralizing antibodies in mice. Immunogenic in mice when delivered orally and parenterally; immunogenic in humans when delivered orally.	Moderately protective against lethal challenge infection in mice.	16, 71
Respiratory syncytial virus (RSV) G and F proteins	Alfalfa mosaic virus in tobacco leaf	High levels of serum antibodies specific for RSV-G. Immunogenic in mice when delivered parenterally.	Protective against challenge with RSV Long strain.	30
Staphylococcus aureus D2 epitope of fibro-nectin-binding protein (FnBP)	Cowpea mosaic virus in cowpea leaf; Potato virus X in tobacco leaf	Elicited FnBP-specific IgA and IgG. Immuno-genic in mice and rats when delivered orally, nasally, or parenterally.	–	85

9.2
Strategies for Vaccine Production in Plants

Three major research strategies have emerged for the production of subunit vaccines in plants (Fig. 9.1). The first, the direct transformation of plants, introduces genes encoding the antigenic proteins of human or animal pathogens stably into the plant genome. In this case, the genes can be inserted into either the nuclear or chloroplast genomes. The resulting plants accumulate the subunit antigens, which can then either be purified from plant tissue for parenteral or mucosal delivery, or the plant tissue can be processed for oral delivery with the antigen as a component of the plant matrix.

In the second research strategy, plant viruses have been utilized as pliable genetic platforms for protein expression. Three formats have been developed, and the one that has undergone the most extensive evaluation is the display of epitopes on the surface of the virus as fusions with the viral coat protein. This epitope-display system

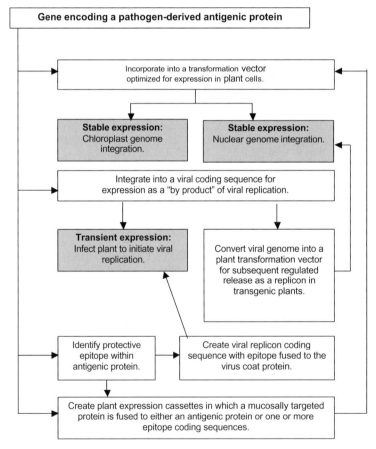

Fig. 9.1 Plant-derived vaccines: Research strategies

usually involves subsequent purification of the viral particle for parenteral delivery. Another viral strategy uses cloned viral genomes to introduce antigenic protein-coding sequences driven by sub-viral promoters, and results in antigen expression as a byproduct of the viral replication cycle. A final variant is the integration of entire viral genomes into plant chromosomes by transformation, with the subsequent transcription of viral RNA as the initiating event in viral replication. A coding sequence for the antigen is included in this viral genome and is thus co-expressed. The potential advantage of viral expression systems compared to stable plant transformation is that viral replication can greatly amplify the template for protein synthesis resulting in high-level protein accumulation.

A third strategy for the production of plant-based vaccines is the design of mucosally targeted fusion proteins. We define this as a separate research strategy to emphasize the goal of creating new mucosal vaccines, even though such fusion proteins can be expressed in transgenic plants or using viral vectors as discussed above. We highlight mucosally targeted proteins because this captures the emerging research emphasis on protein engineering to obtain more effective vaccines, and the potentially unique role of plants for the production and accumulation of novel vaccine products. See Table 9.3 for details of fusion proteins expressed in transgenics.

9.3
The Biomanufacture of Vaccines

Current vaccine technology frequently requires the purification of subunit proteins from mammalian cell cultures or tissues, yeast, fertilized eggs or bacterial fermentation systems to produce the immunogen. Typically, the product requires refrigeration during transport to its final point of use, adding significant cost to the vaccination program. One solution may be to use dried plant extracts containing subunit vaccines, since this provides a product that is stable at ambient temperatures, i.e. equivalent in storage and transport characteristics to dehydrated food products. In addition, the production of antigens in plants may improve product safety by removing animal cell-related contaminants and pathogens.

9.3.1
Advantages of Plants

Plant-derived pharmaceuticals (PDPs) are proteins or organic compounds produced in plants via recombinant DNA technology, which are used to improve human or animal health. Subunit vaccines represent one category of PDPs that have been validated in a variety of studies, including human clinical trials. Current efforts in product formulation utilize food-processing technology to convert transgenic plant material into dried samples that can be delivered in unit doses with assured product uniformity and quality. Plant-derived vaccines appear to offer several product advantages, including oral delivery, heat stability, lower manufacturing costs for the active ingredient and suitability of the manufacturing technology for use in developing countries.

9.3.2
Oral Delivery and Mucosal Immune Responses

Orally delivered, non-replicating subunit vaccines have not yet achieved commercial success using any means of manufacture. Possible hurdles facing the use of orally delivered immunogenic proteins include the likelihood that some proteins will be degraded after ingestion and that some immunogens may not be recognized efficiently at mucosal immune effector sites in the gut. As a result, higher concentrations of immunogen may be required in oral versus parenteral delivery. Although this is a potential limitation, the use of plants as a protein biomanufacturing system offers advantages in that the cost of obtaining the end product is comparatively low. In addition, empirical evidence suggests that encapsulation of the immunogenic protein within the plant cell matrix during administration provides some protection from degradation by gastric enzymes and acids.

The best candidates oral subunit vaccines for production in transgenic plants are the primary antigens of infectious pathogens that aggregate in forms that are recognized at mucosal sites where an immune response is triggered. These include viral surface proteins that co-assemble to form virus like particles (VLPs), and bacterial toxins that naturally aggregate to form mucosally targeted multimeric complexes. In addition, in several laboratories there are also ongoing efforts to produce a variety of fusion proteins that target immunoresponsive mucosal sites [9,10].

9.3.4
Examples of Antigens Produced in Plants

Tables 9.1 and 9.2 shows the range of different plant and vector systems that have been used for the expression of antigens. The immunogenicity of PDPs was demonstrated in a number of studies, including those using hepatitis B surface antigen (HBsAg) [11–13], the S protein of transmissible gastroenteritis coronavirus (TGEV) [14,15] and human immunodeficiency virus 1 (HIV-1) epitopes [16–18]. Furthermore, successful challenge trials resulted after immunization with *Pseudemonas aeruginosa* epitopes [19,20], the FP1 epitope of foot and mouth disease virus (FMDV) [21–23] and *Escherichia coli* labile enterotoxin B subunit (LTB) [24–29]. Additional antigens expressed in plants include the respiratory syncytial virus (RSV) G and F proteins [30,31], the rotavirus VP6 protein [32–35] and VP7 protein [36], the measles virus hemagglutinin protein [37–39] and an epitope from the major surface antigen of *Plasmodium falciparum* (PfMSP1) [40]. Although the immunogenicity of the plant-derived RSV G and F proteins and the measles virus hemagglutinin protein were demonstrated in animal trials, the immunogenicity of the VP6 antigen and the PfMSP1 epitope were not tested.

9.3.5
Targeting Antigens to Specific Tissues

Tissue-specific targeting of antigens can greatly facilitate the harvesting of proteins. Streatfield *et al.* [28] expressed antigens specifically in maize kernels, Sandhu *et al.*

[31] in tomato fruit, and Wright *et al.* [41] in tobacco seeds. Targeting the major glycoprotein (gB) of human cytomegalovirus (hCMV) to transgenic tobacco seeds was previously described [42], and further investigations showed that recombinant gB was almost exclusively deposited in protein storage vesicles in mature tobacco seeds [41]. Sojikul *et al.* [43] and Huang & Mason [44] reported the targeting of a fusion molecule to the endoplasmic reticulum (see **Sect. 9.3.8**).

9.3.6
Expression Systems

The achievement of significant expression levels for recombinant proteins produced in plants often depends on the tailoring and optimization of the expression system. Unfortunately, except for a few rules of thumb, insights gained while studying one protein are not always readily transferable to others, and low levels of expression are found in some of these systems. A few studies (e.g. [11] have attempted to address the low recombinant protein expression levels that may sometimes occur in transgenic plants. The issues of how, where and when to express a transgene in order to achieve maximal accumulation of a functional protein remain priorities in this field. It is especially important in the case of plant-derived vaccines (where the goal is to achieve sufficient dosage with minimal processing of antigenic proteins in edible plant tissue) to ensure that sufficient quantities survive passage through the stomach to elicit a clinically relevant immune response.

As discussed above, vaccine antigens can be produced in plants using two different systems: stable genetic transformation and transient expression. Stable transformation produces a genetic line that can be propagated either by vegetative (stem) cuttings or by seeds, and was the method used in published edible vaccine clinical trials to date [26,45, 69, 70]. Transient expression involves the use of a recombinant plant virus that carries the vaccine gene and, by systemic infection, causes the plant to express the antigen [4]. The main advantage of transient expression with a plant virus system is that virus replication amplifies the gene copy number, typically resulting in a much higher level of expression than with stable transformation. However, the plant virus systems might in some cases suffer from instability and loss of foreign genes larger than 1 kb. In addition, the need to inoculate each plant individually makes large-scale production rather labor intensive.

Recombinant DNA is integrated into the nuclear or chloroplast genome through stable transformation. *Agrobacterium tumefaciens*, a plant pathogen, is typically used for nuclear transformation, and strains have been engineered with deletions of the virulence genes that cause tumor growth in plants. *A. tumefaciens* efficiently transports DNA into cells and then promotes nuclear chromosomal integration at random sites [46]. Integrated nuclear transgenes are expressed and inherited in typical Mendelian fashion.

Some agronomically important plant species (e.g. soybean and most cereal grains) are recalcitrant to *Agrobacterium* transformation, and a biolistic method (microprojectile bombardment) is frequently used for these plants [47]. DNA coated on micron-sized gold particles is propelled into plant cells using compressed helium gas and becomes incorporated into chromosomal DNA. Foreign DNA that integrates into the nuclear genome yields stable transformants that show Mendelian inheritance of the transgenes. The biolistic method, as traditionally carried out, usually results in higher-copy-number plants compared to those generated by *Agrobacterium*, which can enhance expression [48,49]. However, excessive copy numbers or very high-level expression of nuclear genes can cause gene silencing, resulting in low protein accumulation [48–50]. Thus it is important to select transgenic lines that carry only between one and three copies of the transgene.

The biolistic method is also used to introduce transgenes into the chloroplast genome [51]. Although such transplastomic plants have been produced for only a few species, the potential for chloroplast transformation in other crop plants is strong. The high chloroplast genome copy number in plant cells enhances recombinant protein expression, as shown for the cholera toxin B (CTB) subunit [52]. DeCosa *et al.* [53] expressed an insecticidal protein in the chloroplasts of mature tobacco leaves at levels up to 45% of total protein. Stable transgene expression was recently described in the chloroplast-derived chromoplasts of tomato fruit [54]. A further advantage is that, as the chloroplast genome is inherited strictly maternally in most plants, there is minimal danger of unintended spread of transgenes by dissemination of pollen. Although chloroplast transformation is not yet a routine procedure, at least 25 different foreign genes have been expressed in transplastomic plants [55]. Chloroplast transformation is not suitable for some eukaryotic proteins that require post-translational modifications like glycosylation. Gene silencing, however, has not been observed in this system.

Seed-specific production of LTB by Streatfield *et al.* [28] resulted in expression levels of up to 1.8% total soluble protein (TSP), and two separate maize breeding programs have increased antigen production by fivefold [28] and tenfold [27]. The investigations of Chikwamba *et al.* [27] regarding the expression of LTB in maize are among the first to include the use of particle bombardment transformation for the production of plant-derived vaccines.

9.3.7
Mucosally-targeted Fusion Proteins

Most soluble proteins are rapidly hydrolyzed in the intestinal tract. Subunit vaccines are subject to the same proteolytic environment, and as a result some candidates may be ineffective at inducing the mucosal immune system. This is due to degradation prior to reaching immune effector sites in the gut-associated lymphoid tissues (GALT). In order to increase subunit vaccine mucosal immunogenicity, a number of investigators have used protein engineering to create fusion molecules consisting of the antigen (or epitopes) linked to a mucosally active carrier.

Two studies have involved the fusion of an antigen to β-glucuronidase (GUS). Gil *et al.* [61] fused GUS to the 2L21 protective epitope from canine parvovirus, and Dus Santos *et al.* [23] fused GUS to the protective epitope from FMDV. Transformants in both cases were selected on the basis of GUS activity, and both proved to be immunogenic. Mice immunized orally or parenterally with the GUS–FMDV epitope fusion were completely protected against challenge with the native virus.

Cholera toxin subunits have also been used to design targeted, plant-derived fusion vaccines. Using a tobacco mosaic virus (TMV) expression system, Nemchinov *et al.* [62] generated a *Vibrio cholerae* CTB fusion to an epitope (HVR1) from the hepatitis C virus (HCV). Tobacco plants inoculated with the recombinant TMV produced the HVR1 epitope fused to a functionally active, pentameric CTB, which reacted with HVR1-specific monoclonal antibodies and sera from individuals infected with virus from four of the major genotypes of HCV. Nasal immunization of mice with a crude plant extract containing the recombinant CTB–HVR1 elicited both anti-CTB serum antibodies and anti-HVR1 serum antibodies that specifically bound to HCV-like particles.

Sojikul *et al.* [43] reported the expression of a fusion protein designed to improve the expression of the hepatitis B surface antigen (HBsAg). This was achieved by mimicking the process of HBsAg targeting to the endoplasmic reticulum of human liver cells during HBV infection. The gene encoded a recombinant HBsAg modified to contain an N-terminal signal peptide from soybean vegetative storage protein A (VSPaS). This signal peptide directed the HBsAg-VSPaS fusion protein to the endoplasmic reticulum and resulted in greater fusion protein stability, enhanced protein accumulation and formation of more VLPs. Moreover, the HBsAg-VSPaS fusion stimulated higher levels of serum IgG when administered parenterally in mice compared to native HBsAg.

9.3.8
Forming Multivalent and Multicomponent Vaccines

Among vaccine products, those that stimulate several facets of the immune system – such as the induction of strong humoral and mucosal responses as well as effective cellular immune responses – are highly desirable. Combination vaccines targeting multiple pathogens in one formulation are likewise preferred. Therefore, developing both multivalent and multicomponent plant-based vaccines provides both efficacious and cost-effective immunization strategies. This can be accomplished through the development of recombinant plants carrying transgenes that encode the antigens of several pathogens, either by direct simultaneous transformation, sexual crosses of individually transformed lines, or the blending of separately transformed plant tissues.

One approach for the production of multicomponent vaccines was taken by Yu and Langridge [63] and Arakawa *et al.* [64]. They fused peptides containing important protective epitopes derived from two enteric pathogens – enterotoxic *E. coli* (ETEC), which causes bacterial traveler's diarrhea, and rotavirus, which causes acute viral gastroenteritis – to the A2 and B subunits of cholera toxin, respectively. The two recombinant cholera toxin subunit fusions were expressed from a single bidirectional promoter, ensuring coordinated expression for the two gene fusions and potentially facilitating the assembly of the chimeric holotoxin. In this approach, cholera toxin provides a scaffold for presentation of the protective epitopes, acts as a mucosal targeting molecule without toxic effects due to use of the nontoxic A2 and B subunits, and is itself a vaccine candidate. The recombinant protein represents a trivalent vaccine that can elicit significant mucosal and humoral responses against *Vibrio cholerae*, ETEC, and rotavirus. Mice, orally immunized with potatoes expressing these recom-

Tab. 9.2 Vaccine antigens expressed in plant systems: subunit antigen expression in transgenics

Antigen	Expression system	Immunogenicity	Protective challenge	Reference
Bacillus anthracis protective antigen	Tobacco leaf	No immunogenicity assays.	–	86
Bovine herpes virus type 1 (BHV1) glycoprotein D (gDc)	Tobacco leaf	Sustained and specific humoral and cellular immune response in mice and cows when delivered once parenterally.	Cows challenged with infectious dose of BHV1 exhibited milder symptoms over a shorter duration.	87
Enteropathogenic *E. coli* pilus subunit A (BfpA)	Tobacco leaf	Immunogenic when delivered orally in mice. Anti-BfpA in fecal matter.	–	88
Enterotoxigenic *E. coli* B subunits of the heat labile toxin (LTB)	Maize seed	Elicited neutralizing antibodies. Immunogenic when administered orally. Serum and secretory immune responses in humans.	Partially protective in mouse gut fluid assay.	27–29, 89, unpublished data
Enterotoxigenic *E. coli* B subunits of the heat labile toxin (LTB)	Potato tuber	Receptor-binding activity. Immunogenic in mice and humans when administered orally.	Partially protective in mouse gut fluid assay.	24–26, 90
Enterotoxigenic *E. coli* B subunits of the heat labile toxin (LTB)	Tomato fruit and leaf	Offspring were passively immunized through oral immunization of parents.	–	91
E. coli O157:H7 Intimin epitope	Tobacco callus	Mucosally immunogenic in mice when primed parenterally and boosted orally.	Mice immunized parenterally and orally boosted showed reduced duration of colonization upon challenge.	92
Foot and mouth disease virus (FMDV) VP1 epitope	Arabidopsis leaf	Serum antibodies reacted strongly with intact FMDV particles. Immunogenic in mice when administered parenterally.	Protective against challenge with virulent FVDM.	93
Foot and mouth disease virus VP1 epitope	Potato	FMDVVP1-specific antibody response. Immunogenic.	Protective against challenge.	22

Tab. 9.2 (continued)

Antigen	Expression system	Immunogenicity	Protective challenge	Reference
Foot and mouth disease virus VP1 epitope	Alfalfa leaf	Mice developed specific antibody response to synthetic peptide, VP1 epitope, and intact FMDV particle. Immunogenic in mice when administered parenterally or orally.	Mice protected against challenge with FMDV virus.	21, 23
Hepatitis B surface antigen (HBsAg)	Cherry tomatillo leaf and fruit	Immunogenic in mice as an oral booster after parenteral priming.	–	13
Hepatitis B surface antigen (HBsAg)	Lettuce leaf	Immunogenic in humans when administered orally; 2 of 3 human subjects developed very modest serum antibody response titers in primary immunization.	–	70
Hepatitis B surface antigen (HBsAg)	Lupin callus	Immunogenic in mice when administered orally.	–	70
Hepatitis B surface antigen (HBsAg) and middle protein	Potato tuber	Immunogenic in mice when administered orally.	–	11, 12, 94
Hepatitis B surface antigen (HBsAg)	Soybean cell culture	No immunogenicity assays performed.	–	95
Hepatitis B surface antigen (HBsAg)	Tobacco leaf	Mice vaccinated parenterally produced all IgG subclasses and IgM.	–	3, 56
Hepatitis E virus (HEV) ORF2 epitope	Tomato leaf and fruit	No immunogenicity assays performed.	–	6
Human cytomegalovirus glycoprotein B	Tobacco seed	gB-specific antigenic activity.	–	42, 96
Human immunodeficiency virus (HIV) type 1 p24 protein	Tomato bushy stunt virus in tobacco leaf	No immunogenicity assays performed.	–	97, 98

Tab. 9.2 (continued)

Antigen	Expression system	Immunogenicity	Protective challenge	Reference
Human immunodeficiency virus (HIV) type 1 gp120 (V3 loop) protein	Alfalfa mosaic virus in tobacco leaf	Elicited specific virus-neutralizing antibodies in mice when delivered parenterally.	–	16
Human immunodeficiency virus (HIV) type 1 gp120 (V3 loop) protein	Tomato bushy stunt virus in tobacco leaf	HIV epitope detected by V3-specific monoclonal antibodies and human sera from HIV positive patients. Immunogenic in mice when delivered parenterally.	–	17
Human papilloma virus type 11 major capsid protein L1	Tobacco leaf, potato tuber	Reacted well with assembled capsids and isolated L1 capsomers. Immunogenic in mice when administered parenterally with subsequent oral boosting.	–	99
Human papilloma virus (HPV) type 16 oncoprotein E7	Potato virus X in tobacco leaf (complete reading frame)	Immunogenic in mice when administered parenterally.	Mice protected moderately when challenged with surrogate (C3 cells).	100
Human papilloma virus (HPV) type 16 major capsid protein L1	Tobacco leaf	Neutralizing monoclonal antibodies bound plant-produced particles.	–	101
Human papilloma virus (HPV) type 16 major capsid protein L1	Tobacco and potato	Weak, transient anti L1 antibody response in 3 of 24 mice when administered orally. Immunogenic when administered orally after parenteral boost.	–	102
Infectious bronchitis virus (IBV) S1 glycoprotein	Potato tuber	Antibodies neutralized IBV from mice and chickens. Immunogenic in chickens when administered orally, nasally, and parenterally.	Protective after 3 immunizations in chickens when challenged with virulent IBV.	103
Measles virus hemagglutinin (H) protein	Carrot leaf and root	IgG1 and IgG2 cross reacted strongly with measles virus and neutralized virus in vivo. Immunogenic in mice when delivered parenterally.	–	39

Tab. 9.2 (continued)

Antigen	Expression system	Immunogenicity	Protective challenge	Reference
Measles virus hemagglutinin (H) protein	Tobacco leaf	Mice produced serum anti-H antibodies that neutralized measles virus in vitro. Immunogenic in mice when delivered orally or parenterally.	–	37, 38
Measles virus hemagglutinin (H) protein	Carrot leaf and root	H-specific antibody isotypes and subclasses in mouse sera; mouse sera neutralized wt virus. Immunogenic in mice when administered parenterally.	–	104
Murine hepatitis virus glycoprotein S 5B19 epitope	Tobacco mosaic virus in tobacco leaf	Mice developed serum IgG and IgA specific for 5B19. Immunogenic in mice when delivered nasally or parenterally.	Protective in mice immunized parenterally or orally when challenged with MHV strain JHM.	105
Norwalk virus (NV) capsid protein	Potato tuber; tobacco leaf	Mice and humans developed serum IgG and secretory IgA specific for recombinant NV. Immunogenic in mice and humans when administered orally.	–	45, 57
Plasmodium falciparum (PfMSP1)	Tobacco leaf	Antigenic characteristics identical to *E. coli* expressed protein. Immunogenic characteristics identical to *E. coli* expressed protein.		40, 106
Porcine epidemic diarrhea virus (PEDV) spike protein	Tobacco leaf	Systemic and mucosal immune response in mice immunized orally.	–	107
Rabbit hemorrhagic disease virus VP60 epitope	Potato leaf	VP60 specific antibodies elicited in rabbits. Immunogenic in rabbits when administered parenterally.	Rabbits protected when challenged with virulent RDHV.	67
Rabies virus glycoprotein (G)	Tomato leaf and fruit	Intact protein forms.		108

Tab. 9.2 (continued)

Antigen	Expression system	Immunogenicity	Protective challenge	Reference
Respiratory syncytial virus fusion protein	Tomato fruit	Mice developed serum and mucosal RSV-F-specific antibodies. Immunogenic in mice when delivered orally.	–	31
Rinderpest virus hemagglutinin (H) protein	Pigeon pea leaf	–	–	110
Rinderpest virus hemagglutinin (H) protein	Tobacco leaf	High titers of neutralizing antibodies detected in immunized mice.		111
Rotavirus VP6 protein	Potato tuber	Anti-GAR antibodies detected. Immunogenic in mice when delivered parenterally.	–	112
Rotavirus VP6 protein	Tomato callus	No immunogenicity assays.	–	32–34
Rotavirus VP7 protein	Potato tuber	Serum IgG and IgA specific for VP7; IgA had VP7-neutralizing activity. Immunogenic in mice when delivered orally.	–	36
Tetanus toxin fragment C	Tobacco leaves	Systemic IgG and local IgA responses. Immunogenic in mice when administered nasally.	–	113
Transmissible gastro-enteritis coronavirus N-terminal domain of the spike glyco-protein (S)	Arabidopsis leaf	gS elicited virus-neutralizing antibodies. Immunogenic when administered parenterally.	–	114

Tab. 9.2 (continued)

Antigen	Expression system	Immunogenicity	Protective challenge	Reference
Transmissible gastroenteritis coronavirus N-terminal domain of the spike glycoprotein (S)	Maize seed	Orally immunized piglets had a memory immune response leading to rapid accumulation of virus-neutralizing antibodies on challenge.	Reduced symptoms in swine immunized orally and challenged with virulent TGEV.	29
Transmissible gastroenteritis coronavirus N-terminal domain of the spike glycoprotein (S)	Potato tuber	Mice developed serum antibodies specific for gS. Immunogenic in mice when delivered parenterally or orally.	—	115, 116
Transmissible gastroenteritis coronavirus N-terminal domain of the spike glycoprotein (S)	Tobacco leaf	TGEV-specific antibodies (in pigs) showed low levels of virus-neutralizing activity. Immunogenic in pigs when administered parenterally.	—	116
V. cholerae B subunits of the cholera toxin (CTB)	Potato	Receptor-binding activity. Immunogenic in mice when delivered orally.	Mice challenged intraileally with CT showed up to 60% reduction in diarrheal fluid accumulation in small intestines.	117, 118
V. cholerae B subunits of the cholera toxin (CTB)	Tobacco leaf chloroplast	No immunogenicity assays performed.	—	119
B subunits of the cholera toxin (CTB)	Tomato leaf and fruit	No immunogenicity assays performed.	—	120

Tab. 9.3 Vaccine antigens expressed in plant systems: fusion proteins expressed in transgenics

Antigen	Expression system	Immunogenicity	Protective challenge	Reference
Canine parvovirus (CPV) VP2 epitope fused to GUS	Arabidopsis leaf	Specific antibodies detected and immunogenic in mice when delivered parenterally.	–	61
Enterotoxigenic E. coli epitope and rotavirus epitope fused to CTB	Potato tuber	Mice developed detectable levels of serum and intestinal antibodies. Immunogenic in mice against ETEC, rotavirus, and V. cholerae when delivered orally.	Symptoms reduced in passively immunized mouse neonates following rotavirus challenge.	63
E. coli B subunit of the heat labile enterotoxin (LTB) fused to an immunocontraceptive epitope	Tomato fruit and leaves	No immunogenicity assays performed.	–	91
Hepatitis B surface antigen (HBsAg) fused to soybean vegetative protein (VSPαS)	Tobacco leaf	VSPαS-HBsAg generated higher levels of serum IgG than native HBsAg in mice.	–	43
Hepatitis C virus (HCV) HVR1 epitope fused to V. cholerae CTB	Tobacco leaf	Reacted with HRV1-specific monoclonal antibodies and sera from individuals infected with HCV. HVR1 epitope formed fused to CTB was immunogenic in mice when administered nasally.	–	62
Mannheimia haemolytica A1 leukotoxin 50 fused to green fluorescent protein	White clover leaf	Rabbits produced antibodies that recognized and neutralized leukotoxin. Immunogenic in rabbits after drying when delivered parenterally.	–	65
RicinB fused to green fluorescent protein (GFP)	Tobacco leaf and hairy root culture	GFP-specific IgG present in serum; IgA in serum and fecal matter.	–	10
Rotavirus enterotoxin protein NSP4 fused to CTB	Potato	Fusion proteins assembled into cholera holotoxin-like structures that retained enterocyte-binding affinity. Orally immunized mice generated levels of serum and mucosal antibodies specific for the native antigen. Induced TH1 immune response.	–	63, 64

binant antigens, developed immune memory B cells as well as helper T cell type 1 (Th1) responses, which are indicators of successful immunization. Further, pups of immunized dams were protected from challenge with rotavirus, with a significantly lower morbidity rate as compared to controls. These results provide robust evidence for a vaccine strategy employing chimeric microbiological proteins expressed in plants and for multicomponent vaccines.

9.3.9
Stability and Processing

Lee *et al.* [65] carried out preliminary stability studies on dehydrated plant tissues. Clover plants expressing the *Mannheimia haemolytica* A1 leukotoxin 50 fusion protein were harvested and allowed to dry at room temperature and ambient humidity for 1–4 days. After 3 days, the clover tissue retained approximately 20% of its initial fresh weight, but no significant degradation of the fusion protein was observed. Hence, the fusion protein did not require refrigeration for stability. The clover-derived fusion protein induced, in injected rabbits, an immune response that recognized and neutralized the native antigen in modified neutral red cytotoxicity assays.

A comprehensive stability study was conducted by Smith *et al.* [66] on HBsAg from soybean cells. Under optimum conditions, antigen stability was maintained for at least 1 month after isolation of the surface antigen, although excess detergent rendered the antigen susceptible to proteolytic degradation by contaminating plant enzymes. This proteolysis was counteracted by the addition of skimmed milk or its protein component, which stabilized the antigen for up to 2 months. Also, by altering the sodium ascorbate concentration or buffer pH, the proportion of HBsAg displaying monoclonal-reactive epitopes increased between 8- and 20-fold.

Although antigen stability will have to be investigated on a case-by-case basis, it is obvious that simple *in vitro* manipulations may prove valuable in increasing the immunogenicity and stability of plant-derived antigens. Castanon *et al.* [67] investigated the minimal processing of a potato-derived rabbit hemorrhagic disease virus (RHDV) vaccine consisting of the VP60 gene product. Harvested potatoes were peeled, cut into pieces, lyophilized, powdered, stored and used in trials within 3 months of collection. Extracts made from the potato powder were used in rabbits as a subcutaneous primer and intramuscular booster. The rabbits immunized with the transgenic potato elicited specific antibody responses and were protected against challenge with virulent RDHV. Freeze-dried tomato fruit expressing LTB was orally immunogenic in mice, indicating the utility of the tomato system [68].

9.4
Clinical Trials with Plant-derived Vaccines

The results of five clinical trials involving orally delivered plant-derived vaccines have been published [26,45,69–71]. With the exception of LTB, each study used different recombinant antigens, and each reported the stimulation of serum antibodies, and in some cases, mucosal antibodies.

9.4.1
Enterotoxic *E. coli* and *Vibrio cholerae*

Two of the most widely spread and well-studied enterotoxigenic forms of bacterial diarrhea are ETEC and *Vibrio cholerae*. The toxins they produce, labile toxin (LT) and cholera toxin (CT) respectively, are very similar in primary sequence, structure, and mechanism of action [72]. They are homologous multi-subunit proteins in which the non-toxic B subunit mediates GM_1 ganglioside binding, and thus are candidates for vaccines that can neutralize toxin activity.

The B-subunits of these toxins, LTB and CTB, are among the most potent oral immunogens known. Oral delivery efficiently causes the accumulation of specific serum (IgG, IgA) and mucosal (sIgA) antibodies [73]. Both LT and CT also function as mucosal adjuvants, stimulating antibody production against co-delivered antigens. The ganglioside-binding activity of the LTB pentamer is required both for its mucosal immunogenicity and for the adjuvanticity of the holotoxin [74]. Tacket *et al.* [26] performed the first human clinical trial with a transgenic plant-derived vaccine. Fourteen volunteers ingested either 100 g of transgenic potato, 50 g of transgenic potato, or 50 g of non-transgenic potato. The LTB content of the tubers varied between 3.7 and 15.7 µg g^{-1} of tuber weight and the doses were given on days 0, 7 and 21. Volunteers reported only a few instances of minor side effects (nausea, cramps or diarrhea), and the raw potato was well tolerated overall. Ten of 11 volunteers who ate the potatoes expressing LTB developed at least 4-fold increases in levels of toxin-neutralizing serum IgG against LTB; none of the volunteers who ate the placebo potatoes showed any increase in anti-LTB antibodies. Five of 10 volunteers showed at least 4-fold increases in anti-LTB IgA, detected in stool samples. These data compared favorably with an earlier study in which volunteers were challenged with 10^9 ETEC cells (Tacket *et al.*, unpublished). This study was significant because it was the first ever to examine an edible plant vaccine in humans, and showed great potential for this new strategy. A more recent clinical study [69] using LTB expressed in processed corn seed produced similar results to the potato study.

9.4.2
Norwalk Virus

Tacket *et al.* [45] reported a clinical trial performed using transgenic potato tubers carrying the gene for Norwalk virus capsid protein (NVCP), which assembled into Norwalk virus like particles (VLPs). Twenty adult volunteers ingested either two or three doses each of 150 g of raw transgenic potato tuber containing 215–750 µg NVCP (expression was variable). Only about 50% of the NVCP subunits assembled into VLPs in the potato cells, possibly as a result of low recombinant protein concentration. Thus, the effective amount of administered potato vaccine was 125–375 µg per dose. Unassembled subunits are likely to be much less stable in the gastrointestinal tract and thus less immunogenic. However, 19 of 20 subjects in the experimental group showed significant increases in the numbers of IgA-antibody forming cells (AFCs), ranging from 6–280 per 10^6 peripheral blood mononuclear cells (PMBC), and 6 of 20 subjects in this group developed increases in IgG AFCs. Four volunteers showed increases in serum IgG anti-NVCP titers, four showed increased serum IgM and six showed increased IgA in their stool samples (a 17-fold mean increase).

Although the antibody responses were less impressive than those obtained with LTB, the study showed that a plant-derived non-replicating antigens other than LTB and CTB could stimulate human immune responses after oral delivery. Insect cell-derived 250-µg doses of purified NVLP showed more effective seroconversion [75]. Thus it is likely that part of the potato-delivered NVCP was unavailable for uptake in the gastrointestinal tract. More recent studies in transgenic tomato fruits with a plant-optimized NVCP gene resulted in higher expression levels and more potent immune responses in mice fed freeze-dried tomatoes (X. Zhang and H.S. Mason, unpublished results). A clinical trial is planned with dried tomato powder formulated in gelatin capsules (D. Kirk, H.S. Mason, and C. Tacket, trial investigators) in order to evaluate safety and immunogenicity.

9.4.3
Hepatitis B Virus

There is one published report concerning the immunogenicity in humans of orally delivered HBsAg expressed in plants [70]. Two of three volunteers who ate two 150-g doses of transgenic lettuce (containing ~1–2 µg HBsAg per dose) developed a modest protective (>10 IU l^{-1}) serum antibody titer after the second dose. The serum antibody titers declined rapidly after 4 weeks, probably due to the very low antigen dosage. However, the study showed that presumably naïve subjects could be seroconverted by the oral delivery of plant-expressed HBsAg. In the US, a clinical trial was carried out at the Roswell Park Cancer Institute, Buffalo, NY, to test the oral immunogenicity of recombinant potatoes expressing HBsAg [12]. The study was limited to volunteers who had previously been vaccinated and seroconverted with the standard injectable yeast-derived HBsAg. The trial involved 33 volunteers who ate either two (days 0 and 28) or three (days 0, 14 and 28) 100-g doses of HBsAg potato tubers containing approximately 1 mg HBsAg per dose, while a group of 10 volunteers ate non-transgenic potatoes only. The potato HBsAg vaccine boosted serum IgG antibody titers in more than half of the volunteers (Thanavala, Mason and Arntzen, unpublished results), which suggests that the oral delivery of plant-produced HBsAg could be a viable delivery system for an HBV boosting vaccine.

9.4.4
Rabies Virus

Yusibov *et al.* [71] delivered to 14 human volunteers spinach expressing epitopes from the rabies virus glycoprotein and nucleoprotein fused to the coat protein of alfalfa mosaic virus (AlMV). Five of the fourteen subjects had previously received a conventional rabies vaccine. Three of those five and all nine of the initially naïve subjects displayed significant antibody responses to the rabies virus. None of the control individuals showed a significant elevation in rabies-specific antibodies.

The conventional rabies vaccine was administered to the nine initially naïve subjects seven days after completing the oral vaccination. Three of these volunteers produced neutralizing antibodies against rabies virus, although none of the five control subjects did. This study showed a clear indication that the orally delivered rabies vaccine has potential as an oral booster for the conventional rabies vaccine.

9.5
Issues and Challenges

9.5.1
Development and Licensing of Plant-derived Vaccines

Conventional pharmaceutical development has been estimated to costs roughly between $100 million and $800 million per product, and takes over 12 years. The development of conventional vaccines may be somewhat less costly, although new recombinant DNA-derived subunit vaccines produced in fermentation-based systems are likely to be similar to protein pharmaceuticals in development costs. At present, none of the major pharmaceutical companies is directing funding towards the development of plant-derived vaccines for infectious diseases. This may reflect:

- doubts about the potential for significant return on investment;
- uncertainties in the regulatory processes;
- limited human clinical trial data that establish required dosages, timing of delivery, and evaluation of possible adverse immunological effects;
- a lack of personnel with sufficient expertise in plant biology.

As a result, the opportunity to produce vaccines in plants represents a classic example in which the reliance on market forces for the development of health products is failing. Participation of both the public sector and the non-profit sector will be essential to provide leadership and investment support to unlock the potential of plant-derived vaccines. A principal justification for the public sector promotion of plant-based vaccines is the significant favorable characteristics of this technology for the manufacture of vaccines against rare and neglected diseases. Developing new pharmaceuticals for these diseases is not a high priority for the major drug companies due to the low profit margins.

9.5.2
Confronting GM Food Issues

The plant-based production of vaccines is a potentially transformative technology, but the use of a similar technology for agricultural biotechnology has stimulated significant public debate, especially focused on genetically modified foods (GM foods).

Knowledgeable public debate is valuable, but the debates about GM foods have not always been based upon scientific considerations, and consequently the debates have become polarized.

Because plant-derived pharmaceuticals are not intended for use as food products, the crops that produce them must have special stewardship to ensure containment. In addition, the proteins they produce will have to be separated and purified in processing facilities dedicated to that purpose. The use of crop species that are currently part of the food supply for the production of oral vaccines will necessitate genetic separation from the food supply as an essential parameter for production of these materials. It is likely that global health organizations such as the Pan-American Health Organization (PAHO) and WHO will plan an essential role in the process of transferring the technology on a global scale as new regulatory frameworks are implemented.

References

[1] Centers for Disease Control (1998) *Preventing Emerging Infectious Diseases: A Strategy for the 21st Century.* Centers for Disease Control, Atlanta, GA. http://medi-smart.com/infect_emerging2.htm

[2] R. Curtiss III, G.A. Cardineau, *US Patent 5,686,079*, **1997**.

[3] H.S. Mason, D.M. Lam, C.J. Arntzen. *Proc. Natl. Acad. Sci. USA* **1992**, *89* (24), 11745–11749.

[4] H. Koprowski, V. Yusibov, *Vaccine* **2001**, *19* (17–19), 2735–2741.

[5] H.S. Mason, H. Warzecha, T. Mor et al., *Trends Mol. Med.* **2002**, *8* (7), 324–329.

[6] J. K.-C. Ma, P.M.W. Drake, P. Christou, *Nature Rev. Genet.* **2003**, *4* (10), 794–805.

[7] S.J. Streatfield, J.A. Howard, *Int. J. Parasitol.* **2003**, *33* (5–6), 479–493.

[8] A.M. Walmsley, C.J. Arntzen, *Curr. Opin. Biotechnol.* **2003**, *14* (2), 145–150.

[9] M.M. Rigano, F. Sala, C.J. Arntzen et al., *Vaccine* **2003**, *21* (7–8), 809–811.

[10] F. Medina-Bolivar, R. Wright, V. Funk et al., *Vaccine* **2003**, *21* (9–10), 997–1005.

[11] L.J. Richter, Y. Thanavala, C.J. Arntzen et al., *Nature Biotechnol.* **2000**, *18* (11), 1167–1171.

[12] Q.X. Kong, L. Richter, Y.F. Yang et al., *Proc. Natl. Acad. Sci. USA* **2001**, *98* (20), 11539–11544.

[13] Y. Gao, Y. Ma, M Li, T. Cheng, et al., *World J. Gastroenterology* **2003**, *9* (5), 996–1002.

[14] N. Gomez, A. Wigdorovitz, S. Castanon et al., *Arch. Virol.* **2000**, *145* (8), 1725–1732.

[15] T. Tuboly, W. Yu, A. Bailey et al., *Vaccine* **2000**, *18* (19), 2023–2028.

[16] V. Yusibov, A. Modelska, K. Steplewski et al., *Proc. Natl. Acad. Sci. USA* **1997**, *94* (11), 5784–5788.

[17] T. Joelson, L. Akerblom, P. Oxelfelt et al., *J. Gen. Virol.* **1997**, *78* (6), 1213–1217.

[18] C. Marusic, P. Rizza, L. Latanzi et al., *J. Virol.* **2001**, *75* (18), 8434–8439.

[19] H.E. Gilleland, L.B. Gilleland, J. Staczek et al., *FEMS Immunol. Med. Microbiol.* **2000**, *27* (4), 291–297.

[20] J. Staczek, M. Bendahmane, L.B. Gilleland et al., *Vaccine* **2000**, *18* (21), 2266–2274.

[21] A. Wigdorovitz, C. Carrillo, M.J. Dus Santos et al., *Virology* **1999**, *255* (2), 347–353.

[22] C. Carrillo, A. Wigdorovitz, K. Trono et al., *Viral Immunol.* **2001**, *14* (1), 49–57.

[23] M.J. Dus Santos, A. Wigdorovitz, K. Trono et al., *Vaccine* **2002**, *20* (7–8), 1141–1147.

[24] T.A. Haq, H.S. Mason, J.D. Clements et al., *Science* **1995**, *268* (5211), 714–716.

[25] H.S. Mason, T.A. Haq, J.D. Clements et al., *Vaccine* **1998**, *16* (13), 1336–1343.

[26] C.O. Tacket, H.S. Mason, G. Losonsky et al., *Nature Med.* **1998**, *4* (5), 607–609.

[27] R.K. CHIKWAMBA, J. CUNNICK, D. HATHAWAY et al., *Transgenic Res.* **2002**, *11* (5), 479–493.

[28] S.J. STREATFIELD, J.M. MAYOR, D.K. BARKER et al., *In Vitro Cell. Dev. Biol. Plant* **2002**, *38* (1), 11–17.

[29] B.J. LAMPHEAR, S.J. STREATFIELD, J.M. JILKA et al., *J. Control. Release* **2002**, *85* (1–3), 169–180.

[30] H. BELANGER, N. FLEYSH, S. COX et al., *FASEB J.* **2000**, *14* (14), 2323–2328.

[31] J.S. SANDHU, S.F. KRASNYANSKI, L.L. DOMIER et al., *Transgenic Res.* **2000**, *9* (2), 127–135.

[32] I.S. CHUNG, C.H. KIM, K.I. KIM et al., *Biotechnol. Lett.* **2000**, *22* (4), 251–255.

[33] G.J. O'BRIEN, C.J. BRYANT, C. VOOGD et al., *Virology* **2000**, *270* (2), 444–453.

[34] C.H. KIM, K.I. KIM, S.H. HONG et al., *Biotechnol. Lett.* **2001**, *23* (13), 1061–1066.

[35] T. MATSUMURA, N. ITCHODA, H. TSUNE-MITSU, *Arch. Virol.* **2002**, *147* (6), 1263–1270.

[36] Y.-Z. WU, J.-T. LI, Z.-R. MOU et al., *Virology*, **2003**, *313* (2), 337–342.

[37] Z. HUANG, I. DRY, D. WEBSTER et al., *Vaccine* **2001**, *19* (15–16), 2163–2171.

[38] D.E. WEBSTER, M.L. COONEY, Z. HUANG et al., *J. Virol.* **2002**, *76* (15), 7910–7912.

[39] E. MARQUET-BLOUIN, F.B. BOUCHE, A. STEINMETZ et al., *Plant Mol. Biol.* **2003**, *51* (4), 459–469.

[40] S. GHOSH, P. MALHOTRA, P.V. LALITHA et al., *Plant Sci.* **2002**, *162* (3), 335–343.

[41] K.E. WRIGHT, F. PRIOR, R. SARDANA et al., *Transgenic Res.* **2001**, *10* (2), 177–181.

[42] E.S. TACKABERRY, A.K. DUDANI, F. PRIOR et al., *Vaccine* **1999**, *17* (23–24), 3020–3029.

[43] P. SOJIKUL, N. BUEHNER, H.S. MASON, *Proc. Natl. Acad. Sci. USA* **2003**, *100* (5), 2209–2214.

[44] Z. HUANG, H.S. MASON, *Plant Biotechnol. J.* **2004**, *2* (3), 241–249.

[45] C.O. TACKET, H.S. MASON, G. LOSONSKY et al., *J. Infect. Dis.* **2000**, *182* (1), 302–305.

[46] P. ZAMBRYSKI, *Annu. Rev. Genet.* **1988**, *22*, 1–30.

[47] J.C. SANFORD, F.D. SMITH, J.A. RUS-SELL, *Methods Enzymol.* **1993**, *217*, 483–509.

[48] S. HOBBS, T.D. WARKENTIN, C.M.O. DE-LONG, *Plant Mol. Biol.* **1993**, *21* (1), 17–26.

[49] J. FINNEGAN, D. MCELROY, *Bio/Technology* **1994**, *12* (9), 883–888.

[50] H. VAUCHERET, C. BECLIN, T. ELMAYAN et al., *Plant J.* **1998**, *16* (6), 651–659.

[51] Z. SVAB, P. MALIGA, *Proc. Natl. Acad. Sci. USA* **1993**, *90* (3), 913–917.

[52] H. DANIELL, S.-B. LEE, T. PANCHAL et al., *J. Mol. Biol.* **2001**, *311* (5), 10001–1009.

[53] B. DE COSA, W. MOAR, S.-B. LEE et al. *Nature Biotechnol.* **2001**, *19* (1), 71–74.

[54] S. RUF, M. HERMANN, I.J. BERGER et al., *Nature Biotechnol.* **2001**, *19* (9), 870–875.

[55] H. DANIELL, M.S. KHAN, L. ALLISON, *Trends Plant Sci.* **2002**, *7* (2), 84–91.

[56] Y. THANAVALA, Y.F. YANG, P. LYONS et al., *Proc. Natl. Acad. Sci. USA* **1995**, *92* (8), 3358–3361.

[57] H.S. MASON, J.M. BALL, J.J. SHI et al., *Proc. Natl. Acad. Sci. USA* **1996**, *93* (11), 5335–5340.

[58] T.S. MOR, M. STERNFELD, H. SOREQ et al., *Biotechnol. Bioeng.* **2001**, *75* (3), 259–266.

[59] F. BOUCHE, E. MARQUET-BLOUIN, Y. YA-NAGI et al., *Vaccine* **2003**, *21* (17–18), 2065–2072.

[60] T.R. GANAPATHI, N.S. HIGGS, P.J. BA-LINT-KURTI et al., *Plant Cell Rep.* **2001**, *20* (2), 157–162.

[61] F. GIL, A. BRUN, A. WIGDOROVITZ et al., *FEBS Lett.* **2001**, *488* (1–2), 13–17.

[62] L.G. NEMCHINOV, T.J. LIANG, M.M. RI-FAAT, et al., *Arch. Virol.* **2000**, *145* (12), 2557–2573.

[63] J. YU, W.H. LANGRIDGE, *Nature Biotechnol.* **2001**, *19* (6), 548–552.

[64] T. ARAKAWA, J. YU, W.H. LANGRIDGE, *Plant Cell Rep.* **2001**, *20* (4), 343–348.

[65] R.W. LEE, J. STROMMER, D. HODGINS et al., *Infect. Immun.* **2001**, *69* (9), 5786–5793.

[66] M.L. SMITH, H.S. MASON, M.L. SHU-LER, *Biotechnol. Bioeng.* **2002**, *80* (7), 812–822.

[67] S. CASTANON, J.M. MARTIN-ALONSO, M.S. MARIN et al., *Plant Sci.* **2002**, *162* (1), 87–95.

[68] A.M. WALMSLEY, D.D. KIRK, H.S. MA-SON, *Immunol. Lett.* **2003**, *86* (1), 71–76.

[69] C.O. TACKET, M.F. PASETTI, R. EDELMAN et al., *Vaccine* (in press).

[70] J. KAPUSTA, A. MODELSKA, M. FIGLERO-WICZ et al., *FASEB J.* 1999, *13* (13), 1796–1799.

[71] V. YUSIBOV, D.C. HOOPER, S.V. SPITSIN et al., *Vaccine* 2002, *20* (25–26), 3155–3164.

[72] T.K. SIXMA, S.E. PRONK, K.H. KALK et al., *Nature* 1991, *351* (6325), 371–377.

[73] J. HOLMGREN, N. LYCKE, C. CZER-KINSKY, *Vaccine* 1993, *11* (12), 1179–1184.

[74] J.J. GUIDRY, L. CARDENAS, E. CHENG et al., *Infect. Immun.* 1997, *65* (12), 4943–4950.

[75] J.M. BALL, D.Y. GRAHAM, A.R. OPEKUN et al., *Gastroenterology* 1999, *117* (1), 40–48.

[76] J.P. LANGEVELD, F.R. BRENNAN, J.L. MARTINEZ-TORRECUADRADA et al., *Vaccine* 2001, *19*, 3661–3670.

[77] B.L.NICHOLAS, F.R. BRENNAN, J.L. MAR-TINEZ-TORRECUADRADA et al., *Vaccine* 2002, *20*, 2727–2734.

[78] M.R. FERNANDEZ-FERNANDEZ, J.L. MAR-TINEZ-TORRECUADRADA, J.I. CASAL et al., *FEBS Lett.* 1998, *427* (2), 229–235.

[79] R. USHA, J.B. ROHLL, V.E. SPALL et al., *Virology* 1993, *197* (1), 366–374.

[80] A. NATILLA, G. PIAZZOLLA, A. NUZZACI et al., *Arch. Virol.* 2004, *149* (1), 137–154.

[81] C. PORTA, V.E. SPALL, J. LOVELAND et al., *Virology* 1994, *202* (2), 949–955.

[82] K. DALSGAARD, A. UTTENTHAL, T.D. JONES et al., *Nat. Biotechnol.* 1997, *15* (3), 248–252.

[83] F.R. BRENNAN, T.D. JONES, L.B. GILLE-LAND et al., *Microbiology* 1999, *145* (1), 211–220.

[84] M.R. FERNANDEZ-FERNANDEZ, M. MOURINO, J. RIVERA et al., *Virology* 2001, *280* (2), 283–291.

[85] F.R. BRENNAN, T.D. JONES, M. LONG-STAFF et al., *Vaccine* 1999, *17* (15–16), 1846–1857.

[86] M.A. AZIZ, S. SINGH, P.A. KUMAR et al., *Biochem. Biophys. Res.* 2002, *299* (3), 345–351.

[87] D.M.P. FILGUEIRA, P.I. ZAMORANO, M.G. DOMINGUEZ et al., *Vaccine* 2003, *21* (27–30), 4201–4209.

[88] J.V. DASILVA, A.B. GARCIA, V.M.Q. FLORES et al., *Vaccine* 2002, *20* (16), 2091–2101.

[89] S.J. STREATFIELD, J.M. JILKA, E.E. HOOD et al., *Vaccine* 2001, *19* (17–19), 2742–2748.

[90] T.G. LAUTERSLAGER, D.E. FLORACK, T.J. VAN DER WALL et al., *Vaccine*, 2001, *19* (17–19), 2749–2755.

[91] A.M. WALMSLEY, M.L. ALVAREZ, Y. JIN et al., *Plant Cell Rep* 2003, *21* (10), 1020–1026.

[92] N.A. JUDGE, H.S. MASON, A.D. O'BRIEN, *Infect. Immun.* 2004, *72* (1), 168–175.

[93] C. CARRILLO, A. WIGDOROVITZ, J.C. OLI-VEROS et al., *J. Virol.* 1998, *72* (2), 1688–1690.

[94] P. EHSANI, A. KHABIRI, N.N. DO-MANSKY, *Gene* 1997, *190* (1), 107–111.

[95] M.L. SMITH, M.E. KEEGAN, H.S. MASON et al., *Biotechnol. Prog.* 2002, *18* (3), 538–550.

[96] E.S. TACKABERRY, F. PRIOR, M. BELL et al., *Genome*, 2003, *46* (3), 521–526.

[97] G. ZHANG, C. LEUNG, L. MURDIN et al., *Molecular Biotechnol.* 2000, *14* (2), 99–107.

[98] G. ZHANG, L. RODRIGUES, B. ROVINSKI et al., *Molecular Biotechnol.* 2002, *20* (2), 131–136.

[99] H. WARZECHA, H.S. MASON, C. LANE et al., *J. Virol.*, 2003, 77 (16), 8702–8711.

[100] R. FRANCONI, P. DI BONITO, F. DIBELLO et al., *Cancer Research* 2002, *62* (13), 3654–3658.

[101] A. VARSANI, A-L. WILLIAMSON, R.C. ROSE et al., *Arch. Virol.*, 2003, 148 (9), 1771–1786.

[102] S. BIEMELT, U. SONNEWALD, P. GALM-BACHER et al., *J. Virol.* 2003, *77* (17), 9211–9220.

[103] F.B. BOUCHE, E. MARQUET-BLOUIN, Y. YANAGI et al., *Vaccine* 2003, *21* (17–18), 2065–2072.

[104] J.Y. ZHOU, J.X. WU, L.Q. CHENG et al., *Virology* 2003, *77* (16), 9090–9093.

[105] M. KOO, M. BENDAHMANE, G.A. LET-TIERI et al., *Proc. Natl. Acad. Sci. USA* 1999, *96* (14), 7774–7779.

[106] T.H. TURPEN, S.J. REINL, Y. CHAROENVIT et al., *Bio/Technology* 1995, 13 (1), 53–57.

[107] J.-L. BAE, J.-G. LEE, T.-J. KANG et al., *Vaccine* **2003**, *21* (25–26), 4052–4058.

[108] P.B. McGARVEY, J. HAMMOND, M.M. DIENELT et al., *Bio/Technology* **1995**, *13* (13), 1484–1487.

[109] A. MODELSKA, B. DIETZSCHOLD, N. SLEYSH et al., *Proc. Natl. Acad. Sci. USA* **1998**, *95* (5), 2481–2485.

[110] V.V. SATYAVATHI, V. PRASAD, A. KHANDELWAL et al., *Plant Cell Rep.* **2003**, *21* (7), 651–658.

[111] A. KHANDELWAL, L. SITA, M.S. SHAILA, *Virology* **2003**, *308* (2), 207–215.

[112] T. MATSUMARA, N. ITCHODA, H. TSUNEMITSU, *Arch Virol*, **2002**, *147* (6), 1263–1270.

[113] J.S. TREGONING, P. NIXON, H. KURODA et al., *Nucleic Acids Res*, **2003**, *31* (4), 1174–1179.

[114] N. GÓMEZ, C. CARRILLO, J. SALINAS et al., *Virology* **1998**, *249* (2), 352–358.

[115] N. GÓMEZ, A. WIGDOROVITZ, S. CASTANON et al., *Arch Virol*, **2000**, *145* (8), 1725–1732.

[116] T. TUBOLY, W. YU, A. BAILEY et al., *Vaccine*, **2000**, *18* (19), 2023–2028.

[117] T. ARAKAWA, D.K. CHONG, J.L. MERRITT et al., *Transgenic Res.* **1997**, *6* (6), 403–413.

[118] T. ARAKAWA, D.K. CHONG, W.H. LANGRIDGE, *Nature Biotechnol.* **1998**, *16* (5), 292–297.

[119] H. DANIELL, S.J. STREATFIELD, K. WYCOFF, *Trends Plant Sci.* **2001**, *6* (5), 219–226.

[120] D. JANI, L.S. MEENA, Q.M. RIZWAN-UL-HAW et al.. *Transgenic Res.* **2002**, *11* (5), 447–454.

10
Production of Secretory IgA in Transgenic Plants

Daniel Chargelegue, Pascal M.W. Drake, Patricia Obregon and Julian K.-C. Ma

10.1
Introduction

Plant biotechnology is a new and expanding area of science and it is becoming apparent that plant systems may be valuable for the expression and production of recombinant proteins such as pharmaceuticals, vaccines and in particular, antibodies. A unique attraction of this approach is the potential to produce these kinds of reagents on an agricultural scale, thereby significantly reducing the costs of production. However there are also many other advantages related to the use of plants. The assembly and expression of the multimeric, complex molecule secretory immunoglobulin A (sIgA) was first described successfully in transgenic plants [1], and plants remain the best system to express this molecule. Indeed, the large-scale production of recombinant sIgA is a very challenging task for two main reasons: (i) the components of this molecule are naturally produced by two distinct cell types (plasma and epithelial cells); and (ii) the final product is a large complex molecule of almost 400 kDa displaying numerous post-transcriptional modifications (intra- and interchain disulfide bonds and glycosylation sites). SIgA has also been produced in CHO cells but the cost of production might be too high to envisage commercialization on a worldwide scale [2]. In this chapter, we review the characteristics of sIgA and describe recent advances in the expression of antibodies in plants.

10.2
Antibodies

Antibodies are glycoproteins that bind specifically to their cognate antigens. There are five antibody or immunoglobulin (Ig) classes – IgG, IgA, IgM, IgD and IgE. Immunoglobulins can exist in polymeric forms, for example IgM exists as a pentamer and secretory IgA consists of two IgA molecules linked by a joining (J) chain and associated with a secretory component (SC). Some antibody classes are further divided into subclasses that differ slightly in structure and function from other members of the same class. Monomeric immunoglobulins are composed of two identical heavy

Molecular Farming. Edited by Rainer Fischer, Stefan Schillberg
Copyright © 2004 WILEY-VCH Verlag GmbH & Co. KGaA, Weinheim
ISBN: 3-527-30786-9

(H) and two identical light (L) chains, which are held together by disulfide bonds and non-covalent interactions. The heavy chains that define the isotypes are designated by their Greek letter counterparts (e.g. γ, α, μ, δ, ε) and all antibody classes contain either λ or κ light chains. Both light and heavy chains are divided into variable (V) and constant (C) domains. The VL and VH domains are responsible for antigen binding whereas the heavy chain constant region mediates effector functions such as complement activation and Fc receptor-mediated phagocytosis.

10.2.1
Mucosal Antibodies

The mucosal system of the human body has a very large surface area (about 400 m^2), which is exposed to invasion by multiple pathogens (bacteria, viruses and parasites). Protection against these pathogens is provided by innate defense mechanisms (the mucosal barrier) and by adaptive immune recognition that includes sIgA. To exert its protective activity on the mucosa, polymeric IgAs (pIgAs, mostly IgA dimers) are transported across the epithelium after binding to the polymeric immunoglobulin receptor (pIgR), which is expressed basolaterally on the epithelial cells. During transport, the pIgR (also known as the transmembrane secretory component) is cleaved, and the secretory component (SC) is released in association with pIgA to form sIgA [3]. Quantitatively, sIgA is the most important antibody class with 40–60 mg kg^{-1} produced every day, whereas the daily production of IgG is only of 30 mg kg^{-1} [4,5]. Moreover, the covalent binding of SC enhances resistance to proteolytic degradation making this the most stable form of antibody in mucosal secretions [3].

10.2.2
Structure and 'Natural' Production of SIgA

The overall structure of monomeric IgA resembles IgG in that it is composed of two identical heavy chains (comprising three constant domains: Cα Cα and Cα) and two identical light chains (kappa or lambda). Polymeric IgA (pIgA) usually exists as 11S dimers, comprising two IgA monomers that are linked by the J chain (Fig. 10.1). However, tetramers are also produced in a lesser amounts. In humans, there are two IgA subclasses, IgA1 and IgA2, that differ only in the hinge region: IgA1 contains a 13-amino-acid, proline-rich sequence which is not present in IgA2. The amino acid composition of the IgA hinge region renders it more resistant to proteases than other immunoglobulins [3,6] However, IgA1 is particularly sensitive to proteases produced by Gram-negative bacteria, whereas IgA2 is relatively more resistant (due to the absence of the proline-rich region). IgA2 exists as two well-characterized allotypes, IgA2m(1) and IgA2m(2), and a third one, IgA2m(3) or IgA2m(n), that has not yet been fully defined. The different forms of human IgA differ in their heavy chain disulfide bonding and glycosylation patterns. In IgA1 and IgA2m(2), a disulfide bond links the heavy and light chain, whereas IgA2m(1) antibodies lack this covalent bond. Human IgA1 displays five O-linked carbohydrates in the hinge region, whereas IgA2 lacks these residues. Furthermore, two N-linked glycans are present on IgA1 (one in Cα2 and one in Cα3) and on IgA2 (one Cα and one on Cα). A third N-glycan is only present on the Cα domain of IgAm(2) [3,7].

The J chain is a 137-amino-acid, 15.6 kDa glycoprotein that is added just before the secretion of pIgA by the plasma cells [6]. The J chain contains a single N-linked glycan (Asn49) and eight cysteine residues, two of which (Cys15 and Cys69) are involved in forming disulfide bonds with the α chain in pIgA, and six of which are involved in intrachain disulfide bridges [8]. Either one of the two disulfide bonds normally present between the J chain and the α chain is sufficient for polymer formation [9]. The J chain is a key protein in the synthesis of sIgA because it promotes polymerization of IgA and because its presence in these polymers is required for their affinity to pIgR/secretory component. Studies in J chain knockout mice indicate that the J chain is required for stable association of pIgA with the secretory component [6].

The secretory component (SC), an 80 kDa glycoprotein, is the extracellular domain of the pIgR synthesized by the mucosal epithelium. The pIgR ensures efficient secretion of pIgA at mucosal surfaces. During basal-to-apical transport across the epithelial cells, the pIgR ectoplasmic domain is cleaved, releasing SC in association with pIgA, thus forming sIgA (Fig. 10.1). The human pIgR comprises a 103-amino-acid cytoplasmic domain, a 23-amino-acid transmembrane domain and a 589-

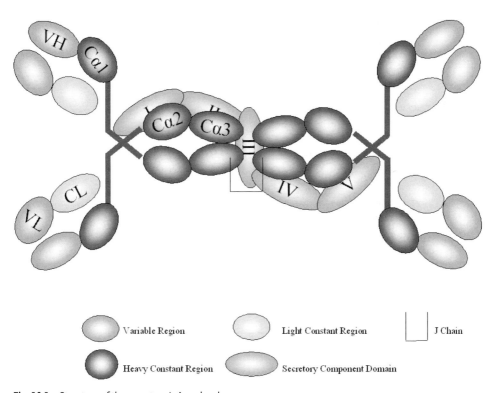

Fig. 10.1 Structure of the secretory IgA molecule.

amino-acid segment (SC) folded into five Ig-like domains stabilized by disulfide bonds (domains I to V) displaying between one and seven N-linked carbohydrates [2]. Domain I of the pIgR is involved in the initial non-covalent interaction with pIgA. Subsequently, during transcytosis, the pIgR becomes covalently attached to pIgA via cysteine residues in domain V. Recently, studies of the mouse secretory component have also revealed a role for domains II and III in covalent binding to IgA [10]. The cytoplasmic domain of pIgR contains a sorting signal to target the pIgR for endocytosis at the basolateral surface and transcytosis to the apical surface. After proteolytic cleavage of the cytoplasmic domain at the apical side of the epithelial cell, the pIgA is released as sIgA (Fig. 10.1).

10.2.3
Passive Immunization with SIgA

Active immunization has been successful in protecting against several infectious diseases. However, vaccines are still not available for numerous pathogens (e.g. human immunodeficiency virus, respiratory syncytial virus, hepatitis C virus). Furthermore, active immunization is generally less effective in immunocompromised individuals. In contrast, passive immunoprophylaxis could provide a high level of protection by neutralizing pathogens at the affected site in all patient groups [11–13] It is important to underline the fact that most infections begin at mucosal surfaces. Therefore, passive immunization with sIgA should provide a better protection level against pathogens than monomeric antibodies (i.e. IgGs or IgAs). It has been shown that the SC is essential for the stability of the whole sIgA molecule when targeting the gastrointestinal (GI) tract [14]. Furthermore, sIgAs have a higher binding avidity for their antigens than monomeric antibodies, because of their four antigen combining sites.

Thus, the tetravalency, anti-inflammatory properties and molecular stability of sIgA make it particularly suitable for protective passive immunity when applied to mucosal surfaces. To date, the clinical evaluation of sIgA protection in humans and animal models has been very limited. Indeed most studies have employed monomeric IgA monoclonal antibodies [3,15]. Hence, differences in IgA and IgG protective activities at the mucosal level have often not been observed [15]. Only a few studies have demonstrated the superior activity of polymeric IgA or sIgA compared with monomeric IgG or IgA [16]. In order to determine the efficacy of sIgA, future animal experiments and clinical trials are needed to compare the activities of IgG monoclonal antibodies and their sIgA counterparts. The ability to engineer sIgAs in plants will allow these comparisons to be made [17].

10.2.4
Production of Recombinant SIgA

As stated in the introduction, the large-scale production of recombinant sIgA is a very challenging task. This is due to the complex post-translational modifications that are required and because two distinct cell types are needed to produce the native molecule in mammals. Thus far, only two expression systems have been successful in producing recombinant sIgA:

(i) Full length sIgA was first expressed in transgenic plants [1] and the methodology employed in this achievement will be reviewed extensively in the second part of this chapter. Briefly, sIgA was produced in tobacco plants with a yield of up to 8% total soluble protein (TSP), which corresponds to 10–80 mg of antibody per kg of fresh material [17]. This figure has been extrapolated to 10–20 kg of therapeutic antibody per acre of tobacco plants [18] or corn [2].

(ii) In the past 5–6 years, several groups have succeeded in expressing recombinant sIgA in mammalian cells [5,19,20]. The use of CHO (Chinese hamster ovary) cells to produce sIgA has been the most successful strategy, and has relied on the successive stable transfection and selection of CHO clones capable of expressing monomeric, dimeric and finally sIgA recombinant molecules [3,5,20]. However, the levels of antibody production in stable transfectomas have generally been lower than in murine hybridomas. Indeed, most transfectomas secrete on average 1–10 μg of sIgA per ml of culture, with a few exceptions where the yield reached 30 μg mL^{-1} [7]. The cost of production of any recombinant protein in CHO cell fermentors is high (estimated to be in excess of $300 per gram of protein and approximately 3000 times higher than estimates for transgenic plants [2].

More recently, recombinant antibodies (mostly IgGs) have been produced in the milk of transgenic animals [21,22]. In particular, one study with transgenic mice has shown that it is possible to produce a porcine chimeric IgA that can form dimers in the presence of the J chain [23]. However, the production of fully assembled sIgA has yet to be reported.

This technology is still in its infancy, and obtaining a large number of animals producing recombinant antibodies is laborious. Indeed, the transgenic birth average is generally only 0.5–5% for cattle, goats and sheep [21]. Furthermore, the presence of host animal IgG and IgA in milk may complicate the purification of recombinant antibodies [22]. Finally, the cost of production is estimated to be 10 times higher than in transgenic plants [2] and the risk of pathogen contamination (e.g. animal viruses or prions) has not yet been fully assessed.

10.3
Production of Recombinant SIgA in Plants

10.3.1
Production of Full-length Antibodies in Plants

Plants have the ability to assemble immunoglobulin heavy chains and light chains to form full-length antibodies very efficiently [24]. In mammalian plasma cells, the assembly mechanism is only partially understood. The immunoglobulin light and heavy chains are synthesized as precursor proteins, and signal sequences direct translocation into the lumen of the endoplasmic reticulum (ER). Within the ER, cleavage of the signal peptides takes place, and stress proteins, such as BiP/GRP78 and GRP94, as well as enzymes such as protein disulfide isomerase (PDI), function as chaperones that bind to unassembled heavy and light chains and direct subsequent folding and assembly. In plants, the passage of immunoglobulin chains through the ER is also necessary, since antibody assembly does not take place in the absence of a signal peptide [24]. However, both plant and non-plant signal sequences from a variety of sources are sufficient for ER-targeting [25,26]. Plant chaperones homologous to mammalian BiP, GRP94 and PDI have been described within the ER [27,28], and the expression of immunoglobulin chains in plants is indeed associated with increased BiP and PDI expression [29]. Recently, binding of plant BiP to IgG heavy chains in transgenic tobacco has been demonstrated and shown to be ATP-sensitive, which is highly suggestive that the interaction with BiP is functional. Co-expression of heavy and light chains resulted in IgG assembly and displacement of BiP from the heavy chain as the amount of light chain increased [29]. Thus, it seems likely that there are broadly similar folding and assembly mechanisms for antibodies in mammals and plants.

Several IgG monoclonal antibodies have been produced in transgenic plants by academic and commercial groups, which may have therapeutic applications in humans or animals (described in Ref. [30]). An example from our group is a murine IgG1 (Guy's 13) that binds to the adhesion protein of *Streptococcus mutans*, the primary cause of dental caries. The strategy used to produce this antibody in plants was to express each immunoglobulin chain separately in different plant lines, and then to stack the two genes in the same plant line by crossing parental plants individually expressing the heavy and light chains. This involved two generations of plants, and using this technique, the yield of recombinant antibody was consistently high (approximately 1% TSP) [24,31]. Guy's 13 IgG is relatively easy to purify in large quantities from tobacco, and functionally there is no discernible difference between the antibody expressed in plants, and that expressed in other systems [31].

Guy's 13 accumulates in the apoplast and recently was shown to be secreted from roots into the surrounding medium in hydroponic cultures giving a yield of 11.7 µg of antibody per gram of root dry mass per day [32]. Guy's 13 can also be expressed in transgenic plants with a transmembrane sequence so that it is retained in the plasma membrane with variable regions protruding into the apoplasm [33]. Transgenic plants expressing antibodies immobilized in such a fashion may have important applications in phytoremediation and phytomining. Indeed, we have recently demonstrated that hydroponic cultures expressing immobilized Guy's 13 are able to take up antigen through the roots, transport it through the vascular system, and form antigen-antibody immune complexes in the leaf tissue [32]. Other groups have expressed IgG antibodies using double transformation techniques [34], or have cloned the light and heavy chain genes together in a single *Agrobacterium* T-DNA vector [25,35]. Both strategies can save time and effort.

10.3.2
Production of Multimeric Antibodies: SIgA

The ability to stack genes in transgenic plants by successive crosses between individually transformed parental plants is a considerable advantage in attempting to construct multimeric protein complexes, such as secretory antibodies. As described earlier, sIgA consists of two basic Ig monomeric units (heavy and light chains) that are dimerized by a joining (J) chain and then associated with a fourth polypeptide, the secretory component (SC) [36].

In order to generate a secretory antibody version of Guy's 13 in plants, the carboxyl terminal domains of the Guy's 13 IgG antibody heavy chain were modified by replacing the Cγ3 domain with the Cα2 and Cα3 domains of an IgA antibody, these being required for binding to the J chain and SC [31]. Four transgenic plants were generated to express independently the Guy's 13 kappa chain, the hybrid IgA-G antibody heavy chain, the mouse J chain and the rabbit SC. A series of sexual crosses was performed between these plants and filial recombinants in order to generate plants in which all four protein chains were expressed simultaneously. In the final, quadruple transgenic plant, three forms of the antibody were detectable by western blot analysis of samples prepared under non-reducing conditions. These bands had approximate molecular masses of 210 kDa (monomeric IgA-G), 400 kDa (IgA-G dimerized with the J chain) and 470 kDa (dimeric IgA-G associated with the SC). The assembly was very efficient, with greater than 50% of the SC being associated with dimeric IgA-G. The sIgA-G yield from fully expanded leaves was in excess of 5% TSP, or 200–500 µg per gram of fresh weight material [1]. Frigerio *et al.* [37] demonstrated that secretion of sIgA-G proceeded at a very slow rate in tobacco leaf cells. After 24 h only about 10% of newly assembled molecules had been secreted with the bulk probably remaining in the ER. In addition, a proportion of the sIgA-G was delivered to the vacuole where it was detected as fragmentation products. Hadlington *et al.* [38] have demonstrated that vacuolar delivery depends on the presence of a cryptic sorting signal in the tailpiece of the IgA heavy chain.

Functional studies of the plant-derived secretory antibody confirmed that the sIgA-G molecule bound specifically to its native antigen and that the binding affinity of each antigen-binding site was no different to that of the native IgG. However, the functional affinity (or avidity) of the sIgA was greater than that of the IgG, which confirmed that a dimeric, tetravalent antibody had been assembled. Finally, in a human trial, the plant-derived secretory Guy's 13 antibody prevented oral colonization by *Streptococcus mutans*, thereby demonstrating for the first time the therapeutic application in humans of a recombinant product derived from plants [17].

The Guy's 13 SIgA-G plantibody technology is licensed to Planet Biotechnology Inc. (USA) and is currently in clinical trials under the product name CaroRx™ [18].

10.3.3
Glycosylation of Antibodies in Transgenic Plants

It has been postulated that differences in glycosylation between mammalian and plant-derived antibodies may result in decreased functional activity or unwanted immunogenicity. Protein modification by glycosylation occurs in all higher eukaryotes, and plant proteins contain N-linked as well as O-linked glycans. The N-linked core high-mannose type glycans have identical structures in plants, mammals and other organisms [39,40], but differences occur between these groups during subsequent modification to complex glycans. Native complex glycans in plant proteins can be quite heterogeneous, but they tend to be smaller than mammalian complex glycans and differ in the terminal sugar residues. For example, a xylose residue linked β(1,2) to the β-linked mannose residue of the glycan core, and an α(1,3)-fucose residue in place of an α(1,6)-fucose linked to the proximal glucosamine, are frequently found in plants, but not in mammals [39]. On the other hand, plants lack the β(1,4)-galactose and terminal α(2,6)N-acetylneuraminic acid (a sialic acid) residues often found on mammalian glycans. However, these complex glycans are not unique to plants. Indeed they are also found in numerous nematodes (e.g. *Schistosoma mansoni*) as well as in baculovirus-derived recombinant glycoproteins.

A structural comparison of the glycans associated with the monoclonal antibody Guy's 13 IgG expressed in plants and in murine hybridoma cells has been performed [41]. The results demonstrated that the same glycosylation sites were utilized in both systems, but that compared to the murine antibody, the glycans on the plant antibody were more heterogeneous. In addition to high-mannose type glycans, approximately two-thirds of the plant antibodies had β(1,2)-xylose and α(1,3)-fucose residues as predicted. The differences in glycosylation patterns had no effect on antigen binding or specificity. Interestingly, it has been demonstrated that antibodies isolated from young transgenic tobacco leaves have a relatively large proportion of high-mannose glycans compared with antibodies from older leaves, which contain a larger number of terminal *N*-acetylglucosamine residues [42].

Glycosylation of IgG antibody C_H2 domain is critical for antibody effector functions, namely Fc receptor binding and complement activation. Indeed, aglycosylated antibodies produced in *E. coli* do not bind to complement C1q or to Fc receptors [43]. The role of N-glycans in sIgA antibodies is not as well defined as it is for IgG. It is known that deletion of the carbohydrates on the α chain does not affect the synthesis and secretion of human IgA1 or interfere with its ability to bind the pIgR or Fcα receptor (CD89) [2]. However, the deletion of glycans on this IgA does affect complement component C3 binding and the alternative pathway.

For systemic applications, it may be necessary to remove either the complex glycans or to alter the heavy chain sequence to remove the sites for N-linked glycosylation (Asn-X-Ser/Thr, where X is any amino acid except proline). An alternative approach is also being developed using mutant plants that lack enzymes involved in the complex glycosylation pathway [44]. In addition, specific fucosyl or xylosyl transferases may also be targeted for gene silencing [18]. Ultimately, it may be possible to generate transgenic plants with a 'humanized' glycosylation pathway. Comparison of plant and mammalian N-glycan biosynthesis indicates that β1,4-galactosyltransferase is the most important enzyme that is missing in plants, and that its addition would allow the conversion of typical plant-N-glycans into mammalian-like N-glycans. Expression of this key enzyme in transgenic tobacco resulted in 15% of proteins expressing terminal β1,4-galactose residues. Backcrossing of β1,4-galactosyltransferase plants with tobacco plants expressing murine Ig heavy and light chains resulted in the expression of an antibody with partially galactosylated N-glycans [45]. A potential problem with humanization is that antibodies expressed in many plant species are not homogeneously glycosylated. For example, transgenic tobacco produces IgG1 in eight different glycoforms. However, antibodies expressed in alfalfa are produced as a single glycoform, and this species appears to be the only system in which homogeneously humanized IgGs may be obtained through *in vivo* or *in vitro* modifications (see chapter 1).

10.3.4
Plant Hosts

There is no consensus as to the choice of plant species for the commercial production of antibodies. Important considerations include intellectual property ownership, ease of transformation, biomass yield, achievable expression levels, purification and storage costs, and processing of valuable co-products such as starch, oil or fiber [46]. For example, the advantages of tobacco for antibody production include the ease of transformation, the high biomass yield per unit area and the production of large numbers of seeds, whereas production in seed systems (e.g. maize) allows the storage of antibodies in a stable form for long periods of time (up to 3 years when refrigerated) and relatively high expression levels due to their high protein content [47]. For a more detailed discussion of the pros and cons of different species for molecular farming, see chapter 13.

10.4
Conclusions

Plants have significant advantages over other expression systems for the expression of antibodies. This is demonstrated by the production of sIgA, which was achieved for the first time in transgenic tobacco [1]. With regards to protein folding and structure, small peptides, polypeptides and even complex proteins can be expressed in plants in a fully assembled and functional form. For larger molecules such as antibodies, this is associated with the presence of ER resident chaperones that are homologous to those involved in protein assembly in mammalian cells. Targeting recombinant proteins for secretion through the ER and Golgi apparatus is achieved using either native or plant-derived leader sequences, and this also ensures that N-glycosylation takes place. Plant glycosylation differs from that in mammals in terms of the structure of complex glycans, but for the recombinant proteins expressed so far, this has not had any adverse effects on structure nor has it enhanced protein immunogenicity. Furthermore, even in mammalian cells, antibodies are not homogeneously glycosylated, and only 10–15% are sialylated.

The storage of immunoglobulin genes and gene products in plants can be very stable. Transgenic plants can be self-fertilized conveniently to produce stable, true breeding lines, propagated by conventional horticultural techniques and stored and distributed as seeds. The expressed recombinant immunoglobulins can be targeted to stable environments within the plant, such as the extracellular apoplastic space. Alternatively, tissue-specific promoters can be used to restrict expression to storage organs such as seeds or tubers. Extraction and purification from these sites is generally a simple process.

One of the most obvious benefits of plants is the potential for production scale up, leading to the production of virtually limitless amounts of recombinant antibody at minimal cost. Plants are easy to grow, and unlike bacteria or animal cells their cultivation is straightforward and does not require specialist media, equipment or toxic chemicals. It has been estimated that plantibodies could be produced at a yield of 10–20 kg per acre at a fraction of the cost associated with production in mammalian cells [2,18] The use of plants also avoids many of the potential safety issues associated with other expression systems, such as contaminating mammalian viruses or prions, as well as ethical considerations involving the use of animals.

References

[1] J. K.-C. MA, A. HIATT, M. HEIN et al., *Science* **1998**, *268* (5211), 716–719.

[2] E. E. HOOD, S. L. WOODARD, M. E. HORN, *Curr. Opin. Biotechnol.* **2002**, *13* (6), 630–635.

[3] B. CORTHESY, *Trends Biotechnol.* **2002**, *20* (2), 65–71.

[4] N. K. CHILDERS, M. G. BRUCE, J. R. McGHEE, *Annu. Rev. Microbiol.* **1989**, *43*, 503–536.

[5] F. E. JOHANSEN, I. N. NORDERHAUG, M. ROE et al., *Eur. J. Immunol.* **1999**, *29* (5), 1701–1708.

[6] K. R. CHINTALACHARUVU, S. L. MORRISON, *Immunotechnology* **1999**, *4* (3–4), 165–174.

[7] E. M. YOO, K. R. CHINTALACHARUVU, M. L. PENICHET et al., *J. Immunol. Methods* **2002**, *261* (1–2), 1–20.

[8] F. E. JOHANSEN, R. BRAATHEN, P. BRANDTZAEG, *Scand. J. Immunol.* **2000**, *52* (3), 240–248.

[9] F. E. JOHANSEN, R. BRAATHEN, P. BRANDTZAEG, *J. Immunol.* **2001**, *167* (9), 5185–5192.

[10] P. CROTTET, B. CORTHESY, *J. Biol. Chem.* **1999**, *274* (44), 31456–31462.

[11] W. D. XU, R. HOFMANN-LEHMANN, H. M. McCLURE et al., *Vaccine* **2002**, *20* (15), 1956–1960.

[12] R. HOFMANN-LEHMANN, J. VLASAK, R. A. RASMUSSEN et al., *J. Virol.* **2001**, *75* (16), 7470–7480.

[13] M. C. GAUDUIN, P. W. H. I. PARREN, R. WEIR et al., *Nature Med.* **1997**, *3* (12), 1389–1393.

[14] B. CORTHESY, A. PHALIPON, *J. Allergy Clin. Immunol.* **2002**, *109* (1), 316.

[15] R. G. FISHER, J. E. CROWE, T. R. JOHNSON et al., *J. Infect. Dis.* **1999**, *180* (4), 1324–1327.

[16] H. STUBBE, J. BERDOZ, J. P. KRAEHENBUHL et al., *J. Immunol.* **2000**, *164* (4), 1952–1960.

[17] J. K.-C. MA, B. Y. HIKMAT, K. WYCOFF et al., *Nature Med.* **1998**, *4* (4), 601–606.

[18] J. W. LARRICK, L. YU, C. NAFTZGER et al., *Biomol. Eng.* **2001**, *18* (3), 87–94.

[19] K. R. CHINTALACHARUVU, S. L. MORRISON, *Proc. Natl Acad. Sci. USA* **1997**, *94* (12), 6364–6368.

[20] J. Berdoz, C. T. Blanc, M. Reinhardt et al., *Proc. Natl. Acad. Sci. USA* **1999**, *96* (6), 3029–3034.

[21] D. P. Pollock, J. P. Kutzko, E. Birck-Wilson et al., *J. Immunol. Methods* **1999**, *231* (1–2), 147–157.

[22] L. M. Houdebine, *Curr. Opin. Biotechnol.* **2002**, *13* (6), 625–629.

[23] I. Sola, J. Castilla, B. Pintado et al., *J. Virol.* **1998**, *72* (5), 3762–3772.

[24] A. Hiatt, R. Cafferkey, K. Bowdish, *Nature* **1989**, *342* (6245), 76–78.

[25] K. During, S. Hippe, F. Kreuzaler et al., *Plant Mol. Biol.* **1990**, *15* (2), 281–293.

[26] M. B. Hein, Y. Tang, D. A. McLeod et al., *Biotechnol. Prog.* **1991**, *7* (5), 455–461.

[27] E. B. P. Fontes, B. B. Shank, R. L. Wroble et al., *Plant Cell* **1991**, *3* (5), 483–496.

[28] J. Denecke, M. H. Goldman, J. De-Molder et al., *Plant Cell* **1991**, *3* (9), 1025–1035.

[29] J. Nuttall, N. Vine, J. L. Hadlington et al., *Eur. J. Biochem.* **2002**, *269* (24), 6042–6051.

[30] K. Peeters, C. De Wilde, G. De Jaeger et al., *Vaccine* **2001**, *19* (17–19), 2756–2761.

[31] J. K.-C. Ma, T. Lehner, P. Stabila et al., *Eur. J. Immunol.* **1994**, *24* (1), 131–138.

[32] P. M. W. Drake, D. Chargelegue, N. D. Vine et al., *FASEB J.* **2002**, *16* (14), 1855–1860.

[33] N. D. Vine, P. M. W. Drake, A. Hiatt et al., *Plant Mol. Biol.* **2001**, *45* (2), 159–167.

[34] M. De Neve, M. De Loose, A Jacobs et al., *Transgenic Res.* **1993**, *2* (4), 227–237.

[35] F. A. VanEngelen, A. Schouten, J. W. Molthoff et al., *Plant Mol. Biol.* **1994**, *26* (6), 1701–1710.

[36] J. Mestecky, J. R. McGhee, *Adv. Immunol.* **1987**, *40*, 153–245.

[37] L. Frigerio, N. D.Vine, E. Pedrazzini et al., *Plant Physiol.* **2000**, *123* (4), 1483–1493.

[38] J. L. Hadlington, A. Santoro, J. Nuttall et al., *Mol. Biol. Cell* **2003**, *14* (6), 2592–2602.

[39] A. Sturm, J. A. Van Kuik, J. F. Vliegenthart et al., *J. Biol. Chem.* **1987**, *262* (28), 13392–13403.

[40] L. Faye, K. D. Johnson, A. Sturm et al., *Physiol. Plant.* **1989**, *75* (2), 309–314.

[41] M. Cabanes-Macheteau, A. C. Fitchette-Laine, C. Loutelier-Bourhis et al., *Glycobiology* **1999**, *9* (4), 365–372.

[42] I. J. W. Elbers, G. M. Stoopen, H. Bakker et al., *Plant Physiol.* **2001**, *126* (3), 1314–1322.

[43] L. C. Simmons, D. Reilly, L. Klimowski et al., *J. Immunol. Methods* **2002**, *263* (1–2), 133–147.

[44] A. Vonschaewen, A. Sturm, J. O'Neill et al., *Plant Physiol.* **1993**, *102* (4), 1109–1118.

[45] H. Bakker, M. Bardor, J. W. Molthof et al., *Proc. Natl Acad. Sci. USA* **2001**, *98* (5), 2899–2904.

[46] E. Stoger, M. Sack, Y. Perrin et al., *Mol. Breeding* **2002**, *9* (3), 149–158.

[47] H. Daniell, S. J. Streatfield, K. Wycoff, *Trends Plant Sci.* **2001**, *6* (5), 219–226.

11
Production of Spider Silk Proteins in Transgenic Tobacco and Potato

Jürgen Scheller and Udo Conrad

11.1
Introduction

11.1.1
Structure and Properties of Spider Silk

Over the last 400 million years, spiders have become highly diverse in the production and use of silks [1], reviewed in Ref. [2]. This diversity is made necessary by the central role silk plays in a spider's life, e.g. prey capture, construction of shelter and reproduction.

Dragline silk is used by spiders for the frames of their webs and as safety lines (Table 11.1) [3–5]. Dragline silk is stronger than high tensile steel and approaches the stiffness and strength of the widely used high-performance p-aramid fiber KEVLAR, which is the raw material for bulletproof vests. One advantage of dragline silk over any synthetic fiber is the unique combination of strength and extensibility before breakage (Table 11.2) [6,7].

Dragline silk fibers are made up of crystalline regions of anti-parallel β-sheets and 'non structured' amorphous regions (coiled coil, preformed β-sheets and elastic β-turn spirals). The crystalline arrays are thought to be responsible for the stiffness of the fiber. The amorphous regions (55 to 60% of dragline spider silk) are more-or-less kinetically free, and can change their shape under the influence of external load and through entropic elasticity [8].

Spider silk fiber has been studied at the protein and genetic levels, and a number of genes from orb-weaving spiders have been partially sequenced (for review see Ref. [9]). Phylogenetic analysis suggests that expansions, contractions and recombination events occurred in orthologous genes from closely related species as well as within sets of alleles in the same species. Such genetic events have been critical for the homogenization of amino acid repeats within spider silk proteins [10]. Here we discuss the structure of dragline silk and flagelliform silk (elastic capture spiral) from the golden orb weaving spider *Nephila clavipes*.

The dragline silk is composed of the two spidroins MaSpI and MaSpII. A 2.4-kb segment from the 3′ end of the original *MaSpI*-mRNA and a 2-kb segment from the

Molecular Farming. Edited by Rainer Fischer, Stefan Schillberg
Copyright © 2004 WILEY-VCH Verlag GmbH & Co. KGaA, Weinheim
ISBN: 3-527-30786-9

Tab. 11.1 Spider silk proteins MaSpI, MaSpII and Flag from *Nephila clavipes*.

Silk	Use	Spinneret	Proteins	Consensus amino acid repeat
Major ampullate dragline	Web frame and radii	Anterior	MaSpI	GGAGQGGYGGLGGQGAGR GGLGGQ(GA)$_2$A$_5$
			MaSpII	(GPGGYGPGQQ)$_2$GPSGPGSA$_8$
Flagelliform	Elastic capture spiral	Posterior	Flag	(GPGGX)$_{43-63}$-(GGX)$_{12}$- flag spacer

Tab. 11.2 Mechanical properties of spider silk from *Nephila clavipes* compared to other structural materials.

Material	Strength (N m^{-2})	Elasticity (%)	Energy to break (J kg^{-1})
Dragline silk	4×10^9	35	1×10^5
Flagelliform silk	1×10^9	>200	1×10^5
Kevlar	4×10^9	5	3×10^4
Rubber	1×10^6	600	8×10^4
Tendon	1×10^9	5	5×10^3
Nylon, type 6	4×10^7	200	6×10^4

3′ end of the original *MaSpII*-mRNA were cloned and sequenced [11,12]. The predicted amino acid sequence of MaSpI is highly repetitive, consisting of polyalanine stretches of 6–9 amino acids followed by Gly-Gly-Xaa repeats with Xaa representing alanine, tyrosine, leucine or glutamine. The total length of the mRNA transcript was estimated to be 12 kb. The predicted amino acid sequence of MaSpII is also highly repetitive. A typical structural element consists of polyalanine stretches of 6–9 amino acids followed by repeating pentapeptides like Gly-Tyr-Gly-Pro-Gly, Gly-Pro-Gly-Gly-Tyr and Gly-Pro-Gly-Gln-Gln.

The elastic capture spiral silk or flagelliform silk is composed of the structural protein Flag. At the current time, *Flag* is the only spider silk gene whose intron-exon structure is completely known. The *Flag* gene produces a 15.5-kb mRNA and the entire *Flag* locus spans > 30 kb and consists of 13 exons [13]. The flagelliform protein is largely composed of iterated sequences with the dominant repeat Gly-Pro-Gly-Gly-Xaa (Xaa: any Amino acid), which appears up to 63 times in tandem arrays. Hayashi and Lewis [13] proposed that the secondary structure of Flag contains a large number of elastic, spring-like helices (β-turn spirals), which explains the greater elasticity of flagelliform silk in comparison to dragline silk (200% *vs.* 35%). This makes the flagelliform silk one of the most extensible known proteins [14,15].

The function of the non-repetitive C-terminal regions of the MaSpI, MaSpII and Flag proteins is still unclear. The Flag protein also contains a non-repetitive N-terminal region of unknown function. It is therefore likely that MaSpI and MaSpII also contain non-repetitive N-terminal regions.

11.1.2
Strategies for the Production of Recombinant Spider Silk Proteins

Genetic engineering has led to the design and cloning of a large variety of synthetic spider silk-like genes (for review see Refs [4] and [16]) and their expression in microorganisms such as *Escherichia coli* [17–21] and lower eukaryotes such as *Pichia pastoris* [22]. It was found that spider silks were expressed in these organisms at low levels, and the proteins were only sparingly soluble in aqueous buffers. Because spider silk proteins consist largely of the hydrophobic amino acids glycine and alanine, a large pool of these two building blocks has to be provided if spider silk proteins are to be produced in rapidly growing microorganisms such as bacteria or yeast.

As an alternative to microbes, fragments of the spider silk proteins MaSpI, MaSpII and Adf3, in the range of 60–140 kDa, have been produced in cultured mammalian cells [23]. These experiments might lead to the development of transgenic animals secreting spider silk proteins into their milk. However, mass production of a structural protein for technical purposes from animal cells or transgenic animals would be very expensive and time consuming in terms of fermentation or animal breeding.

Another approach for the production of spider silk proteins is genetic engineering of the silkworm *Bombyx mori*. Silkworm fibers are not as strong as dragline silk, but in the future the silkworm silk genes could be replaced with spider silk genes. Initial experiments showed that it is possible to generate stable silkworm transformants using baculovirus-based gene targeting [24]. Transformants were constructed carrying a fibroin light chain gene connected to a truncated human type III procollagen gene [25]. The chimeric gene was expressed in the posterior silk gland, and the gene product spun into the cocoon layer. The authors claim that the introduction of foreign genes downstream of a powerful promoter, like the fibroin gene promoter, would allow large-scale production of recombinants protein in the silkworm. Therefore, the creation of silkworms capable of spinning spider silk might be a realistic goal.

We have explored the possibility of producing recombinant spider silk proteins in plants. We constructed stable transgenic tobacco and potato plants containing various synthetic spider silk genes ranging in size from 420–3600 bp. The synthetic genes were >90% identical to the native spider genes. Spider silk protein accumulated in the plants and in the best cases represented 2% of the total soluble protein [26]. In addition, we have combined the synthetic spider silk protein SO1 (51.2 kDa) with the elastic biopolymer 100xELP (100 repeats of the pentapeptide Val-Pro-Gly-Xaa-Gly, where Xaa can be Gly, Val or Ala) [27] making a protein of 94.2 kDa (Fig. 11.1). The best performing plants accumulated spider silk-elastin fusion proteins at levels of up to 4% total soluble protein [28]. Natural elastin fibers provide elasticity to many tissues that need to deform repetitively and reversibly [29]. Synthetic elastin-like polypeptides consist of oligomeric repeats of the pentapeptide Val-Pro-Gly-Xaa-Gly (where Xaa is any amino acid except proline) and can undergo inverse temperature transition [30]. Even as a fusion protein, elastin-like polypeptides become reversibly insoluble if the temperature rises above their transition temperature ("inverse transition cycling" [27]). The proposed secondary structure of elastin is the β-turn spiral. By combining a spider silk protein that exhibits a high tensile

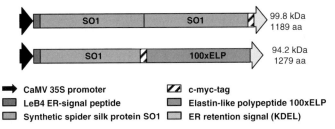

Fig. 11.1 Synthetic spidroin and spidroin-ELP plant expression cassettes.

strength with an elastic biopolymer, such as the elastin-like polypeptide, we have taken a first step towards the development of novel biomaterials for industrial and medical purposes.

11.1.3
Applications of Spider Silk Proteins

11.1.3.1 Synthetic Spider Silk Fibers: 'Natural' *vs* Artificial Spinning Strategies
The first spinning experiments were performed with dissolved raw silk of the silk-worm *Bombyx mori* [31] and the golden orb-weaver *Nephila clavipes* [32]. For this purpose a minimized wet-spinning apparatus was constructed [31]. The apparatus was capable of spinning fibers from solutions containing 10 mg of soluble protein.

Raw silk was dissolved in hexafluoro-iso-propanol (HFIP) [17, 33]. A typical working concentration for spinning was 2.5% (w/v) silk fibroin in HFIP. The spinning solution was pressed through a small needle ($\varnothing$ 80–250 μm) into a precipitation bath (methanol for *Bombyx mori* silk proteins and acetone for *Nephila clavipes* silk proteins) and the silk solution immediately precipitated as a fiber. The best performing fibers approached the maximum strength measured for native fibers of *Bombyx mori*, but did not achieve the mechanical properties of natural spider silk.

In contrast to the harsh organic solvents and high pressures needed for current artificial spinning of solubilized silk, spiders have developed mechanisms for spinning fibers from a highly-concentrated, aqueous protein solution. In the spinning dope, the secreted silk proteins form so-called liquid crystals, where the single protein molecules are already aligned approximately parallel to one another, but are not fully packed [34]. The spider can process this highly viscous solution (> 50% protein) with a slow flow rate under low pressure [35]. In a process called "internal tapedown taper" the perpendicular solutions are rapidly extended and the thread is immediately detached from the walls of the duct. The higher stress forces during this extension might join together the dope molecules with hydrogen bonds to give the antiparallel beta-conformation of the final thread. During this separation phase, the silk becomes more and more hydrophobic and loses its water. The slight acidification of the duct solution further enhances the process of silk formation [36].

The spinning of silk monofilaments from a concentrated aqueous solution (>20% protein) of recombinant spider silk protein might be the best way to generate stress-resistant fibers. This technique has been developed by Lazaris *et al.* [23] for the spinning of recombinant spider silk proteins from human cell culture. The water-insoluble fibers were tough, and modulus values were comparable to those of native dragline silk but with lower tenacity.

In a recent paper, Jin und Kaplan [37] described the *in vitro* recapitulation of silkworm fibroin silk. To increase solubility in water, single fibroin proteins were assembled into micelles, with larger-chain terminal hydrophilic blocks in contact with the surrounding aqueous solution. 'Globule' formation was driven by increased fibroin concentration and lower water content, further hydrophobic interactions, and at the final stages the presence of polyethylene oxide (PEO). The naturally occurring glue-like protein sericin was substituted by PEO during the solubilization of raw silk. Finally, elongation and alignment of globule structures and interactions among globules promoted by physical shear forces produced a fibrillar structure. The next step will be to control the features that promote solubility of the hydrophobic silk proteins in water at high concentrations while avoiding premature crystallization.

11.1.3.2 Synthetic Spider Silk Proteins for the *In Vitro* Proliferation of Anchorage-dependent Cells

Silk fibers or monolayers of silk proteins have a number of potential biomedical applications. Biocompatibility tests have been carried out with scaffolds of fibers or solubilized silk proteins from the silkworm *Bombyx mori* (for review see Ref. [38]). Some biocompatibility problems have been reported, but this was probably due to contamination with residual sericin. More recent studies with well-defined silkworm silk fibers and films suggest that the core fibroin fibers show *in vivo* and *in vivo* biocompatibility that is comparable to other biomaterials, such as polyactic acid and collagen. Altmann *et al.* [39] showed that a silk-fiber matrix obtained from properly processed natural silkworm fibers is a suitable material for the attachment, expansion and differentiation of adult human progenitor bone marrow stromal cells. Also, the direct inflammatory potential of silkworm silk was studied using an *in vitro* system [40]. The authors claimed that their silk fibers were mostly immunologically inert in short and long term culture with murine macrophage cells.

Plant-produced recombinant spider silk-elastin fusion proteins were used for the proliferation of anchorage-dependent chondrocytes [28]. The chondrocytes showed similar growth behavior and a rounded phenotype on plates coated with either collagen or recombinant plant-produced spider silk-elastin, and in each case the rate of cell proliferation was significantly greater than was possible using untreated polystyrene plates. This fusion protein might be useful for coating the surfaces of implants to suppress the immune rejection response, or the generation of 3D scaffolds for tissue engineering.

11.1.4
Molecular Farming: Plants as Biofactories for the Production of Recombinant Proteins

In our opinion, the production of recombinant proteins for technical and medical applications should fulfill two main prerequisites. First, it should be possible to produce large amounts of proteins without the need to build new facilities/fermenters. Second, the production and downstream processing costs should be as low as possible.

Plants are grown over large areas of the world, not only for food production but also to provide resources for industrial purposes, e.g. cellulose. Mass production is common and plant production and harvesting technologies are well established. Plants have been successfully used for the production of different transgenic protein products (for review see Ref. [41]). The storage organs of crop plants (e.g. seeds and tubers) have been used for the stable expression of foreign proteins [42–45]. In several cases, stable accumulation of functional proteins to high levels has been achieved by retention in the endoplasmic reticulum of plant cells [46] (reviewed in Ref. [42]). Synthetic spider silk proteins were successfully produced by this approach in transgenic tobacco and potato. Furthermore, the extreme heat stability of these plant-produced synthetic spider silk proteins has been exploited for the development of simple purification procedures.

11.2
Spider Silk and Spider Silk-ELP Fusion Proteins from Plants: Expression, Purification and Applications

11.2.1
Spider Silk-ELP Expression in Transgenic Tobacco and Potato

Synthetic spider silk genes closely related to the natural spidroin sequences were constructed and cloned in plant expression vectors allowing ubiquitous expression in transgenic plants (Fig. 11.1). The constructs were designed so that the spider silk protein was expressed with an N-terminal signal sequence and a C-terminal KDEL signal [47], allowing high-level and stable accumulation in the ER of tobacco and potato cells to a level of up to 2% total soluble protein. The accumulation level of the transgenic silk proteins did not depend on size [26]. In further experiments, spider silk-elastin fusion proteins were expressed in transgenic tobacco and potato plants. For this purpose, we combined the synthetic spider silk protein SO1 (51.2 kDa) [26] with the elastic biopolymer 100xELP (100 repeats of pentapeptide Val-Pro-Gly-Xaa-Gly, where Xaa can be Gly, Val or Ala) [27] coding for a protein of 94.2 kDa (Fig. 11.1). The expression level of this protein was compared to that of a synthetic spider silk protein, SO1SO1 (Fig. 11.2) [28]. The best plants accumulated the spider silk-elastin fusion protein at a level of up to 4% total soluble protein. Thus, the expression level could be almost doubled through the use of ELP fusions. The protein was stable in tobacco leaves and in potato leaves and tubers, as indicated by the clearly defined bands in western blots (Fig. 11.2).

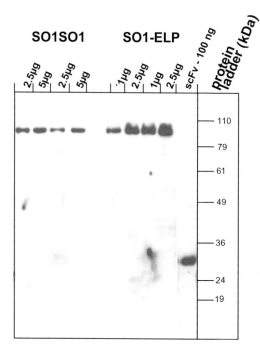

Fig. 11.2 SO1-100xELP fusion proteins in transgenic tobacco and potato plants detected by western blot analysis. Plant material was ground in liquid nitrogen, extracted with SDS-sample buffer and loaded onto SDS-polyacrylamide gels. Recombinant proteins were detected using an anti-c-myc antibody, anti-mouse-IgG-POD (peroxidase) conjugate and an ECL western blotting detection system (ECL-kit: chemiluminescent, non-radioactive method to detect antigens that have been immobilized onto membranes).
1: SO1-ELP – 1 µg of total soluble leaf protein (JS154/50, tobacco);
2: SO1-ELP – 2.5 µg of total soluble leaf protein (JS154/50, tobacco);
3: SO1-ELP – 1 µg of total soluble leaf protein (JS158/4, potato);
4: SO1-ELP – 2.5 µg of total soluble leaf protein (JS158/4, potato);
5: 100 ng scFv standard.

11.2.2
Purification of Spider Silk-Elastin Fusion Proteins by Heat Treatment and Inverse Transition Cycling

Spider silk proteins from plants remain soluble at high temperatures, allowing them to be enriched by boiling [26]. In order to enrich the spider silk-ELP fusion protein, we therefore exposed tobacco leaf extracts to heat treatment at 95°C for 60 min and then cleared the supernatant by centrifugation. In further steps, the reversible precipitation behavior of ELP fusion proteins was exploited to develop a suitable purification strategy. For the selective precipitation of SO1–100xELP, NaCl was added at a final concentration of 2 M and the temperature was increased to 60 °C. In this man-

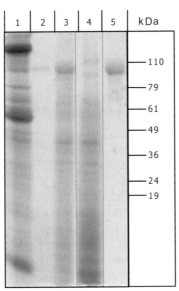

Fig. 11.3 Purification of SO1–100xELP-proteins from transgenic tobacco plants by inverse transition cycling and analysis by SDS-PAGE. 1: 15 µg of total soluble leaf protein extracted in raw extract buffer; 2: cleared supernatant of original 15 µg total soluble leaf protein after heat treatment (60 min, 95 °C); 3: cleared supernatant of original 300 µg leaf protein after heat treatment; 4: cleared supernatant of original 300 µg leaf protein after heat treatment (60 min, 60 °C) with 2 M NaCl; 5: redissolved spider silk-elastin protein pellet from original 300 µg of total soluble leaf protein after heat treatment (60 min, 60 °C) with 2 M NaCl.

ner, the recombinant spider silk-elastin fusion proteins could then be precipitated by centrifugation, whereas cellular proteins remained in the supernatant. The precipitated recombinant proteins were homogenous and relatively pure, and could be dissolved to a final concentration of up to 1 mg mL^{-1} (Fig. 11.3). Dialysis against water followed by drying produced storable membranes. Extraction of 1 kg tobacco leaves produced 80 mg of pure recombinant spider silk-elastin protein. The detailed experimental protocol is described in Ref. [28].

11.2.3
Applications of Spider Silk-ELP Fusion Proteins in Mammalian Cell Culture

Mammalian cells need extracellular matrix (ECM) proteins for attachment and proliferation both in vivo and in vitro. Recombinant ECM-like proteins for medical applications must be biocompatible and they should enhance cell growth in cell culture, but inhibit differentiation. Deriving such proteins from plants would be beneficial because this would reduce the risk of contamination with mammalian viruses. In a first attempt to test spider silk-ELP fusion proteins for this purpose, human chondrocytes (HCH-371) and CHO cell lines were grown using either the plant–derived fusion protein or collagen as a substrate, or using uncoated dishes in the presence of fetal calf serum. The performance of both substrates was comparable [28]. In addition to rapid proliferation, an important criterion for the successful cultivation of chondrocytes is the correct morphological phenotype. The cells should be rounded, as they are *in vivo*, and fibroblastoid morphology must be avoided. In the presence of SO1-100xELP or fetal calf serum, the cells had a rounded shape (Fig. 11.4). The effect of SO1-100xELP on the quality of the chondrocytes will be

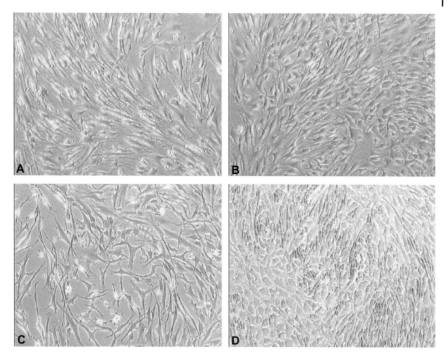

Fig. 11.4 Morphology of human HCH 386 chondrocytes in mono-
layer culture, 8 days after initial seeding at 1×10^4 cells cm^{-2} [28].
A. Collagen substrate, 0% FCS (fetal calf serum); B. Spider silk-ELP
substrate, 0% FCS; C. No substrate, 0% FCS; D: No substrate, 10%
FCS. In A and C, the cells have an elongated, fibroblastoid morphology, which characterizes dedifferentiation. The spider silk-elastin
substrate (B) or the presence of 10% FCS in dishes without a substrate (D) produce a more favorable morphology.

tested in further experiments. Plant-derived spider silk-elastin fusion proteins appear to be an effective, biocompatible matrix to promote the growth of mammalian
cells.

11.3
Discussion

Recombinant proteins with unique properties can potentially generate new markets
and penetrate into existing markets if they can be supplied on a large scale. An ideal
system would produce the safest biologically active material at the lowest cost, and
would be used in combination with an inexpensive and simple purification process.
So far, there have been several examples of the high-yield production of recombinant
proteins in transgenic crop plants, mainly in the area of molecular medicines such
as antibodies, enzymes and vaccines [45, 48–50]. Modern agricultural practices offer

rapid scale-up, harvesting and processing of large quantities of leaves, tubers or seeds. Health risks arising from the contamination of recombinant proteins with potential human pathogens or toxins are minimized [51]. Proteins can be produced in plants at 2–10 % of the cost of those produced in *E. coli*. The relative costs could be reduced further if it would be possible to produce high-quality and high-value proteins in plants.

The development of material for superior fibers or as scaffolds for artificial organs could result in the development of high-value plant products. The ability of our spider silk-elastin fusion protein to support the proliferation of chondrocyte monolayer cultures makes it an attractive alternative to collagen-coated cell culture plates. Recently, cell culture studies showed that chondrocytes cultured in ELP coacervate maintain a rounded shape and their normal chondrocytic phenotype [52]. Additionally, the fusion protein used in our studies might be useful for future *in vivo* and *in vitro* applications, e. g. coating the surface of implants to suppress the immune rejection response or the generation of three-dimensional scaffolds for the growth of artificial organs. The simple and scalable purification procedure and the use of transgenic plants will make spider silk-elastin proteins a serious competitor to collagen proteins produced in pigs and cattle.

The development of a by-production system could further minimize the cost of such plant-derived products. For example, potatoes are the raw material for the production of technical grade starch. During this process, soluble proteins are separated by heat treatment and sold as animal fodder. Recombinant proteins could be produced in transgenic potato tubers as a by-product of starch extraction, and this would be useful for proteins produced in large amounts with a low commercial impact, like structural fiber proteins.

References

[1] W. A. Shear, J. M. Palmer, J. A. Coddington et al., *Science* 1989, *246* (4929), 479–499.

[2] M. B. Hinman, Z. Dong, M. Xu et al., *Results Probl. Cell Differ.* 1992, *19*, 227–254.

[3] D. A. Tirrell, *Science* 1996, *271* (5245), 39–40.

[4] M. B. Hinman, A. J. Jones, R. V. Lewis, *Trends Biotechnol.* 2000, *18* (9), 374–379.

[5] F. Vollrath, D. P. Knight, *Nature* 2001, *410* (6828), 541–548.

[6] M. W. Denny, *J. Exp. Biol.* 1976, *65*, 483–506.

[7] S. A. Wainwright, W. D. Biggs, J. D. Currey et al., *Mechanical Design in Organisms*, Princeton University Press, New Jersey, 1982.

[8] J. M. Gosline, M. W. Denny, M. E. DeMont, *Nature* 1984, *309* (5968), 551–552.

[9] J. Gatesy, C. Hayashi, D. Motriuk et al., *Science* 2001, *291* (5513), 2603–2605.

[10] C. Y. Hayashi, *EXS.* 2002, *92*, 209–223.

[11] M. Xu, R. V. Lewis, *Proc. Natl Acad. Sci. USA* 1990, *87* (18), 7120–7124.

[12] M. B. Hinman, R. V. Lewis, *J. Biol. Chem.* 1992, *267* (27), 19320–19324.

[13] C. Y. Hayashi, R. V. Lewis, *Science* 2000, *287* (5457), 1477–1479.

[14] C. Y. Hayashi, R. V. Lewis, *J. Mol. Biol.* 1998, *275* (5), 773–784.

[15] C. Y. Hayashi, R. V. Lewis, *BioEssays* 2001, *23* (8), 750–756.

[16] J. M. Gosline, P. A. Guerette, C. S. Ortlepp et al., *J. Exp. Biol.* 1999, *202* (23), 3295–3303.

[17] S. R. Fahnestock, International Patent WO 94/29450, **1994**.

[18] J. T. Prince, K. P. McGrath, C. M. Di-Girolamo et al., *Biochemistry* **1995**, *34* (34), 10879–10885.

[19] R. V. Lewis, M. Hinman, S. Kothakota et al., *Protein Expr. Purif.* **1996**, *7* (4), 400–406.

[20] S. Arcidiacono, C. Mello, D. Kaplan et al., *Appl. Microbiol. Biotechnol.* **1998**, *49* (1), 31–38.

[21] S. R. Fahnestock, Z. Yao, L. A. Bedzyk, *J. Biotechnol.* **2000**, *74* (2), 105–119.

[22] S. R. Fahnestock, L. A. Bedzyk, *Appl. Microbiol. Biotechnol.* **1997**, *47* (1), 33–39.

[23] A. Lazaris, S. Arcidiacono, Y. Huang et al., *Nature* **2002**, *295* (5554), 472–476.

[24] M. Yamao, N. Katayama, H. Nakazawa et al., *Genes Dev.* **1999**, *13* (5), 511–516.

[25] M. Tomita, H. Munetsuna, T. Sato et al., *Nature Biotechnol.* **2003**, *21* (1), 52–56.

[26] J. Scheller, K. H. Gührs, F. Grosse et al., *Nature Biotechnol.* **2001**, *19* (4), 573–577.

[27] D. E. Meyer, A. Chilkoti, *Nature Biotechnol.* **1999**, *17* (11), 1112–1115.

[28] J. Scheller, D. Henggeler, A. Viviani et al., *Transgenic Res.* **2004**, *13* (1), 51–57.

[29] J. Rosenbloom, W. R. Abrams, R. Mecham, *FASEB J.* **1993**, *7* (13), 1208–1218.

[30] D. W. Urry, *J. Protein Chem.* **1988**, *7* (1), 1–34.

[31] O. Liivak, A. Blye, N. Shah et al., *Macromolecules* **1998**, *31* (9), 2947–2951.

[32] A. Seidel, O. Liivak, L. W. Jelinski, *Macromolecules* **1998**, *31* (19), 6733–6736.

[33] R. L. Lock, US Patent 5,252,285, **1993**.

[34] K. Kerkam, C. Viney, D. Kaplan et al., *Nature* **1991**, *349* (6310), 596–598.

[35] D. P. Knight, F. Vollrath, *Proc. Roy. Soc. B.* **1999**, *26*, 1–5.

[36] D. P. Knight, F. Vollrath, *Naturwissenschaften* **2001**, *88* (4), 179–182.

[37] H. J. Jin, D. L. Kaplan, *Nature* **2003**, *424* (6952), 1057–1061.

[38] G. H. Altman, F. Diaz, C. Jakuba et al., *Biomaterials* **2003**, *24* (3), 401–416.

[39] G. H. Altman, R. L. Horan, H. H. Lu et al., *Biomaterials* **2002**, *23* (20), 4131–4141.

[40] B. Panilaitis, G. H. Altman, J. Chen et al., *Biomaterials* **2003**, *24* (18), 3079–3085.

[41] G. Giddings, *Curr. Opin. Biotechnol.* **2001**, *12* (5), 450–454.

[42] U. Conrad, U. Fiedler, *Plant Mol. Biol.* **1998**, *38* (1–2), 101–109.

[43] Y. Perrin, C. Vaquero, I. Gerrad et al., *Plant Mol. Biol.* **2000**, *6* (4), 345–352.

[44] I. Saalbach, M. Giersberg, U. Conrad, *J. Plant Physiol.* **2001**, *158* (4), 529–533.

[45] O. Artsaenko, B. Kettig, U. Fiedler et al., *Mol. Breeding* **1998**, *4* (4), 313–319.

[46] C. I. Wandelt, M. R. I. Khan, S. Craig et al., *Plant J.* **1992**, *2* (2), 181–192.

[47] S. Munro, H. R. Pelham, *Cell* **1987**, *48* (5), 899–907.

[48] E. Stoger, C. Vaquero, E. Torres et al., *Plant Mol. Biol.* **2000**, *42* (4), 583–590.

[49] J. W. Larrick, D. W. Thomas, *Curr. Opin. Biotechnol.* **2001**, *12* (4), 411–418.

[50] L. J. Richter, Y. Thanavala, C. J. Arntzen et al., *Nature Biotechnol.* **2001**, *18* (11), 1167–1171.

[51] H. Daniell, S. J. Streatfield, K. Wycoff, *Trends Plant Sci.* **2001**, *6* (5), 219–226.

[52] H. Betre, L. A. Setton, D. E. Meyer et al., *Biomacromolecules* **2002**, *3* (5), 910–916.

12
Gene Farming in Pea Under Field Conditions

Martin Giersberg, Isolde Saalbach and Helmut Bäumlein

12.1
Introduction

Plant seeds have evolved to synthesize and accumulate reserve compounds in large quantities. Therefore, transgenic seed crops, particularly the large seeds of cereals and legumes, offer an effective and economical system for the production of biomolecules. Plant systems produce recombinant proteins that are free from animal viruses and other pathogens that could potentially infect humans. This ability to produce safe recombinant proteins opens new perspectives in agriculture using conventional techniques for harvesting and downstream processing. However, public concerns about large-scale, field-grown transgenic plants require the investigation of practical issues such as biosafety and containment (e.g. [1–8]). Seeds of fodder pea varieties (*Pisum sativum* L.) appear to be especially suitable as safe bioreactors. Pea plants are self-fertile, with a rate of less than 1% cross-pollination because the flowers remain closed. Pea cultivars are sexually incompatible with most of the wild pea species, and spring cultivars do not survive frost conditions.

Although all plant organs can in principle be used for the expression of foreign genes, seeds appear to be a preferable choice. Therefore, strong and seed-specific gene promoters are essential tools for efficient and controlled transgene expression. The broad bean (*Vicia faba* L.) 'unknown seed protein' (USP) promoter meets most of the requirements for molecular farming in pea [9]. The molecular basis of seed specificity for this promoter is sufficiently well understood [10,11] and it has been used to control the expression of various transgenes under greenhouse conditions [12].

In this chapter we describe the use of pea seeds to express the bacterial enzyme α-amylase. Bacterial exoenzymes like the heat stable α-amylase from *Bacillus licheniformis* are important for starch hydrolysis in the food industry. The enzymatic properties of α-amylase are well understood [13,14], it is one of the most thermostable enzymes in nature and it is the most commonly used enzyme in biotechnological processes. Although fermentation in bacteria allows highly efficient enzyme production, plant-based synthesis allows *in situ* enzymatic activity to degrade endogenous reserve starch, as shown in experiments with non-crop plants performed under greenhouse conditions [12,15]. Finally, the quantitative and sensitive detection of α-amylase activ-

Molecular Farming. Edited by Rainer Fischer, Stefan Schillberg
Copyright © 2004 WILEY-VCH Verlag GmbH & Co. KGaA, Weinheim
ISBN: 3-527-30786-9

ity makes it a suitable reporter gene product to study the strength and specificity of different promoters. This chapter includes results obtained during a two-year field trial, which – to our knowledge – was the first field trial with transgenic pea in Europe. The experiment was performed to test the applicability of a transformation and expression system comprising a vector cassette with a strong, seed-specific promoter, the transformation of a commercial fodder pea variety and stable transgene expression under field conditions during two vegetation periods.

12.2
Procedures for Foreign Protein Expression in Transgenic Pea Seeds

12.2.1
Plant Material, Transformation and Field Growth

The experiments described in this chapter involved fodder pea (Pisum sativum L.) varieties "Erbi", "Eiffel" and "Power". The transformation method we used was suitable for gene transfer to several different commercial fodder pea varieties and is based on the protocol of Schroeder *et al.* [16]. Essential modifications have been described previously [17]. Explants for transformation were cut from the embryonic axis of immature seeds. For explant preparation, the root end of each segment was cut off and the epicotyl and apical meristem regions were sliced transversely into 3–5 segments. The slices were co-cultivated with *Agrobacterium tumefaciens* (see below) for 3–4 days at 21 °C with a 16-h photoperiod. After co-cultivation, explants were washed 3–4 times with sterile water. After callus induction, developing shoots (20 mm in length), were selected on 10 mg L^{-1} phosphinothricin for several days. The resistant shoots were grafted to a rootstock of "Erbi" seedlings *in vitro*. After 6–10 days, the plants were adapted to soil and grown up in a climatic chamber. When primary transformants reached the 4–6 node stage, leaves were painted with a 0.5 % solution of BASTA herbicide (Hoechst). Inheritance of the transgene was also monitored by herbicide application.

Plants from the first six generations after transformation were grown in growth chambers and under greenhouse conditions. The field trial was performed with a homozygous line from generations T_7 and T_8 in spring/early summer 2000 and 2001, on a field of approximately 100 m^2. One hundred transgenic plants and one hundred wild type control plants were grown. To prevent uncontrolled seed distribution by birds, the field was covered with a net. The environmental conditions of the field trial were as follows: the field was 111.5 m above sea level, the average precipitation was 492.1 mm, the average temperature was 8.5 °C, and the field comprised highly fertile black earth surrounded by a protective strip planted with *Avena sativa* L. var. *nigra*, *Fagopyrum esculentum* and *Trifolium resupinatum*. The field experiment was performed according to German law with permission from the state of Saxony-Anhalt (No. 6786-01-114). To protect plants against pests, compounds such as "Ripcoord" (SHELL AGRAR), "Karate" (ZENECA AGRO) and "Perfecthion" (BASF) were applied during the field trial. Plants were harvested by cutting the stem and whole plants were bagged. Mature seeds were separated from the plants in the laboratory to avoid seed loss during harvesting. To ensure that no transgenic pea plants remained after the field trial, the area was left unused for two further years. During this period, the area was inspected every two months. To date, three plants have been found.

12.2.2
Transformation Vectors and Analysis of Transgenic Plants

The expression cassette contained the 638-bp *V. faba* USP promoter, the *Bacillus licheniformis* α-amylase gene and the *Agrobacterium tumefaciens* octopine synthase (*ocs*) terminator [12]. The cassette was isolated as an *Xba*I fragment and inserted into the binary vector pGPTV-*bar* [18]. Subcloning procedures were carried out according to Sambrook *et al.* [19]. A modified protocol (ICC Standard No. 303, Megazyme) was used to determine the α-amylase activity in transgenic pea seeds [20]. All enzyme activities are presented in Ceralpha Units (CU). Pea seeds were milled under liquid nitrogen and 500 mg flour was extracted with 10 ml of extraction buffer B (0.1 M maleic acid, 5.8 g NaCl, 2 mM $CaCl_2 \times 2 H_2O$, 0.01% sodium azid, pH 6.5) for 20 min at 42 °C. One milliliter of the extract was incubated at 80 °C for 30 min and centrifuged at 13000 x g for 10 min in an Eppendorf bench centrifuge. The α-amylase activity was determined in the supernatant. Extracts of seeds younger than 13 days after pollination (DAP) were measured without dilution, whereas seed extracts from later developmental stages were diluted 5 or 10 fold before measurement. Agar plates containing 0.5% starch were used for a qualitative test of α-amylase activity. Ten microliters of extract was spotted on the surface of the plate, incubated overnight at 37 °C and the remaining non-hydrolyzed starch was detected with Lugol solution.

12.3
Expression of α-Amylase in Transgenic Pea Seeds

Three commercial fodder pea varieties ("Erbi", "Eiffel" and "Power") were transformed using the transgene construct shown in Fig. 12.1. The transgene copy number was determined by Southern blot hybridization and segregation analysis. The characteristics of each line, in terms of transgene copy numbers and enzyme activities, are summarized in Table 12.1. Doubling the copy number by self-crossing and achieving homozygosity resulted in a rather precise doubling of the enzyme activity.

The expression profile of the USP promoter was monitored under field conditions using α-amylase activity as a reporter. The earliest sign of α-amylase activity was observed 8 DAP. Although the absolute level remained low, a ~10-fold increase in promoter activity was detected between 12 and 13 DAP. The activity steadily increased during seed development reaching maximum levels of about 6000 CU kg^{-1} in mature seeds (Fig. 12.2).

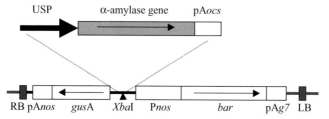

Fig. 12.1 Structure of the expression construct. USP, USP-gene promoter; pA*ocs*, polyadenylation site from the *Agrobacterium tumefaciens* octopine synthase gene; RB, right T-DNA border; LB, left T-DNA-border; pA*nos*, polyadenylation site from the *A. tumefaciens* nopaline synthase gene; P*nos*, nopaline synthase promoter; pA*g7*, polyadenylation site from the *g7* gene; *bar*, phosphinothricin resistance gene; *gusA*, gene for β-glucuronidase.

Tab. 12.1 Number of independent transgenic lines, number of single insertion lines and the range of enzyme activities in pea seeds of three commercial varieties obtained under greenhouse conditions.

Commercial variety	Independent transgenic lines	Single insertion lines	Enzyme activity (CU kg^{-1} seeds)
Erbi	3	1	4000–9000
Eiffel	4	1	4000–10,000
Power	6	4	4000–6000

Evaluation of alpha amylase activity

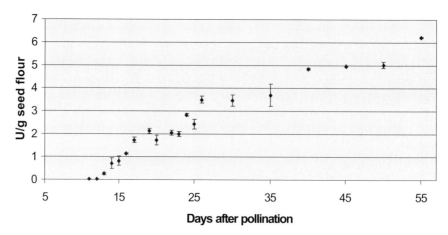

Fig. 12.2 Enzyme activity during pea seed development under field conditions.

One homozygous genotype with a single transgene insertion was selected for the field experiment. As shown in Table 12.2, the enzyme activity in transgenic seeds remained constant over several years without any herbicide selection for the presence of the transgene. No clear difference in α-amylase expression could be detected in plants grown under greenhouse, growth chamber and field conditions. Wild type controls and transgenic lines showed similar yields under all growth conditions. In both 2000 and 2001, the yield (measured as pods and seeds per plant), was found to be approximately 10-fold higher under field conditions (Table 12.3). As expected from previous reports, the USP promoter fragment was predominantly seed-specific. The only exception was a low level of transgene expression in pollen, but this was only 1–2% of the level observed in seeds.

Tab. 12.2 Transgene expression is stable over eight generations and is independent of growth conditions.

Generation	Year	Growth conditions	Enzyme activity (CU kg^{-1} seeds)
T1	1996	Growth chamber	n. d.
T2	1996/97	Growth chamber	7000
T3	1997	Growth chamber	6800
T5	1997/98	Growth chamber	7800
T6	1998	Growth chamber	7000
T6	1998	Greenhouse	6500
T7	1999	Growth chamber	6900
T7	2000	Greenhouse	6300
T7	2000	Field trial	7800
T8	2001	Greenhouse	6500
T8	2001	Field trial	6200

Tab. 12.3 Transformed (TG) and non-transformed (WT) plants do not differ in yield. Field conditions result in approximately 5–10 fold higher yields.

	Pods per plant		Seeds per plant	
	WT	TG	WT	TG
T$_7$/2000				
Growth chamber	8.0	7.8	30.4	34.1
Greenhouse	7.6	8.0	22.5	20.4
Field trial	43.4	43.0	220.8	216.0
T$_8$/2001				
Greenhouse	8.1	6.5	25.3	23.2
Field trial	64.0	61.0	253.0	251.0

12.4
Conclusions

The data presented in this chapter demonstrate the potential of pea seeds as bioreactors under field conditions. The chosen transgene was stably inherited and expressed without selection pressure. The expression level remained unchanged over several generations. Apart from presence and expression of the transgene, there was no observable difference between transgenic plants and wild type controls. With the level of transgene expression per seed unchanged, the seed yield per plant was about 5–10 fold higher under field conditions than under greenhouse or growth chamber conditions. One reason for this could be the choice of an established commercial fodder pea variety instead of an experimental laboratory line. No transgenic plants survived over winter, thus providing a built-in biosafety feature.

Based on a seed biomass yield of approximately $4 \, t \times ha^{-1}$ for pea in Saxony-Anhalt in the year 1999 [21] and an enzyme activity of 7000 CU kg^{-1} (average of the two field trials), a theoretical yield of 28×10^6 CU ha^{-1} would be expected. This amount clearly exceeds the theoretical enzyme yield calculated for the grain legume *Vicia narbonensis* (about 10^4 CU ha^{-1} [12]). However, even given the 100-fold improvement in yields achieved when switching from *V. narbonensis* to pea, plants cannot yet match the efficiency of enzyme production in recombinant bacteria. Therefore, enzymes like α-amylase should not be expressed in seeds with the intention of purifying them, since this would not make economic sense. Rather, the potential future advantage of plant-based enzyme production is likely to be the digestion of starch and other substrates *in planta*, as has been shown in *Vicia narbonensis*. This situation is likely to be different for more valuable recombinant proteins, like antibodies, although the yields reported thus far are difficult to compare. Using the seed-specific legumin A promoter, transgenic pea plants produced up to 9 µg per gram fresh weight of a functional single-chain Fv fragment antibody [7].

There are several approaches that could be used to increase transgene expression levels in seeds. One possibility is the co-expression of a second transgene using another seed-specific gene promoter with a different developmental expression profile. The rather early developmental profile of the USP promoter would be complemented best by promoters whose activity peaked later on in seed development, such as the legumin A promoter of *Pisum sativum* [22] or the legumin B promoter of *Vicia faba* [23]. Alternative promoters have been suggested previously [7]. Another possibility is the use of a longer version of the USP promoter, since extension of the promoter fragment used in this study was found to increase the expression level under greenhouse conditions by a factor of at least two. Finally it would be tempting to use the lines described in this chapter in a conventional breeding program, aiming to a increase the expression level still further.

Based on previous experiments in tobacco and *Arabidopsis*, the USP promoter fragment we used was considered to be highly seed specific. However, thorough examination of the field-grown pea plants revealed some promoter activity in pollen, albeit 100-fold lower than the activity seen in seeds (Fig. 12.3). To test whether or not the pollen expression is restricted to pea, we also examined pollen from transgenic

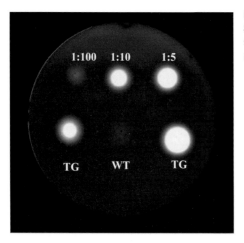

Fig. 12.3 Transgene activity in pollen, 8 days after pollination, as detected on a starch plate. WT, wild type; TG, transgenic pollen, undiluted and diluted extracts.

tobacco plants carrying the USP promoter and other seed-specific promoters. Anther activity was observed for the promoters of several seed-specific genes, including those encoding phaseolin, legumin and sucrose binding protein. Anther activity may therefore be a more general property of seed-specific promoters, although the reasons behind this phenomenon are unknown. It may reflect a common regulatory program in somatic cells competent for embryogenesis [24] or a common regulatory program is tissues that undergo desiccation. In any case, the phenomenon needs further investigation, since even the low promoter activity in anthers might lead to a reduction of pollen fertility with transgene products other than the α-amylase.

In summary, we provide an initial technology package for biofarming in pea seeds consisting of a seed-specific promoter cassette for stable transgene expression, an efficient procedure suitable for the transformation of commercial fodder pea varieties and the safe handling of transgenic peas under field conditions. The package was validated by expressing the bacterial exoenzyme α-amylase.

12.5
Acknowledgements

The work described in this chapter was supported by a grant from the Sachsen-Anhalt Ministry of Research and Education. For excellent technical support we would like to thank Ursula Hebenstreit, Andreas Czihal, Petra Hoffmeister, Enk Geyer, Peter Schreiber and Katrin Menzel.

References

[1] R. FISCHER, N. EMANS, *Transgenic Res.* 2000, *9* (4–5), 279–299.

[2] G. GIDDINGS, G. ALLISON, D. BROOKS et al., *Nature Biotechnol.* 2000, *18* (11), 1151–1155.

[3] O. J. M. GODDIJN, J. PEN, *Trends Biotechnol.* 1995, *13* (9), 379–387.

[4] K. HERBERS, U. SONNEWALD, *Curr. Opin. Biotechnol.* 1999, *10* (2), 163–168.

[5] E. E. HOOD, J. M. JILKA, *Curr. Opin. Biotechnol.* 1999, *10* (4), 382–386.

[6] J. K. C. MA, P. M. W. DRAKE, P. CHRISTOU, *Nature Rev. Genet.* 2003, *4* (10), 794–805.

[7] Y. PERRIN, C. VAQUERO, I GERRAND et al., *Mol. Breeding* 2000, *6* (4), 345–352.

[8] E. ROSCHE, D. BLACKMORE, M. TEGEDER et al., *Plant J.* 2002, *30* (2), 165–175.

[9] H. BÄUMLEIN, W. BOERJAN, I. NAGY et al., *Mol. Gen. Genet.* 1991, *225* (3), 459–467.

[10] U. FIEDLER, R. FILISTEIN, U. WOBUS et al., *Plant Mol. Biol.* 1993, *22* (4), 669–679.

[11] W. REIDT, T. WOHLFARTH, M. ELLERSTRÖM et al., *Plant J.* 2000, *21* (5), 401–408.

[12] A. CZIHAL, B. CONRAD, P. BUCHNER et al., *J. Plant Physiol.* 1999, *155* (2), 183–189.

[13] B. CONRAD, V. HOANG, A. POLLEY et al., *Eur. J. Biochem.* 1995, *230* (2), 481–490.

[14] B. SVENSSON, M. SOGAARD, *Biochem. Soc. Trans.* 1992, *20* (1), 34–41.

[15] J. PEN, L. MOLENDIJK, W. J. QUAX et al., *Bio/Technol.* 1992, *10* (3), 292–296.

[16] H. E. SCHROEDER, A. H. SCHOTZ, T. WARDLEY-RICHARDSON et al., *Plant Physiol.* 1993, *101* (3), 751–757.

[17] I. SAALBACH, M. GIERSBERG, U. CONRAD, *J. Plant Physiol.* 2001, *158* (4), 529–533.

[18] D. BECKER, E. KEMPER, J. SCHELL et al., *Plant Mol. Biol.* 1992, *20* (6), 1195–1197.

[19] J. SAMBROOK, E. F. FRITSCH, T. MANIATIS, *Molecular Cloning: A Laboratory Manual, 2nd ed.*, Cold Spring Harbor Laboratory Press, Cold Spring Harbor NY, 1989.

[20] B. V. MCCLEARY, H. SHEEHAN, *J. Cereal Sci.* 1989, *6* (3), 237–251.

[21] J. BISCHOFF, Anbauempfehlung Körnererbsen und Ackerbohnen. Ministerium für Raumordnung, Landwirtschaft und Umwelt des Landes Sachsen-Anhalt, Magdeburg, 2000. (http://lsa-st23.sachsen-anhalt.de/llg/acker_pflanzenbau/produktionstechnik/anbau/erbsen_ackerbohnen00.pdf)

[22] G. W. LYCETT, R. R. D. CROY, A. H. SHIRSAT et al., *Nucleic Acids Res.* 1984, *12* (11), 4493–4506.

[23] H. BÄUMLEIN, W. BOERJAN, I. NAGY et al., *Mol. Gen. Genet.* 1991, *225* (1), 121–128.

[24] Y. V. CHESNOKOV, A. MEISTER, R. MANTEUFFEL. *Plant Sci.* 2002, *162* (1), 59–77.

13
Host Plants, Systems and Expression Strategies for Molecular Farming

RICHARD M. TWYMAN

13.1
Introduction

Plants have numerous advantages over traditional expression technologies for the large-scale production of recombinant proteins. The major benefits of whole plants include the comparatively low cost of large-scale production, the inherent scalability of agricultural systems and the convenience of existing infrastructure for harvesting, processing and distribution [1–4]. The start-up and running costs for molecular farming in plants are significantly lower than those of cell-based production systems because there is no need for fermenters or the skilled operators to run them. It has been estimated that proteins can be produced in plants at 2–10% of the cost of microbial fermentation systems and at 0.1% of the cost of mammalian cell cultures or transgenic animals, as long as adequate yields can be achieved [5].

All plant expression platforms, including whole plants and cell/tissue culture systems, also have safety benefits over microbial and animal cells. This is because they lack the endotoxins produced by bacterial cells and they do not harbor human pathogens [1]. Proteins produced in mammalian cells and transgenic animals must be screened for viruses, oncogenic DNA sequences and prions, which adds significantly to processing costs. Indeed, regardless of the production system, over 85% of the costs associated with recombinant protein production reflect extraction and downstream processing steps rather than the production phase itself [6]. For many proteins, however, these steps can be circumvented or eliminated if plants are used as production hosts. For example, recombinant subunit vaccines produced in plants can be administered orally as raw or partially processed fruits and vegetables [7], therapeutic proteins designed for topical application can be applied as crude extracts or pastes, and industrial enzymes used in food and feed processing can be expressed in the plant that needs to be processed [8–11]. Even for those proteins that must be extracted and purified using conventional methods, plant systems can provide practical advantages to facilitate processing and make it more economical. One example is the oleosin fusion method, which is being developed as a commercial platform technology by SemBioSys Genetics Inc., Calgary, Canada [12]. This allows recombinant proteins to be concentrated in the oil bodies of oilseed crops and extracted using a

Molecular Farming. Edited by Rainer Fischer, Stefan Schillberg
Copyright © 2004 WILEY-VCH Verlag GmbH & Co. KGaA, Weinheim
ISBN: 3-527-30786-9

simple and economical method. Another example is the rhizosecretion of recombinant proteins from the roots of tobacco plants, which allows the protein to be collected continually from root exudates [13,14]. This approach is being developed as a commercial platform technology by Phytomedics Inc., Dayton, NJ.

Given the multitude of benefits listed above, it is not surprising that many of our crop species have been investigated as potential hosts for molecular farming (Table 13.1). More than 30 species of plants have now been transformed for the sole purpose of expressing and exploiting recombinant proteins, and it is becoming increasingly difficult to choose which expression host is the most suitable for particular products. Further complications are added by the diversity of expression systems available for each crop species. These may include transgenic plants, transplastomic plants, virus-infected plants, transiently transformed leaves, hairy roots and suspension cell cultures. There is also a wide choice of expression strategies, including leaf expression, seed expression, fruit expression, inducible expression, targeting to different subcellular compartments and secretion.

The choice of host species, expression system and expression strategy must be evaluated carefully on a case-by-case basis, depending on many inter-related factors [15]. Strategic choices may be made for geographical reasons, such as the site intended for production and the local availability of labor, processing infrastructure, storage facilities, transport and distribution networks. The value of the recombinant protein is also important, since it would make economic sense to produce a high-value protein such as a recombinant antibody in an expensive expression system if other benefits were gained, while bulk products with a low market value would better be produced in a less expensive expression system. The widest choice is generally available where the recombinant protein needs to be purified to homogeneity, while there are greater constraints for proteins intended to be delivered in unprocessed or partially processed plant material. For example, proteins expressed for the purpose of oral vaccination in humans need to be expressed in the edible parts of food crops (e. g. tomato or banana fruits) while those intended for the oral vaccination of animals need to be expressed in fodder crops (e. g. alfalfa or clover leaves). Additional factors that may need to be taken into consideration include the degree of containment afforded by the crop (see Chapter 16) and the extent to which the structure and homogeneity of N-linked glycans can be controlled (see Chapter 15). Different host species also vary in the expediency and convenience of transformation and regeneration, the availability of useful regulatory elements to control transgene expression, the extent to which endogenous compounds interfere with downstream processing, and in the absolute yields of recombinant proteins that can be achieved.

This chapter provides an overview of the different expression hosts, systems and strategies available in molecular farming, and discusses their advantages and disadvantages for the production of different types of recombinant proteins.

Tab. 13.1 Plant expression hosts used for molecular farming – advantages and disadvantages.

Species	Advantages	Disadvantages
Model plants		
Arabidopsis thaliana	Range of available mutants, accessible genetics, ease of transformation	Not useful for commercial production (low biomass)
Simple plants		
Physcomitrella patens *Chlamydomonas reinhardtii* *Lemna minor*	Containment, clonal propagation, batch consistency, secretion of proteins into medium, regulatory compliance, homologous recombination in *Physcomitrella*	Scalability
Leafy crops		
Tobacco	High yield, established transformation and expression technology, rapid scale-up, non-food/feed	Low protein stability in harvested material, presence of alkaloids
Alfalfa, clover	High yield, useful for animal vaccines, clonal propagation, homogenous N-glycans (alfalfa)	Low protein stability in harvested material, presence of oxalic acid
Lettuce, spinach	Edible, useful for human vaccines	Low protein stability in harvested material
Cereals		
Maize, rice	Protein stability during storage, high yield, easy to transform and manipulate	Food crops
Wheat, barley	Protein stability during storage, low producer price	Food crops, lower yields, more difficult to transform and manipulate
Legumes		
Soybean	High seed protein content, seed coat expression, low producer price	Lower expression levels
Pea, pigeon pea, peanut	High protein content	Low expression levels
Fruits and vegetables		
Potato, carrot	Edible, proteins stable in storage tissues	Must be cooked before consumption (potato), high starch content (potato)
Tomato	Edible, containment in greenhouses	More expensive to grow, must be chilled after harvest
Banana	Edible, staple in developing countries, eaten by adults and children	Transformation and regeneration are currently difficult and time-consuming
Oilcrops		
Oilseed rape, safflower, *Camelina sativa*	Oil-body purification, sprouting system	Lower biomass yields, oil bodies incompatible with glycosylation

13.2
Host Species for Molecular Farming

13.2.1
Leafy Crops

The two major leafy crops used for the production of recombinant proteins are tobacco and alfalfa, both of which have high leaf biomass yields in part because they
can be cropped several times every year. The main limitation of such crops is that
the harvested leaves tend to have a restricted shelf life. The recombinant proteins exist in an aqueous environment and are therefore relatively unstable, which can reduce product yields [16]. For proteins that must be extracted and purified, the leaves
need to be dried or frozen for transport, or processed immediately after harvest at
the production site. This adds considerably to the processing costs.

13.2.1.1 **Tobacco (*Nicotiana tabacum*)**
Cultivated tobacco has a long and successful history in molecular farming. The po-
pularity of this species reflects its dual status as a model plant and a cultivated crop,
providing many practical advantages for the large-scale production of recombinant
proteins. As a model plant, tobacco benefits from well-established gene transfer and
regeneration methodologies, and the availability of many robust expression cassettes
for the control of transgene expression. The practical advantages of tobacco include
its high biomass yield (up to 100 tonnes of leaf biomass per hectare each year), the
wide range of available expression systems (transgenic plants, transplastomic plants,
virus-infected plants, transient expression in leaves, transformed suspension cell cultures, hairy roots and shooty teratomas; see Sect. 13.3), and the fact that tobacco is
neither a food nor a feed crop, thus reducing the likelihood of transgenic material
contaminating the food or feed chains. The first recombinant human therapeutic
protein to be produced in transgenic plants was expressed in tobacco leaves [17,18][1],
as was the first plant-derived recombinant antibody [19], the first plant-derived vaccine candidate [20], the first plant-derived industrial enzyme [21] and the first plant-
derived synthetic biopolymer [22]. However, there are several drawbacks to molecular
farming in tobacco leaves, particularly for pharmaceutical proteins. Many tobacco
cultivars have high contents of nicotine and other alkaloids, which must be removed
during the downstream processing steps. Even where low-alkaloid cultivars are used,
processing is still necessary to remove these toxic components [23, 24]. It has also
been shown that glycoproteins produced in tobacco are very heterogeneous in terms
of their N-glycan structures, which could make tobacco unsuitable for the bulk production of certain pharmaceutical proteins [25,26].

---

1) Human serum albumin was the first full-length human protein to
be expressed in the leaves of a transgenic plant (both tobacco and
potato leaves were shown to express the protein). This report was
published in 1990 [18]. Four years earlier, Barta and colleagues had
demonstrated the expression of human growth hormone in tobacco
and sunflower *callus*, and the protein was expressed as a fusion
with the *Agrobacterium tumefaciens* nopaline synthase enzyme [17].

13.2.1.2 Tobacco (*Nicotiana benthamiana*)

N. benthamiana is a non-cultivated tobacco species whose main advantage as a host plant for molecular farming is that it supports the systemic replication of many different viruses. This has two major applications. First, *N. benthamiana* is a suitable host species when the aim is to produce recombinant proteins using viral vectors, such as tobacco mosaic virus (TMV) and potato virus X (PVX) (Sect. 13.3.3). Second, transgenic *N. benthamiana* plants are often used to produce antibodies that are active against plant viruses, allowing the transgenic plant to be used as a viral assay system. While this is not strictly molecular farming (in that the aim is to confer disease resistance on the plant rather than to exploit the recombinant protein as a pharmaceutical or industrial product), the expression of antibodies in plants is of general interest because strategies to improve antibody expression levels or target them to specific compartments are also applicable to pharmaceutical proteins. *N. benthamiana* is occasionally used for molecular farming without the involvement of viruses, as in the production of a V_H domain recognizing the neuropeptide substance P [27].

13.2.1.3 Alfalfa (*Medicago sativa*)

The leaf biomass produced by alfalfa is somewhat lower than that of tobacco (12 tonnes per hectare per year), but it has several advantageous agronomic characteristics compared to tobacco including the fact that it is a perennial plant (vegetative growth can be maintained for many years), it can be clonally propagated by stem cutting, and it fixes its own nitrogen thus eliminating the need for fertilizer input. The technology for gene transfer and transgene control in alfalfa is not so well established as it is in tobacco, but due mainly to the efforts of researchers at Medicago Inc., Québec City, Canada, much progress has been made in the development of constitutive and inducible expression cassettes for molecular farming, and alternative expression systems such as agroinfiltrated leaves and cell suspension cultures (see Chapter 1). While alfalfa lacks toxic alkaloids, the leaves do contain high levels of oxalic acid, which can in some cases interfere with downstream processing. The major advantage of alfalfa for the production of pharmaceutical proteins is that glycoproteins expressed in alfalfa leaves have homogeneous glycan structures [28,29]. This, together with the ease of clonal propagation, provides enormous production benefits in terms of batch-to-batch reproducibility. Since alfalfa is a fodder crop, the other major application of this species in molecular farming is the delivery of vaccines to domestic animals, as has been demonstrated in the case of a foot-and-mouth-disease vaccine [30]. Alfalfa has also been used for the production of three industrial enzymes: 1,4-β-D-endoglucanase and cellobiohydrolase [11], and phytase [31].

13.2.1.4 White clover (*Trifolium repens*)

Like alfalfa, the clover family (*Trifolium* spp.) consists of perennial legume plants that fix their own nitrogen. However, they suffer several limitations as general production crops for molecular farming, such as restricted perenniality, low protein content and the presence of high levels of condensed tannins which interfere with protein extraction. Despite these disadvantages, clovers are forage crops and are therefore useful for the production of vaccines against animal diseases. Thus far, white clover has been used to produce a *Mannheimia haemolytica* A1 leukotoxin fusion protein antigen to protect cattle against pneumonic pasteurellosis, otherwise known as shipping fever. The antigen remained stable for over one year in dried clover leaves that had been baked in the oven at 50 °C immediately after harvest [32].

13.2.1.5 Lettuce (*Lactuca sativa*)

Lettuce has been used for the production of a hepatitis B surface antigen and was chosen essentially because it is an edible salad crop that can be used in human clinical trials [33,34]. It does have a higher biomass yield than alfalfa (30 tonnes per hectare per year) but it has a much higher producer price and a very high water content (98%) which reduces protein yields and stability. Human volunteers, fed with transgenic lettuce plants expressing hepatitis B virus surface antigen, developed specific serum-IgG responses to the plant-derived vaccine. The investigators who published this original report are now producing further pharmaceutical proteins in lettuce, including anthrax protective protein and antibodies against rabies and colorectal cancer [35]. Other investigators are considering the use of lettuce for the production of further antigens, including most recently the SARS virus spike glycoprotein.

13.2.1.6 Spinach (*Spinacia oleracea*)

Spinach, like lettuce, has been used for the production of edible vaccines. In the first report, Yusibov and colleagues used alflafla mosaic virus (AlMV) to produce rabies fusion epitopes on the virus surface in infected spinach leaves. This resulted in the development of anti-rabies (as well as anti-AlMV) antibodies when the spinach leaves were fed to mice [36]. Vaccines against HIV gp120 and Tat have been produced in spinach, and a construct of gp120 with the CD4 receptor is now being adapted for this plant [35]. Spinach is also being used to make an anthrax vaccine [37].

13.2.1.7 Lupin (*Lupinus* spp.)

Lupin (*Lupinus luteus*) has been used to express the same hepatitis B surface antigen as produced in lettuce, and the transgenic leaves were used in pre-clinical trials in an attempt to promote an immune response in mice following oral administration [34]. Narrow-leaf lupin (*Lupinus angustifolius*) has been used to express a gene encoding a plant allergen (sunflower seed albumin), in a successful attempt to suppress experimentally-induced asthma [38]. This was the first study involving the production of a heterologous vaccine in plants to provide protection against allergic diseases.

13.2.2
Dry Seed Crops

The dry seed crops that have been used as host plants for molecular farming include the cereals maize, rice, wheat and barley (see also Chapter 4), and the grain legumes soybean, pea, pigeon pea and peanut. Maize, rice, wheat, barley, soybean and pea have been investigated as general production platforms, while pigeon pea and peanut have been used solely for the expression of animal vaccine candidates. The major advantage of all seed crops is that recombinant proteins can be directed to accumulate specifically in the desiccated seeds [15]. Therefore, although seed biomass yields are smaller than the leaf biomass yields of tobacco and alfalfa, this is offset by the increased stability of the proteins. Seeds are natural storage organs, with the optimal biochemical environment for the accumulation of large amounts of protein. In the best cases, recombinant proteins expressed in seeds have been shown to remain stable and active after storage at room temperature for over three years. The accumulation of proteins in the seed rather than vegetative organs also prevents any toxic effects on the host plant. Finally, the extraction of proteins from seeds is facilitated because the target protein is concentrated in a small volume, most cereal seeds lack the phenolic compounds that are often found in leaves and which interfere with processing, and the seed proteome is fairly simple, which reduces the likelihood that contaminating proteins will co-purify with the recombinant protein during downstream processing [39].

Several factors need to be weighed up when choosing an appropriate dry seed expression host, including geographical considerations, the ease of transformation and regeneration, the annual yield of seed per hectare, the yield of recombinant protein per kilogram of seed, the producer price of the crop, the percentage of the seed that is made up of protein and, inevitably, intellectual property issues [15]. Together, these determine the overall cost of producing the recombinant protein in the chosen seed crop.

Maize (*Zea mays*)

Maize was chosen as a platform expression host by ProdiGene Inc., College Station, TX, and has the honor of being the only crop thus far to be used commercially for the production of plant-derived recombinant proteins (avidin and β-glucuronidase were first produced in maize seeds on a commercial basis in 1997 [40,41]). Maize was chosen over the other cereals because it has the highest annual grain yield (8300 kg ha^{-1}), and the seeds have a relatively high seed protein content (10%) resulting in potentially the highest recombinant protein yields per hectare. Maize is also relatively easy to transform and manipulate in the laboratory, while in the field it has the greatest capacity of all the cereals for rapid scale up. Maize is the most widely cultivated crop in North America, so the complex infrastructure for growing, harvesting, processing, storing and transporting large volumes of corn is already in place. Prodigene is developing maize for the production of a range of pharmaceutical and technical proteins, including recombinant antibodies, vaccine candidates and enzymes [5,42,43].

Rice (*Oryza sativa*)

Rice is the most important staple food crop in the world and has also emerged as the model cereal species (it is the only terrestrial plant other than *Arabidopsis thaliana* to benefit from a completed genome sequence, and extensive EST resources are also available). Rice has a lower annual grain yield than maize (6600 kg ha^{-1}) and the grain has a lower protein content (8%), but like maize it is easy to transform and manipulate in the laboratory, a range of useful expression cassettes have been developed, and field populations can be scaled up rapidly. The major disadvantage of rice compared to maize is that the producer price is significantly higher, so it may be excluded as a major expression host in the West simply on economic grounds. However, the story may be different in Asia and Africa, where rice is traditionally grown. Several pharmaceutical proteins have been expressed in rice grains, including α-interferon [44], recombinant antibodies [39] and most recently a cedar pollen allergen [45]. Rice suspension cell cultures have also been used for the production of pharmaceutical proteins (Sect. 13.3.8).

Wheat (*Triticum aestivium*)

The main advantages of wheat for molecular farming are the low producer price and the high protein content of the grain (> 12%) [15]. The adoption of this species as a production crop will depend, however, on the development of better transformation procedures and stronger expression cassettes, since at the current time only low expression levels have been achieved in transgenic wheat grains [15]. The annual grain yield per hectare (2800 kg) is also much lower in wheat than in maize or rice, so larger areas of land would need to be cultivated to produce the same amounts of protein.

Barley (*Hordeum vulgare*)

Barley, like wheat, has a low producer price and a high protein content in the grain (about 13%). Also like wheat, gene transfer and regeneration techniques are not so well developed as they are in maize and rice. A significant advantage of barley, however, is that while few molecular farming experiments have been carried out using this species, the results have generally been very encouraging in terms of recombinant protein yields. In early reports, recombinant glucanase and xylanase were expressed at very low levels in barley [46,47]. More recently, however, a recombinant diagnostic antibody (SimpliRED, for the detection of HIV) was expressed in transgenic grains at levels exceeding 150 µg g^{-1} [48], and a recombinant cellulase enzyme was expressed at levels exceeding 1.5% total seed protein [49]. Further recombinant proteins will need to be expressed in barley before its performance can be compared meaningfully with maize and rice.

Soybean (*Glycine max*)

The advantage of soybean as a production crop is that while it has a relatively low annual grain yield compared to maize and rice (about 2500 kg ha^{-1}), the protein content of the seed is very high (> 40%) which means that it has the highest potential recombinant protein yields per hectare of any seed crop. This, combined with its relatively low producer price, makes it one of the least expensive crops for recombinant protein production. However, these advantages are balanced by the more difficult transformation and regeneration procedures. Therefore, only one report of molecular farming in soybean has been published, that of a humanized antibody against herpes simplex virus, and this was produced constitutively in the plant rather than in the seeds alone [50]. Although the use of soybean as a production host has been limited, the anti-HSV2 antibody produced in soybean is one of the few plant-derived antibodies in clinical development [51]. Another potential advantage of soybeans is the ability to express proteins in the seed coat [52]. This structure contains very few complex soluble proteins, which makes isolation of the target protein simple and inexpensive. Furthermore, the seed coats are easily removed and separated in the milling process, which makes them readily attainable and free of contamination from other components of the seed.

Pea (*Pisum sativum*)

Pea has a similar annual grain yield and seed protein content to soybean, and therefore has the same potential in terms of high recombinant protein yields per hectare. However, the producer price is about 50% higher than that of soybean, so proportionately higher yields would be necessary to make this a more competitive crop. To date, only a single pharmaceutical protein has been expressed in pea, a recombinant scFv antibody recognizing a cancer antigen. This was expressed under the control of the seed-specific legumin A promoter, and the maximum expression level achieved was 9 µg g^{-1} of seed [53]. Trials have also been carried out with field pea varieties producing the enzyme α-amylase from *Bacillus licheniformis* under the control of the bean USP (unknown seed protein) promoter. In this study, which is described in detail in Chapter 12, a yield of 7000 enzyme units per kg of seed was achieved.

Pigeon pea (*Cajanus cajan*)

Pigeon pea is often used in Africa, the Middle East and South Asia as a fodder crop, and is therefore suitable as an expression host for animal vaccines against diseases endemic in those areas. In one report thus far, pigeon pea plants have been generated expressing Rinderpest virus hemagglutinin at a level just below 0.5% TSP, with the aim of protecting ruminants against the disease caused by this virus [54]. As with many crops envisaged as delivery vehicles for vaccines, this is a very specialized application and the crop is unlikely to be used as a general production system for recombinant proteins.

Peanut (*Arachis hypogaea*)

Peanuts are used as fodder in much of Africa and Asia, making this species also a suitable delivery vehicle for vaccines against animal diseases. As is the case for pigeon pea, transgenic peanut plants expressing Rinderpest virus hemagglutinin have been generated, but more general applications are unlikely [55,56].

13.2.3
Fruit and vegetable crops

Most fruit and vegetable crops that have been considered as hosts for molecular farming have been chosen not because they are particularly advantageous for bulk production, but because they might serve as vehicles for the delivery of edible vaccines. Therefore, the choice of a fruit or vegetable production crop usually reflects either geographical or cultural preferences that suit the production and distribution of a given vaccine, rather than any intrinsic advantages in terms of yields or stability that each species may exhibit. The exceptions to this general rule are potato and to-mato, both of which have merits as general production platforms.

Potato (*Solanum tuberosum*)

The potato is the fourth most important food crop (after rice, wheat and maize) and is therefore widely grown throughout the world. It also has a very high tuber bio-mass yield (about 125 tonnes per hectare annually) which makes it eminently suita-ble for bulk protein production. Like cereals and grain legumes, potato plants have specialized storage organs (tubers) that are adapted for the accumulation of large amounts of protein. Targeting recombinant proteins to these organs therefore en-hances protein stability in a manner similar to seed endosperm-specific expression in cereals. The first example of molecular farming in potato was the expression of human serum albumin in 1990, although in this case the protein was expressed in leaves [18]. Potato leaves have also been used to express an industrial cellulase [9]. Pharmaceutical proteins that have been expressed in potato tubers include human glutamic acid decarboxylase [57,58], human interferons [59], human interleukins [60,61] as well as various diagnostic and therapeutic antibodies [62,63]. Potato has also been used to express nutrition-enhancing proteins, such as human milk casein, and antibacterial lysozyme as potential additives to baby food [64,65]. The most wide-spread use of this species, however, has been in the production of vaccine candi-dates. At least 10 different vaccine subunits have been expressed in potato [66], and in three cases thus far these have been used in clinical trials [67–69]. Recently, potato was used simultaneously to express three vaccine antigens: cholera toxin B and A2 subunits, rotavirus enterotoxin and enterotoxigenic *Escherichia coli* fimbrial antigen for protection against several enteric diseases [70]. The only disadvantage of using potatoes for the delivery of vaccines is that potatoes are cooked before eating in nor-mal domestic settings, and heating may denature the recombinant proteins and ren-der them inactive in terms of eliciting an appropriate immune response. This limita-tion does not apply to the other fruit and vegetable crops discussed below. A disad-vantage of potatoes for the production of other pharmaceutical proteins, which need to be isolated and purified, is that the large tuber starch content may interfere with downstream processing.

Tomato (*Lycopersicon esculentum*)

Tomato plants have a high fruit biomass yield (about 60 tonnes per hectare per year) and offer other advantages in terms of containment, because they are grown in greenhouses. Tomato fruits have therefore been investigated as a general production system in molecular farming, and have been used to express one recombinant anti-body (an scFv recognizing carcinoembryonic antigen) [15] and several potentially pharmaceutical proteins (e.g. angiotensin-converting enzyme). As with potatoes, however, the most widespread use of tomato fruits thus far has been in the expres-sion of vaccine candidates. The first such report involved the expression of rabies surface glycoprotein, which achieved the relatively high expression level of 1% TSP [71]. Other vaccines that have been expressed in tomato include cholera toxin B sub-unit [72] respiratory syncytial virus-F protein [73] and, most recently, hepatitis E virus partial OFR2 [74] and the B subunit of *E. coli* heat-labile enterotoxin [75]. On the negative side, tomatoes must be chilled after harvest to keep the recombinant proteins stable, and the yields of recombinant proteins in tomatoes are generally relatively low at the present time. However, this could be addressed in some cases by transforming the fruit chromoplasts rather than the nuclear genome (Sect. 13.3.2).

Carrot (*Daucus carota*)

The carrot taproot is a useful site for protein accumulation since this is both a natural storage organ and the edible portion of the plant. Unlike potato, carrots do not need to be cooked prior to consumption, so vaccines are more likely to remain in their native conformation. Several antigens have already been expressed in carrot, including the *Mycobacterium tuberculosis* MPT64 protein [76], a diabetes-associated autoantigen (an isoform of glutamic acid decarboxylase) [77] and various derivatives of measles hemagglutinin [78,79].

Banana (*Musa* spp.)

Vaccine production in transgenic plants began with the idea that transgenic bananas could be used to administer oral vaccines to adults (consuming whole fruits) and children (fed with banana paste or puree). Banana plants are cheap to grow and the fruits are a staple food source in many developing countries where vaccination campaigns are needed the most. One limitation of bananas, however, is that the transgenic plants take nearly two years to produce. Transgenic banana plants have been produced expressing vaccines against measles virus and hepatitis B virus. In order to prevent vaccine-containing bananas entering the food chain, attempts are in progress to introduce a marker gene that turns the banana flesh blue, enabling them to be distinguished from wild type fruits.

13.2.4
Oilcrops

Oilcrops are potentially advantageous for molecular farming because the oil bodies in developing seeds can trap the recombinant proteins and can be used to facilitate extraction and processing. Oil bodies are seed-specific organelles whose function is to accumulate triacylglycerides. Each oil body comprises a triacylglyceride core surrounded by a phospholipid membrane, which is peppered with oil-body-specific proteins termed oleosins. Recombinant protein can be trapped in the oil bodies by expressing them as oleosin-fusion proteins. A proprietary system developed by the biotechnology company SemBioSys Inc., Calgary, Canada, can then be used to extract the oil bodies and isolate the recombinant protein by endoproteolytic cleavage. It may also be possible to exploit the natural oils normally extracted from such crops as byproducts, which can be used to offset production costs.

Rapeseed/Canola (*Brassica napus*)

Brassica napus is a widely grown crop used primarily for the production of oil, which is classed as either rapeseed oil or canola oil depending on its quality and content.

The seeds of this plant also have a high protein content (25%), which makes them suitable for molecular farming. Transgenic rapeseed/canola plants can be produced relatively easily, and scale-up is rapid due to the large number of seeds produced by each plant. Although the overall biomass yields from this crop are comparatively low (about 1 tonne per hectare) significant cost savings can be achieved during downstream processing steps by targeting recombinant proteins to the oil bodies and using these organelles to facilitate protein extraction and purification. This was first achieved in the case of leech hirudin by fusing the *Hirudo medicinalis* hirudin cDNA to the plant's oleosin gene. The recombinant hirudin accumulated to 1% total seed protein, and following extraction of the oil bodies and purification of the fusion protein, the leech protein could be removed from its fusion partner by endoproteolytic cleavage in vitro [80]. As well as providing a simple extraction and purification process, the expression of pharmaceutical proteins as fusions renders them inactive, and thus poses less risk to both the developing plant and any other organisms that come into adventitious contact with it. Enzymes have also been produced in rapeseeds, including a bacterial xylanase and *Aspergillus* phytase [81]. Seeds expressing recombinant phytase are commercially available as a feed additive. One disadvantage of the oil body system is that proteins directed to oil bodies do not pass through the secretory pathway and therefore are not glycosylated. For this reason, the oil body system is not useful for the production of glycoproteins in cases where the N-glycans are necessary for function or activity.

Another useful feature of *B. napus* is the rapid sprouting of the seeds, which is the basis of a distinct platform technology being developed by UniCrop Ltd, Helsinki, Finland. In this method, proteins are expressed in the developing sprouts using cotyledon-specific Rubisco small subunit promoters, and proteins are extracted from the sprouts which are grown in an airlift tank. Although lower yields are produced in this method compared to agricultural scale production, the added advantage of containment may be useful for the production of pharmaceutical proteins under defined conditions. The technology is described in more detail in Chapter 3.

Falseflax (*Camelina sativa*)

Falseflax is a self-pollinating oilseed crop, rarely used for food, which originates from the Fertile Crescent. This is also being developed as a sprout-based production platform by UniCrop Ltd, Helsinki, Finland (Chapter 3).

Safflower (*Carthamus tinctorius*)

Safflower has many qualities in common with rapeseed including the suitability for oleosin fusion technology, and has been chosen as a platform crop by SemBioSys Inc., Calgary, Canada, for the following reasons: Safflower plants are readily transformed, they grow counter-seasonally, and they can be contained easily (there are no weedy relatives in the West, and seeds show minimal dormancy). The system is easily scalable and can produce clinical quantities of pure protein with an unprecedented manufacturing capacity. In addition, the seed-based system offers seasonally-independent availability of raw materials and improved inventory management since the recombinant proteins are stable in transgenic seeds for extended periods.

13.2.5
Unicellular Plants and Aquatic Plants Maintained in Bioreactors

Single-celled plants and aquatic plants can be maintained in bioreactors, which offers two critical advantages over molecular farming in terrestrial plants. First, the growth conditions can be controlled precisely, which means that optimal growth conditions can be maintained, batch-to-batch product consistency is improved and the growth cycle can conform to good manufacturing practice (GMP) procedures. Second, growth in bioreactors offers complete containment, thus sidestepping the environmental biosafety issues associated with all transgenic terrestrial plants, whether or not they are used for molecular farming (Chapter 16). Although more expensive than agricultural molecular farming, the use of simple plants in bioreactors is not as expensive as cultured animal cells because the media requirements are generally very simple. Added to this, the proteins can be secreted into the medium, which reduces the downstream processing costs and allows the product to be collected in a non-destructive manner. A final, major advantage is the speed of production. The time from transformation to first product recovery is on the scale of days to weeks because no regeneration is required, and stable producer lines can be established in weeks rather than months to years because there is no need for crossing, seed-collection and the testing of several filial generations to check transgene stability. Three major bioreactor-based systems are currently under commercial development: algae, moss and duckweed.

Chlamydomonas reinhardtii
Although the alga *Chlamydomonas reinhardtii* is a model organism which has been instrumental in the study of photosynthesis and light-regulated gene expression, it has only recently been explored as a potential host for molecular farming. A single report discusses the production of monoclonal antibodies in algae [82], and shows that the production costs are similar to those of recombinant proteins produced in terrestrial plants, mainly due to the inexpensive media requirements (the medium does not cost very much to start with, and in any case can be recycled for algal cultures grown in continuous cycles). Aside from the economy of producing recombinant proteins in algae, there are further attributes that make alga ideal candidates for recombinant protein production. First, transgenic algae can be generated quickly, requiring only a few weeks between the generation of initial transformants and their scale up to production volumes. Second, both the chloroplast and nuclear genome of algae can be genetically transformed, providing scope for the production of several different proteins simultaneously. In addition, algae have the ability to be grown on various scales, ranging from a few milliliters to 500,000 liters in a cost-effective manner. These attributes, and the fact that green algae fall into the GRAS (generally regarded as safe) category, make *C. reinhardtii* a particularly attractive alternative to other plants for the expression of recombinant proteins. The production technology has been reviewed recently [83].

Physcomitrella patens

The moss *Physcomitrella patens* is a haploid bryophyte which can be grown in bioreactors in the same way as algae, suspension cells and aquatic plants. Like these other systems, it has the advantages of controlled growth conditions, synthetic growth media, and the ability to secrete recombinant proteins into the medium [84]. The unique feature of this organism, relative to all other plants, is that it is amenable to homologous recombination [85]. This means that not only can it be transformed stably with new genetic information, but that endogenous genes can be disrupted by gene targeting. The major application of gene targeting in molecular farming is the modification of the glycosylation pathway (by knocking out enzymes that add non-human glycan chains to proteins) thus allowing the production of humanized glycoproteins [84, 86].

The *P. patens* system is being developed by the German biotechnology company Greenovation Biotech GmbH, which is based in Freiburg. The company has developed transient expression systems that allow feasibility studies, and stable production strains that can be scaled up to several thousand liters.

Duckweed (*Lemna minor*)

The *Lemna* System developed by the US biotechnology company Biolex Inc. has a number of significant advantages for the production of recombinant pharmaceutical proteins [87]. Unlike transgenic terrestrial plants, this aquatic plant is cultured in sealed, aseptic vessels under constant growth conditions (temperature, pH and artificial light). Only very simple nutrients are required (water, air and completely synthetic inorganic salts) and under these conditions, the plant proliferates vegetatively and doubles its biomass every 36 hours. This provides the optimal production environment for batch-to-batch consistency. Duckweed constitutes about 30% dry weight of protein, and recombinant proteins can either be extracted from wet plant biomass or secreted into the growth medium. Biolex Inc. has reported the successful expression of 12 proteins in this system, including growth hormone, interferon-α, and several recombinant antibodies and enzymes [87].

13.2.6
Non-cultivated Model Plants

Rather than study every different species in the world, researchers focus on model organisms to learn general principles that can be applied to a wider range of life forms. Model organisms are often chosen for historical reasons, or because they are particularly easy to handle and manipulate in the laboratory. Model organisms may be chosen because they have short generation intervals or other features that make them amenable to genetic analysis. They may be chosen because they are particularly suitable for genetic manipulation or surgical procedures, or they may have a small, compact genome compared to related species. Often, a combination of the above features is present. Model organisms are rarely, if ever, chosen because of their commercial value. In terms of molecular farming, the advantage of model plants is that they are very well-characterized, which means that there is a disproportionately large amount of genetic, biochemical and physiological data available about them. This makes them suitable test systems for the design of novel expression systems. Three model plants, rice, tobacco and *C. reinhardtii*, have already been discussed in this section. The most important model plant of all, however, is *Arabidopsis thaliana*.

Arabidopsis thaliana

Arabidopsis thaliana is a small dicot plant of the mustard family. It has a number of features that make it ideal as a model organism: e.g. its size, short life cycle, the fact that it produces large numbers of seeds, and its relatively small genome of 125 Mb. The genome sequence of *A. thaliana* was completed in 2000, extensive EST resources are available and in terms of functional annotation, more is known about this higher plant than any other. The availability of such large amounts of information, together with the ease of transformation by methods such as floral dipping, make *A. thaliana* a convenient host species to use as a test system for molecular farming. Various pharmaceutical and industrial proteins have been expressed in the leaves, seeds and undifferentiated callus of this plant, including an *Acidothermus* endoglucanse which accumulated to 26% of total soluble protein (TSP) [10]. Despite the encouraging results in terms of protein expression levels, however, *A. thaliana* is not suitable for cultivation on an agricultural scale. In this setting, its small size, weediness and low biomass yield are disadvantages and the plant has no commercial value. Therefore, while useful as a test system, it is unlikely this species will ever be used for commercial molecular farming.

13.3
Expression systems for molecular farming

While the range of available expression hosts for molecular farming is impressive, the choice becomes even more varied when the different expression systems are considered (Table 13.2). In this chapter, an expression host and an expression system are considered as separate entities: an *expression host* is defined as a particular species while an *expression system* is defined as a transformation, propagation and expression strategy. The combination of host and system results in an *expression platform*, e.g. stably transformed rice seeds, transiently transformed alfalfa leaves, stably transformed tobacco suspension cells, virus-infected spinach leaves etc. Note that this is not a universal nomenclature, and in the literature the terms host, system, platform, strategy etc. appear to be used interchangeably. By far the predominant system used in molecular farming is the nuclear transgenic plant with the recombinant protein accumulating within the plant tissues [4]. However, biosafety concerns and the disadvantage of long development phases have driven researchers to look at alternative systems based either on transient expression, extra-nuclear transgenesis, or the use of cultured plants, organs or cells. These systems are discussed at great length in other chapters and will be discussed only briefly here.

Tab. 13.2 Comparison of different plant-based production systems.

System	Advantages	Disadvantages
Transgenic plants, accumulation within plant	Economy, biomass yields, scalability, establishment of permanent lines	Production timescale, regulation, biosafety, product yields
Transgenic plants, secretion from roots or leaves	Containment, purification	Scale, cost of production facilities
Transplastomic plants	Yield, multiple gene expression, low toxicity, containment	Absence of glycosylation
Virus infected plants	Yield, timescale, mixed infections, epitope presentation systems	Construct size limitations
Agroinfiltration	Timescale	Cost, scalability
Cell or tissue culture	Timescale, containment, secretion into medium (purification), regulatory compliance	Cost, scalability

13.3.1
Transgenic plants

Transgenic plants usually contain foreign DNA incorporated into the nuclear genome. Such plants have a combination of valuable attributes for molecular farming, but in terms of commercialization the most important of these are the low overall cost of production, the inherent scalability of agricultural systems, the fact that stable transgenic lines are a permanent resource, and the variety of production hosts suitable for different applications. Scalability is probably the most important advantage because the cost of recombinant proteins produced in field plants is inversely proportional to the production scale. In a market which can see demand rapidly increase and decrease, fermenter-based production systems are often unable to cope or left with surplus capacity. In contrast, the scale of plant-based production can be modulated rapidly simply by using more or less land as required. It can take several years to achieve e. g. a tenfold scale-up or scale-down in fermenter systems or in transgenic animals, but a field of transgenic plants can be scaled up or down more than 1000-fold in a single generation by planting greater or smaller numbers of seeds [88].

The two major disadvantages of transgenic plants are the development timescales and the increasing importance of biosafety and regulatory compliance. The 'gene-to-protein' time for transgenic plants encompasses the preparation of expression constructs, transformation, regeneration and the production and testing of several generations of plants. The testing phase is necessary to ensure transgene and expression stability and the biochemical activity of the product, as well as the absence of adverse phenotypic changes in the host plant. These processes take up to two years depending on the plant species although milligram amounts of protein might be available after several months for initial testing [5]. Biosafety concerns include the risk of transgenic plants becoming naturalized outside their intended production sites due to human or animal activities, the risk of transgene spread by outcrossing or horizontal gene transfer, and the risk to human health and the environment caused by the presence of potentially toxic recombinant proteins. These issues are discussed at more length in Chapter 16. One further limitation is the low yields that are obtained in many transgenic plants. In some cases, this reflects the poor performance of the expression construct and can be addressed by improved construct design (see Sect. 13.4). However, in other cases the problem is intrinsic to the transgenic plant, and reflects rate limiting processes of transgene expression, protein synthesis or protein turnover.

13.3.2
Transplastomic plants

Transplastomic plants are transgenic plants generated by introducing DNA into the chloroplast genome, usually by particle bombardment [89,90]. The plants are grown in the same way as nuclear transgenic plants and therefore suffer the same disadvantages in terms of production timelines. However, the advantages of chloroplast transformation are many: the transgene copy number is high because of the many chloroplasts in a typical photosynthetic cell, there is no gene silencing, multiple genes can be expressed in operons, the recombinant proteins accumulate within the chloroplast thus limiting toxicity to the host plant, and the absence of functional chloroplast DNA in the pollen of most crops provides natural transgene containment. The high transgene copy numbers and the absence of silencing have resulted in extraordinary expression levels, e.g. 25% TSP for a tetanus toxin fragment [91], 11% TSP for human serum albumin [92] and 6% TSP for a thermostable xylanase [93]. The ability to express multiple genes in operons means that chloroplast expression will be particularly amenable to the production of multisubunit proteins such as antibodies. However, while proteins expressed in the chloroplast have been shown to fold properly and form appropriate disulfide bonds, glycosylation does not occur, so this system has limited use for the production of therapeutic glycoproteins. The biosafety advantages of the chloroplast system are particularly noteworthy since the inability of functional chloroplast DNA to reach the egg in most commercial crop species means that transgene spread by outcrossing is strongly inhibited. However, other biosafety concerns such as seed spread by human and animal activities are just as prevalent in transplastomic crops as they are for standard transgenic crops. At the current time, chloroplast transformation is a routine procedure only in tobacco and *C. reinhardtii* (see Sect. 13.2.5.1). However, plastid transformation has been achieved in a growing number of other species, including carrot and tomato [94]. The ability to transform the chromoplasts of fruit and vegetable crops has obvious advantages for the expression of subunit vaccines [94]. The chloroplast transgenic system is discussed in more detail in Chapter 8.

13.3.3
Virus-infected plants

Recombinant plant viruses have been used as expression vectors because the infections are rapid and systemic, leading to high levels of recombinant protein production soon after inoculation [95]. Compared to transgenic and transplastomic plants, the development phase is significantly reduced, but the scalability of virus-mediated expression is equivalent to that of other field plant expression systems. No known plant viruses integrate into the genome, so the genetic modification of plants is entirely avoided. Two major strategies have been developed with viral vectors: the expression of full-length recombinant proteins and the presentation of foreign epitopes on the surface of viral particles. Both TMV and PVX have been used in the context of the first strategy to produce antibodies, vaccine candidates and some other pharmaceutical proteins. Cowpea mosaic virus (CPMV) and AlMV are the most popular epitope presentation systems, but TMV and PVX have also been used for this purpose.

Perhaps the most significant use of viral expression systems for molecular farming was described by McCormick and colleagues [96]. These investigators used TMV vectors in *N. benthamiana* to produce a scFv antibody based on the idiotype of malignant B-cells from the murine 38C13 B-lymphoma cell line. When administered to mice, the recombinant protein stimulated the production of anti-idiotype antibodies capable of recognizing 38C13 cells, providing immunity against lethal challenge with the lymphoma. This strategy could be used to develop personalized therapies for diseases such as non-Hodgkin's lymphoma. Antibodies capable of recognizing unique markers on the surface of any malignant B-cells could be produced for each patient. The necessity for speed in the derivation of such prophylactic antibodies is recognized by the use of viral vectors rather than transgenic plants. These antibodies are now undergoing Phase I clinical trials. Another important development in the use of viral vectors was the production of full-size immunoglobulins in *N. benthamiana* plants using two TMV vectors [97]. One of the vectors expressed the heavy chain and one the light chain. This study showed that viral coexpression was compatible with the correct assembly and processing of multimeric recombinant proteins. Many epitopes from human and animal pathogenic viruses and bacteria have been expressed as coat protein fusions in CPMV and AlMV, and in most cases mice either injected with plant extracts or administered the extracts intranasally developed suitable immune responses. The use of plant virus expression vectors, with a focus on vaccine development, is discussed in detail in Chapter 6.

13.3.4
Transiently transformed leaves

Transient expression assays are often used to evaluate expression constructs or test the functionality of a recombinant protein before committing to the long term goal of transgenic plants. However, transient expression can also be used as a routine molecular farming method if enough protein can be produced to make the system economically viable. An example of a transient expression system is the agroinfiltration method, where recombinant *Agrobacterium tumefaciens* are infiltrated into leaf tissue and genes carried on the T-DNA are expressed for 2–5 days without integration. Agroinfiltration was developed in tobacco [98], but there appears to be no intrinsic limitation to the range of species that can be used, since preliminary data obtained at the RWTH Aachen demonstrates the successful infiltration of over 20 different plant species (Markus Sack, personal communication). Although originally considered difficult to scale up, agroinfiltration is now known to be suitable for the routine production of milligram amounts of protein in a timescale of weeks. Scientists at Medicago Inc., Québec, Canada, regularly process up to 7500 infiltrated alfalfa leaves per week (see Chapter 1) and similarly we have shown that up to 100 kg of wild type tobacco leaves can be processed by agroinfiltration, resulting in the production of 50–150 mg of protein per kg (Stefan Schillberg, personal communication).

A number of different antibodies and their derivatives have been produced by agroinfiltration, including the full-size IgG T84.66 (along with its scFv and diabody derivatives) [99], and a chimeric full-size IgG known as PIPP which recognizes human chorionic gonadotropin [100]. Recently, several reports have described how agroinfiltration can be scaled-up more efficiently. Baulcombe and colleagues have shown that the loss of protein expression seen a few days after agroinfiltration is predominantly caused by gene silencing. They managed to increase the expression levels of several proteins at least 50-fold by co-expressing the p19 protein from tomato bushy stunt virus, a known inhibitor of gene silencing [101].

13.3.5
Hydroponic cultures

In most cases where nuclear transgenic plants have been used for the production of recombinant protein, the proteins have been extracted from plant tissues. An alternative is to attach a signal peptide to the recombinant protein thus directing it to the secretory pathway. In this way, the protein can be recovered from the root exudates or leaf guttation fluid, processes known respectively as rhizosecretion and phyllosecretion [13,14]. Although not widely used, the secretion of recombinant proteins into hydroponic culture medium is advantageous because no cropping or harvesting is necessary. The technology is being developed by the US biotechnology company Phytomedics Inc. In an exciting recent development, a monoclonal antibody was shown to be secreted into hydroponic culture medium resulting in a yield of 11.7 µg antibody per gram of dry root mass per day [102].

13.3.6
Hairy roots

Hairy roots are neoplastic structures that arise following transformation of a suitable plant host with *Agrobacterium rhizogenes*. If the plant is already transgenic, or if the transforming *A. rhizogenes* strain is transgenic and transfers the foreign gene to the host plant during the process of transformation, then hairy root cultures can be initiated which will produce recombinant proteins and secrete them into the growth medium [103]. Hairy roots grow rapidly and can be propagated indefinitely in liquid medium. Thus far, hairy root cultures have been used to produce a relatively small number of antibodies [104–106] mainly because of the relative ease with which multi-subunit proteins can be produced. The cultures can be initiated from transgenic plants already carrying multiple transgenes, wild type plants can be infected with multiple *A. rhizogenes* strains, or established hairy root cultures can be super-transformed with *A. tumefaciens*.

13.3.7
Shooty teratomas

Shooty teratomas are differentiated cell cultures produced by transformation with certain strains of *A. tumefaciens* [107]. Thus far, there has been only one report of pharmaceutical protein production in teratoma cultures, and the levels of antibody were very low [106].

13.3.8
Suspension cell cultures

Suspension cell cultures are usually derived from callus tissue by the disaggregation of friable callus pieces in shake bottles or fermenters of liquid medium. Recombinant protein production is achieved by using transgenic explants to derive the cultures, or transforming the cells after disaggregation, usually by co-cultivation with *A. tumefaciens*. Suspension cultures have the same advantages as the simple plants discussed in Sect. 13.2.5, i.e. controlled growth conditions, batch-to-batch reproducibility, containment and production under GMP procedures. The main disadvantage is the scale of production, although tobacco suspension cells have been cultivated at volumes of up to 100,000 liters [108]. Many foreign proteins have been expressed successfully in suspension cells, including antibodies, enzymes, cytokines and hormones [108–112]. Tobacco cultivar Bright Yellow 2 (BY-2) is the most popular source of suspension cells for molecular farming, since these proliferate rapidly and are easy to transform. However, rice suspension cells have been used to produce several biopharmaceutical proteins [113–117] and soybean suspension cells have been used to produce a hepatitis B vaccine candidate [118]. Proteins expressed in suspension cells can either be extracted from the wet biomass or secreted into the culture medium for continuous, non-destructive recovery. The advantages and disadvantages of these approaches in terms of product yield and quality, plus a discussion of the optimization of culture conditions, can be found in Chapter 2.

13.4
Expression strategies and protein yields

Finally in this chapter, we consider some general strategies for the control of gene expression and protein accumulation in plants. These strategies play an important role in defining the overall yields of recombinant proteins, but they also have wider impact, e.g. on biosafety and product authenticity. To achieve high yields, all stages of gene expression must be optimized, including transcription, mRNA stability, mRNA processing, protein synthesis, protein modification, protein accumulation and protein stability. Since expression constructs are chimeric structures in which the transgene is enclosed by various regulatory elements, the considered choice of these regulatory elements is an essential component of the development phase in molecular farming.

For high-level transcription, the two most important elements are the promoter and the polyadenylation site. In dicot plants, the cauliflower mosaic virus (CaMV) 35S promoter is the most popular choice because it is strong and constitutive, and therefore drives high-level transgene expression in leaves, fruits, tubers, roots and any other relevant organs [119,120]. The promoter can be made even more active by various modifications, such as duplicating the enhancer region [121]. In monocots, where seed expression is the normal strategy, the CaMV 35S promoter has lower activity and is generally replaced either with the maize ubiquitin-1 promoter (which is constitutive, but drives high level transgene expression in seeds [122]), or with a seed-specific promoter from a seed storage protein gene (e.g. maize zein, rice glutelin, bean unknown seed protein (USP), pea legumin). The seed-specific *arc5-1* promoter from the common bean (*Phaseolus vulgaris*) has been used to express a single chain antibody in *Arabidopsis thaliana*, and this resulted in the accumulation of 36 times more recombinant protein than when the transgene was driven by the CaMV 35S promoter [123]. Other tissue-specific promoters that are useful in molecular farming include fruit specific promoters for tomato and tuber-specific promoters for potato. In each case, restriction of transgene expression to the target tissue prevents expression in vegetative organs, therefore reducing any negative impact of the recombinant protein on normal plant growth and development, and limiting the exposure of non-target organisms such as pollinating insects or microbes in the rhizosphere.

Inducible promoter systems are also valuable assets in molecular farming because transgene expression can be controlled externally [124]. One example is the mechanical gene activation (MeGA) system that was developed at CropTech Corp., VA. This utilizes a tomato hydroxy-3-methylglutaryl CoA reductase 2 (HMGR2) promoter, which is inducible by mechanical stress. Transgene expression is activated when harvested tobacco leaves are sheared during processing, which leads to the rapid induction of protein expression, usually within 24 hours. Another potentially useful inducible promoter that has been described recently is the peroxidase gene promoter from sweet potato (*Ipomoea batatas*). When linked to the *gusA* reporter gene, this promoter produced 30 times more GUS activity than the CaMV 35S promoter following exposure to hydrogen peroxide, wounding or ultraviolet light [125].

Other parts of the expression construct are also important. A strong polyadenylation signal is required for transcript stability and those from the CaMV 35S transcript, the *Agrobacterium tumefaciens nos* gene and the pea *ssu* gene are popular choices. In monocots, the presence of an intron in the 5' untranslated region of the expression construct has been shown to improve transgene expression [126]. The structure of the 5' and 3' untranslated regions should also be inspected for AU-rich sequences, which can act as cryptic splice sites and instability elements. Some sequences, such as the 5' leader of the petunia chalcone synthase gene, have been identified as translational enhancers and these can be incorporated into the expression construct to boost protein synthesis. Other important factors that influence translation include the presence of a consensus Kozak sequence, the absence of multiple AUG codons, and the disparity of codon bias between the transgene donor and the host species [127].

One of the most important considerations for the improvement of protein yields is subcellular protein targeting, because the compartment in which a recombinant protein accumulates strongly influences the interrelated processes of folding, assembly and post-translational modification. All of these contribute to protein stability and hence help to determine the final yield [88].

Comparative targeting experiments with full size immunoglobulins and single chain Fv fragments have shown that the secretory pathway is often a more suitable compartment for folding and assembly than the cytosol, and is therefore advantageous for high-level protein accumulation [128,129]. Because many plant-derived recombinant proteins under development are human proteins that normally pass through the endomembrane system, this principle can be applied not only to antibodies but also more generally. However, it is not a universal rule, since there are examples of proteins that are more abundant when directed to the cytosol than the secretory pathway, e.g. α-galactosidase A (see Chapter 6). Antibodies targeted to the secretory pathway using either plant or animal N-terminal signal peptides usually accumulate to levels that are several orders of magnitude greater than those of antibodies expressed in the cytosol. Even with antibodies, however, there are occasional exceptions, and this suggests that intrinsic features of each antibody might also contribute to overall stability [130,131]. The endoplasmic reticulum provides an oxidizing environment and an abundance of molecular chaperones, while there are few proteases. These are likely to be the most important factors affecting protein folding and assembly. It has been shown recently that antibodies targeted to the secretory pathway in transgenic plants interact specifically with the molecular chaperone BiP [132].

In the absence of further targeting information, the expressed protein is secreted to the apoplast. The stability of antibodies in the apoplast is lower than in the lumen of the ER. Therefore, antibody expression levels can be increased up to ten times higher if the protein is retrieved to the ER lumen using an H/KDEL C-terminal tetrapeptide tag [133]. Again, although the principles of ER-retention in molecular farming have been established using antibodies, it is likely that they will also apply to many other proteins.

13.5
Conclusions

In this chapter, we have looked at the properties of different expression hosts and expression systems, and considered some of the available strategies to control transgene expression and protein accumulation. When all these variations are combined, there exists a very diverse range of potential expression platforms that can be used to produce recombinant proteins. The choice depends on many factors, some intrinsic to the plant species or expression system, some dependent on the recombinant protein and its intended use, and some determined by external factors such as regional, economic and regulatory constraints. There is no ideal production platform for molecular farming, and each of the host plants and systems described in this chapter has its merits and drawbacks. As ever, the choice of production platform therefore should be determined empirically, and on a case-by-case basis.

References

[1] R. FISCHER, N. EMANS, *Transgenic Res.* 2000, *9* (3–4), 279–299.

[2] G. GIDDINGS, G. ALLISON, D. BROOKS et al., *Nature Biotechnol.* 2000, *18* (11), 1151–1155.

[3] J.K.-C. MA, P.M.W. DRAKE, P. CHRISTOU, *Nature Rev. Genet.* 2003, *4* (10), 794–805.

[4] R.M. TWYMAN, E. STOGER, S. SCHILLBERG et al., *Trends Biotechnol.* 2003, *21* (12), 570–578.

[5] E.E. HOOD, S.L. WOODARD, M.E. HORN et al., *Curr. Opin. Biotechnol.* 2002, *13* (6), 630–635

[6] R.L. EVANGELISTA, A.R. KUSNADI, J.A. HOWARD et al., *Biotechnol. Prog.* 1998, *14* (4), 607–614

[7] H.S. MASON, H. WARZECHA, T. MOR et al., *Trends Mol. Med.* 2002, *8* (7), 324–329.

[8] D.M. DENBOW, E.A. GRABAU, G.H. LACY et al., *Poultry Sci.* 1998, *77* (6), 878–881.

[9] Z.Y. DAI, B.S. HOOKER, D.B. ANDERSON et al., *Mol. Breeding* 2000, *6* (3), 277–285.

[10] M.T. ZIEGLER, S.R. THOMAS, K.J. DANNA, *Mol. Breeding* 2000, *6* (1), 37–46.

[11] T. ZIEGELHOFFER, J.A. RAASCH, S. AUSTIN-PHILLIPS, *Mol. Breeding* 2001, *8* (2), 147–158.

[12] M. MOLONEY, J. BOOTHE, G. VAN ROOIJEN, US Patent 6,509,453, 2003.

[13] N.V. BORISJUK, L.G. BORISJUK, S. LOGENDRA et al., *Nature Biotechnol.* 1999, *17* (5), 466–469.

[14] S. KOMARNYTSKY, N.V. BORISJUK, L.G. BORISJUK et al., *Plant Physiol.* 2000, *124* (3), 927–933.

[15] E. STOGER, M. SACK, Y. PERRIN et al., *Mol. Breeding* 2000, *9* (3), 149–158.

[16] U. CONRAD, U. FIEDLER, *Plant Mol. Biol.* 1998, *38* (1–2), 101–109.

[17] A. BARTA, K. SOMMERGRUBER, D. THOMPSON et al., *Plant Mol. Biol.* 1986, *6* (5), 347–357.

[18] P.C. SIJMONS, B.M.M. DEKKER, B. SCHRAMMEIJER et al., *Bio/Technology* 1990, *8* (3), 217–221.

[19] A. HIATT, R. CAFFERKEY, K. BOWDISH, *Nature* 1989, *342* (6245), 76–78.

[20] H.S. MASON, D.M.K. LAM C.J. ARNTZEN, *Proc. Natl Acad. Sci. USA* 1992, *89* (24), 11745–11749.

[21] J. PEN, L. MOLENDIJK, W.J. QUAX et al., *Bio/Technology* 1992, *10* (3), 292–296.

[22] X. ZHANG, D.W. URRY, H. DANIELL, *Plant Cell Rep.* 1996, *16* (3–4), 174–179.

[23] S. W. MA, D. L. ZHAO, Z. Q. YIN et al., *Nature Med.* 1997, *3* (7), 793–796.

[24] R. MENASSA, V. NGUYEN, A. JEVNIKAR et al., *Mol. Breeding* 2001, *8* (2), 177–185.

[25] M. CABANES-MACHETEAU, A. C. FITCHETTE-LAINE, C. LOUTELIER-BOURHIS et al., *Glycobiology* 1999, *9* (4), 365–372.

[26] H. BAKKER, M. BARDOR, J. W. MOLTHOFF et al., *Proc. Natl Acad. Sci. USA* 2001, *98* (5), 2899–2904.

[27] E. BENVENUTO, R.J. ORDAS, R. TAVAZZA et al., *Plant Mol. Biol.* 1991, *17* (4), 865–874.

[28] H. KHOUDI, S. LABERGE, J. M. FERULLO et al., *Biotechnol. Bioeng.* 1999, *64* (2), 135–143.

[29] M. BARDOR, C. LOUTELIER-BOURHIS, T. PACCALET et al., *Plant Biotech. J.* 2003, *1* (6), 451–462.

[30] A. WIGDOROVITZ, C. CARRILLO, M. J. DUS SANTOS et al., *Virology* 1999, *255* (2), 347–353.

[31] S. AUSTIN-PHILLIPS, R. G. KOEGEL, R. J. STRAUB et al., US patent application 5900525, 1999.

[32] R.W.H. LEE, A.N. POOL, A. ZIAUDDIN et al., *Mol. Breeding* 2003, *11* (4), 259–266.

[33] J KAPUSTA, A MODELSKA, M FIGLERO-

WICZ et al., *FASEB J.* **1999**, *13* (13), 1796–1799.

[34] J. KAPUSTA, A. MODELSKA, T. PNIEWSKI et al., *Adv. Exp. Med. Biol.* **2001**, *495*, 299–303.

[35] H. KOPROWSKI, *Arch. Immunol. Ther. Exp.* **2002**, *50* (6), 365–369.

[36] V. YUSIBOV, D.C. HOOPER, S. SPITSIN et al., *Vaccine* **2002**, *20* (25–26), 3155–3164,

[37] H.E. SUSSMAN, *Drug Discov. Today* **2003**, *8* (10), 428–430.

[38] V. SMART, P.S. FOSTER, M.E. ROTHEN-BERG et al., *J. Immunol.* **2003**, *171* (4), 2116–2126.

[39] E. STOGER, C. VAQUERO, E. TORRES et al., *Plant Mol. Biol.* **2000**, *42* (4), 583–590.

[40] E.E. HOOD, D.R. WITCHER, S. Maddock et al., *Mol. Breeding* **1997**, *3* (4), 291–306.

[41] D.R. WITCHER, E.E. HOOD, D. PETERSON et al., *Mol. Breeding* **1998**, *4* (4), 301–312.

[42] E.E. HOOD, *Enzyme & Microbial Technol.* **2002**, *30* (3), 279–283.

[43] E.E. HOOD, in: P. Christou. H. Klee (eds) *Handbook of Plant Biotechnology* **2004**, Wiley-VCH, NY, 791–800.

[44] Z. ZHU, K. HUGHES, L. HUANG et al., *Plant Cell Tiss. Org. Cult.* **1994**, *36* (2), 197–204.

[45] A. OKADA, T. OKADA, T. IDE et al., *Mol. Breeding* **2003**, *12* (1), 61–70.

[46] L.G. JENSEN, O. OLSEN, O. KOPS et al., *Proc. Natl Acad. Sci. USA* **1996**, *93* (8), 3487–3491.

[47] M. PATEL, J.S. JOHNSON, R.I.S. BRETTELL et al., *Mol. Breeding* **2000**, *6* (1), 113–123.

[48] P.H.D. SCHUNMANN, G. COIA, P.M. WATERHOUSE, *Mol. Breeding* **2002**, *9* (2), 113–121.

[49] G.P. XUE, M. PATEL, J.S. JOHNSON et al., *Plant Cell Rep.* **2003**, *21* (11), 1088–1094.

[50] L. ZEITLIN, S.S. OLMSTED, T.R. MOENC et al., *Nature Biotechnol.* **1998**, *16* (13), 1361–1364.

[51] R. FISCHER, R.M. TWYMAN, S. SCHILLBERG, *Vaccine* **2003**, *21* (7–8), 820–825.

[52] R.A. VIERLING, J.R. WILCOX, *Seed Sci. Technol.* **1996**, *24* (3), 485–494.

[53] Y. PERRIN, C. VAQUERO, I. GERRARD et al., *Mol. Breeding* **2000**, *6* (4), 345–352.

[54] V.V. SATYAVATHI, V. PRASAD, A. KHANDELWAL et al., *Plant Cell Rep.* **2003**, *21* (7), 651–658.

[55] A. KHANDELWAL, G.L. SITA, M.S. SHAILA, *Vaccine* **2003**, *21* (23), 3282–3289.

[56] A. KHANDELWAL, K.J.M. VALLY, N. GEETHA et al., *Plant Sci.* **2003**, *165* (1), 77–84.

[57] T. ARAKAWA, D.K.X. CHONG, J. Yu et al., *Transgenics* **1999**, *3* (1), 51–60.

[58] S.W. MA, A.M. JEVNIKAR, *Adv. Exp. Med. Biol.* **1999**, *464*, 179–194.

[59] K. OHYA, T. MATSUMURA, K. OHASHI et al., *J. Interf. Cytok. Res.* **2001**, *21* (8), 595–602.

[60] B. ZHANG, Y.H. YANG, Y.M. LIN et al., *Biotechnol. Lett.* **2003**, *25* (19), 1629–1635.

[61] Y. PARK, H. CHEONG, *Protein Express. Purif.* **2002**, *25* (1), 160–165.

[62] O. ARTSAENKO, B. KETTIG, U. FIEDLER et al., *Mol. Breeding* **1998**, *4* (4), 313–319.

[63] C. DE WILDE, K. PEETERS, A. JACOBS et al., *Mol. Breeding* **2002**, *9* (4), 271–282.

[64] D.K.X. CHONG, W.H.R. LANGRIDGE, *Transgenic Res.* **2000**, *9* (1), 71–78.

[65] D.K.X. CHONG, W. ROBERTS, T. ARAKAWA et al., *Transgenic Res.* **1997**, *6* (4), 289–296.

[66] A.M. WALMSLEY, C.J. ARNTZEN, *Curr. Opin. Biotech.* **2003**, *14* (2), 145–150.

[67] C.O. TACKET, H.S. MASON, G. LOSONSKY et al., *Nature Med.* **1998**, *4* (5), 607–609.

[68] C.O. TACKET, H.S. MASON, G. LOSONSKY et al., *J. Infect. Dis.* **2000**, *182* (1), 302–305.

[69] L.J. RICHTER, Y. THANAVALA, C.J. ARNTZEN et al., *Nature Biotechnol.* **2000**, *18* (11), 1167–1171.

[70] J. YU, W. LANGRIDGE, *Transgenic Res.* **2003**, *12* (2), 163–169.

[71] P.B. MCGARVEY, J. HAMMOND, M.M. Dienelt et al., *Bio/Technology* **1995**, *13* (13), 1484–1487.

[72] D. JANI, L.S. MEENA, Q.M. RIZWAN-ul-Haw et al., *Transgenic Res.* **2002**, *11* (5), 447–454.

[73] J.S. SANDHU, S.F. KRASNYANSKI, L.L. DOMIER et al., *Transgenic Res.* **2000**, *9* (2): 127–135.

[74] Y. Ma, S.Q. Lin, Y. Gao et al., *World J. Gastroenterol.* **2003**, *9* (10), 2211–2215.

[75] A.M. Walmsley, M.L. Alvarez, Y. Jin et al., *Plant Cell Rep.* **2003**, *21* (10), 1020–1026.

[76] L.J. Wang, D.A. Ni, Y.N. Chen et al., *Acta Bot. Sin.* **2001**, *43* (2), 132–137.

[77] A. Porceddu, A. Falorni, N. Ferradini et al., *Mol. Breeding* **1999**, *5* (6), 553–560.

[78] F.B. Bouche, E. Marquet-Blouin, Y. Yanagi et al., *Vaccine* **2003**, *21* (17–18), 2065–2072.

[79] E. Marquet-Blouin, F.B. Bouche, A. Steinmetz et al., *Plant Mol. Biol.* **2003**, *51* (4), 459–469.

[80] D.L. Parmenter, J.G. Boothe, G.J.H. vanRooijen et al., *Plant Mol. Biol.* **1995**, *29* (6), 1167–1180.

[81] J.H. Liu, L.B. Selinger, K.J. Cheng et al., *Mol. Breeding* **1997**, *3* (6), 463–470.

[82] S.P. Mayfield, S.E. Franklin, R.A. Lerner, *Proc. Natl Acad. Sci. USA* **2003**, *100* (2), 438–442.

[83] S.E. Franklin, S.P. Mayfield, *Curr. Opin. Plant Biol.* **2004**, *7* (2), 159–165.

[84] E.L. Decker, R. Reski, *Curr. Opin. Plant Biol.* **2004**, *7* (2), 166–170.

[85] D.G. Schaefer, *Annu. Rev. Plant Biol.* **2002**, *53*, 477–501.

[86] V. Gomord, L. Faye, *Curr. Opin. Plant Biol.* **2004**, *7* (2), 171–181.

[87] J.R. Gasdaska, D. Spencer, L. Dickey, *Bioprocess J.* **2003**, *2*, 49–56.

[88] S. Schillberg, N. Emans, R. Fischer, *Phytochem. Rev.* **2002**, *1* (1), 45–54.

[89] P. Maliga, *Curr. Opin. Plant Biol.* **2002**, *5* (2), 164–172.

[90] H. Daniell, M.S. Khan, L. Allison, *Trends Plant Sci.* **2002**, *7* (2), 84–91.

[91] J.S. Tregoning, P. Nixon, H. Kuroda et al., *Nucleic Acids Res.* **2003**, *31* (4), 1174–1179.

[92] A. Fernandez-San Milan, A. Mingo-Castel, M. Miller et al., *Plant Biotechnol. J.* **2003**, *1* (2), 77–79.

[93] S. Leelavathi, N. Gupta, S. Maiti et al., *Mol. Breeding* **2003**, *11* (1), 59–67.

[94] S. Ruf, M. Hermann, I.J. Berger et al., *Nature Biotechnol.* **2001**, *19* (9), 870–875.

[95] C. Porta, G.P. Lomonossoff, *Biotechnol. & Genet. Eng. Rev.* **2002**, *19*, 245–291.

[96] A.A. McCormick, M.H. Kumagai, K. Hanley et al., *Proc. Natl Acad. Sci. USA* **1999**, *96* (2), 703–708.

[97] T. Verch, V. Yusibov, H. Koprowski, *J. Immunol. Methods* **1998**, *220* (1–2), 69–75.

[98] J. Kapila, R. De Rycke, M. Van Montagu et al., *Plant Sci.* **1997**, *122* (1), 101–108.

[99] C. Vaquero, M. Sack, F. Schuster et al., *FASEB J.* **2002**, *16* (1), 408–410.

[100] S. Kathuria, R. Sriraman, R. Nath et al., *Hum. Reprod.* **2002**, *17* (8), 2054–2061.

[101] O. Voinnet, S. Rivas, P. Mestre et al., *Plant J.* **2003**, *33* (5), 949–956.

[102] P.M.W. Drake, D.M. Chargelegue, N.D. Vine et al., *Plant Mol. Biol.* **2003**, *52* (1), 233–241.

[103] J. M. Sharp, P. M. Doran, *Biotechnol. Prog.* **2001**, *17* (6), 979–992.

[104] R. Wongsamuth, P. M. Doran, *Biotechnol. Bioeng.* **1997**, *54* (5), 401–415.

[105] J. M. Sharp, P. M. Doran, *Biotechnol. Bioprocess Eng.* **1999**, *4* (4), 253–258.

[106] J. M. Sharp, P. M. Doran, *Biotechnol. Bioeng.* **2001**, *73* (5), 338–346.

[107] M. A. Subroto, J. D. Hamill, P. M. Doran, *J. Biotechnol.* **1996**, *45* (1), 45–57.

[108] R. Fischer, N. Emans, F. Schuster et al., *Biotechnol. Appl. Biochem.* **1999**, *30* (2), 109–112.

[109] S. Y. Hong, T. H. Kwon, J. H. Lee et al., *Enzyme & Microbial Tech.* **2002**, *30* (6), 763–767.

[110] E. James, J. M. Lee, *Adv. Biochem. Eng. Biotechnol.* **2001**, *72*, 127–156.

[111] T. H. Kwon, J. E. Seo, J. Kim et al., *Biotechnol. Bioeng.* **2003**, *81* (7), 870–875.

[112] N. S. Magnuson, P. M. Linzmaier, R. Reeves et al., *Protein Express. Purif.* **1998**, *13* (1), 45–52.

[113] M. Terashima, Y. Ejiri, N. Hashikawa et al., *Biochem. Eng. J.* **1999**, *4* (1), 31–36.

[114] J. Huang, T. D. Sutliff, L. Wu et al., *Biotechnol. Prog.* **2001**, *17* (1), 126–133.

[115] M. M. Trexler, K. A. McDonald, A. P. Jackman, *Biotechnol. Prog.* **2002**, *18* (3), 501–508.

[116] E. Torres, C. Vaquero, L. Nicholson et al., *Transgenic Res.* **1999**, *8* (6), 441–449.

[117] J.M. HUANG, L. WU, D. YALDA et al., *Transgenic Res.* **2002**, *11* (3), 229–239.

[118] M. L. SMITH, M. E. KEEGAN, H. S. MASON et al., *Biotechnol. Prog.* **2002**, *18* (3), 538–550.

[119] J. T. O'DELL, F. NAGY, N. H. CHUA, *Nature* **1985**, *313* (6005), 810–812.

[120] M.A. LAWTON, M.A. TIERNEY, I. NAKAMURA et al., *Plant Mol. Biol.* **1987**, *9* (4) 315–324.

[121] R. KAY, A. CHAN, M. DALY et al., *Science* **1987**, *236* (4806), 1299–1302.

[122] A. H. CHRISTENSEN, P. H. QUAIL, *Transgenic Res.* **1996**, *5* (3), 213–218.

[123] G. DE JAEGER, S. SCHEFFER, A. JACOBS et al., *Nature Biotechnol.* **2002**, *20* (12), 1265–1268.

[124] M. PADIDAM, *Curr. Opin. Plant Biol.* **2003**, *6* (2), 169–177.

[125] K.Y. KIM, S.Y. KWON, H.S. LEE et al., *Plant Mol. Biol.* **2003**, *51* (6), 831–838.

[126] P. VAIN, K.R. FINER, D.E. ENGLER et al., *Plant Cell Rep.* **1996**, *15* (7), 489–494.

[127] R. FISCHER, N.J. EMANS, R.M. TWYMAN et al., in: M. Fingerman, R. Nagabhushanam (eds) *Recent Advances in Marine Biotechnology, Volume 9 (Biomaterials and Bioprocessing)*, Science Publishers Inc., Enfield NH, pp 279–313, **2003**.

[128] S. ZIMMERMANN, S. SCHILLBERG, Y.-C. LIAO et al., *Mol. Breeding* **1998**, *4* (4), 369–379.

[129] S. SCHILLBERG, S. ZIMMERMANN, A. VOSS et al., *Transgenic Res.* **1999**, *8* (4), 255–263.

[130] G. DE JAEGER, E. BUYS, D. EECKHOUT et al., *Eur. J. Biochem.* **1998**, *259* (1–2), 1–10.

[131] A. SCHOUTEN, J. ROSSIEN, J. BAKKER et al., *J. Biol. Chem.* **2002**, *277* (22), 19339–19345.

[132] J. NUTTALL, N. VINE, J.L. Hadlington et al., *Eur. J. Biochem.* **2002**, *269* (24), 6042–6051.

[133] U. CONRAD, U. FIEDLER, *Plant Mol. Biol.* **1998**, *38* (1–2), 101–109.

14
Downstream Processing of Plant-derived Recombinant Therapeutic Proteins
Juergen Drossard

14.1
Introduction

It is common practice to divide biotechnological processes into *upstream* and *downstream* sections. In a standard process based on microbial or animal cell fermentation, the starting point for the upstream section is generally the transfer of the contents of a vial from a cell bank into liquid growth medium for expansion and subsequent inoculation of a seed fermenter. The process continues with sequential scale-up into larger-volume vessels and into the final production-size fermenter. Here, after the activation of inducible promoters, the target product is expressed by the host cells and is either released into the medium or retained within the cells. When the production-scale cultivation is complete, harvesting of either the cell mass or the fermentation broth is the linking step between upstream and downstream stages. The downstream part of the process focuses on the treatment of the crude fermentation broth or the harvested cells to obtain the product of interest in a suitable form and quality, which is very dependent on both the nature of the product and its intended use. The downstream section ends with the formulation of a bulk product ready for final testing and packaging. It is crucially important that these upstream and downstream processes are carefully designed and synchronized to insure a smooth transition between them, and that they are bracketed by a system of monitoring and quality control to help identify critical steps and avoid suboptimal results due to problems at the interface.

This description of a generic production process excludes all the early-stage research and development work such as genetic engineering, transformation and selection, expression analysis, media and process optimization etc. Although these activities form the basis of bioprocessing and have a great influence over product yield and quality, they are discrete tasks that should be completed before the onset of a production process. The above description also excludes non-product-related aspects of the process like waste treatment, environmental protection and legal requirements for recombinant DNA technology.

The above definition of classical biotechnological processes can easily be adapted to the concept of molecular farming. With plant cell fermentation the analogy is obvious, but processes starting with greenhouse- or field-cultivated plants can be viewed in the same way, with the harvesting of plant biomass (leaves, seed or fruit) and its transport to the processing facility representing the link between upstream and downstream.

The versatility of the molecular farming approach in biotechnology has been discussed extensively in other chapters of this book. It is this technological versatility, allowing expression of a broad range of products in different species ranging from algae and mosses to higher plants, and in different organs or cellular compartments, that makes the downstream processing of plant-derived recombinant proteins a challenging task. The broad range of products that can be generated by the genetic engineering of plants indicates that different procedures are required to turn a constituent of a plant cell into a finished product ready for its intended use. Industrial enzymes, for example, may require only minimal purification [1], while certain pharmaceutical products, such as orally-administered subunit vaccines, may even be administered as whole unprocessed plant tissues, purees or juices, without any purification [2]. Other products, in particular recombinant pharmaceutical proteins such as antibodies, require extensive purification treatment under a strict regime of quality assurance and quality control to achieve the approval of regulatory agencies [3]. Any alternative production system for active pharmaceutical ingredients will, from a regulatory point of view, have to deliver a product that fulfills the requirements for product safety, quality, potency and efficacy equally well or better than the established comparator product. In the EU, by May 2004, regulatory compliance with the principles of good manufacturing practice (GMP) in the manufacture of medicinal products (Table 14.1) had been extended to the production of clinical trial material [4], an area of interest and activity for many researchers in the molecular farming community. This chapter discusses the technological and regulatory challenges encountered in the downstream processing of plant-derived recombinant proteins, with the emphasis on products designed for pharmaceutical use.

14.2
Similarities and Differences in the Processing of Pharmaceutical Proteins from Different Sources

Looking at the downstream processing of recombinant pharmaceutical proteins from different sources as a whole, there are more common steps than operations addressing expression system-specific problems or requirements. One of the most important common features is that a given end product must meet the same standards and specifications in terms of safety, quality, potency and efficacy, regardless of the production host. Furthermore, the physicochemical properties of such end products should be identical, so that the intrinsic features used for purification (affinity, hydrophobicity etc.) are the same. Well-established procedures and protocols should therefore be utilized, and should be adapted to the special requirements of the source material only when absolutely necessary. This is particularly true in the case of pharmaceuticals, since the tendency in this field is to stick to established methods

Tab. 14.1 The role of GMP (good manufacturing practice) in the production and processing of APIs (active pharmaceutical ingredients) from difference sources. It is not yet clear how biotechnology-derived plants fit into this scheme. Modified from the Good Manufacturing Practice Guide for Active Pharmacuetical Ingreedients, ICH (2000).

API (product type and process)	Stages of the process. Those subject to GMP guidelines are shown in grey.				
Chemical	Production of starting material	Introduction of API starting material into process	Production of intermediates	Isolation and purification	Physical processing and packaging
Derived from animal sources	Collection of tissue or fluid	Cutting, mixing, initial processing	Introduction of API starting material into process	Isolation and purification	Physical processing and packaging
Extracted from plant sources	Collection of plants	Cutting and initial extraction	Introduction of API starting material into process	Isolation and purification	Physical processing and packaging
Herbal extracts	Collection of plants	Cutting and initial extraction		Further extraction	Physical processing and packaging
Comminuted or powdered herbs	Collection of plants/ Cultivation and harvesting	Cutting/ comminuting			Physical processing and packaging
Biotechnology (fermentation/ cell culture)	Establishment of master and working cell banks	Maintenance of working cell bank	Cell culture and fermentation	Isolation and purification	Physical processing and packaging
Classical fermentation	Establishment of a cell bank	Maintenance of the cell bank	Cell culture and fermentation	Isolation and purification	Physical processing and packaging

as far as possible due to the immense efforts, both financially and in terms of labor and time input, required to implement new processes in a regulated environment.

The genetic engineering of plants is a mature technology, and the food and feed industry has the capacity to process any quantity of plant material, generating virtually unlimited amounts of juice, puree, flour etc. The utilization of plants and their ingredients for pharmaceutical purposes is also nothing new, with a product range including teas and herbal extracts, well-established chemicals like digitalis and morphine, and new potent drugs derived from secondary plant metabolites such as taxol. The novel and challenging task in molecular farming is the combination of genetic engineering, protein extraction, and the development of adequate manufacturing and processing technology.

Another important issue is process economy. Some general statements on this topic can be made that are valid for all production systems:

- downstream processing contributes significantly to the overall costs of a biotechnological process, in particular if the target is a medicinal product;
- as stated above, the majority of product-specific requirements for downstream processing are not associated with the particular expression system, leading to the conclusion that the potential economic advantages of plant production systems lie in the upstream rather than the downstream part of the process;
- the large contribution of the downstream costs to the overall cost of a process will always put pressure on the downstream side to meet the specifications as economically as possible.

Several key issues have to be addressed in the downstream processing of biopharmaceuticals regardless of the expression system. The removal of host cell proteins and nucleic acids, as well as other product- or process-related or adventitious contaminants, is laid down in the regulations and will not differ between the individual expression hosts. The identity, activity and stability of the end product has to be demonstrated regardless of the production system. The need for pharmaceutical quality assurance, validation of processes, analytical methods and cleaning procedures are essentially the same.

There are, however, expression system-specific risks for product quality and safety that must be adequately taken into consideration. For example, rodent cell culture systems (e. g. CHO cells, which are widely used for monoclonal antibody production) are susceptible to inherent or adventitious contamination with human pathogenic viruses, and therefore require rigorous virus inactivation or removal procedures to be included in the purification process. In bacterial systems, endotoxins are a major concern. The content of the host cells in terms of proteinases, oxidizing agents, allergens, toxins and other unwanted by-products will, to a large extent, be species-specific. It is therefore necessary to break down every downstream process into individual, well-defined unit operations and carefully analyze the individual steps for their efficiency, robustness and reproducibility in the context of the expression system. In contrast to established host systems like *Escherichia coli* and mammalian cells, the regulatory requirements for plant-based therapeutics are not yet fully defined. However, both the US Food and Drug Administration (FDA) and the European Agency for the Evaluation of Medicinal Products (EMEA) have recently published draft guidance documents addressing this issue [5,6], so it has become clearer what the future focus of the agencies' concerns and activities in molecular farming will be.

14.3
Process Scale

If a recombinant protein is developed and expressed for research purposes, the most labor- and cost-intensive part of the project normally lies with the upstream tasks, i. e. cloning, expression vector design, sequencing, transformation and selection. Once a suitable expresser strain or plant line is available, it is often sufficient to visualize expression of the recombinant protein, verify its activity, study its biological effects on the host or follow transmission of the transgene to the progeny. If purification is necessary, isolation of low milligram quantities of the target protein is performed from a small amount of source material for biochemical characterization or structural analysis. Recovery (expressed as a percentage of the total content of the target molecule in the raw material) is usually not a major concern. When working on a laboratory scale, well-established procedures are available to reduce the problems of initial protein purification. Selective salting-out by ammonium sulfate or other structure-forming salts is often used to separate proteins from cell debris and reduce contamination with nucleic acids, lipids and small organic or inorganic compounds. This process also reduces the eluent volume by redissolving the sedimented precipitate in a small amount of a suitable buffer. The addition of nucleases for viscosity reduction, protease inhibitors for protection against proteolytic cleavage, and stabilizing agents to counteract oxidation and other adverse environmental effects are common practices. The resulting conditioned extract, usually having a volume of a few milliliters, can then be further prepared as necessary for chromatography by centrifugation, filtration and dialysis.

This situation changes dramatically when scaling-up production. Although few data are publicly available concerning the process economics of commercial recombinant protein production [7], it can be estimated that the downstream part of the process may, in the case of a therapeutic protein, account for more than 80% of the total production cost [8]. Under laboratory conditions, the use of expensive buffers and additives (e.g. protease inhibitors) may be acceptable when balancing benefits against their purchase costs. In large-scale production, such agents have to be replaced by inexpensive substances like acetate or phosphate salts. Other factors affecting the design of a purification scheme include the influence of percentage recovery on the unit price of the final product and the high degree of purity required for therapeutic proteins. Finally, many of the standard procedures for laboratory-scale protein extraction and purification require a lot of hands-on operation and will, for technical or cost reasons, not be applicable in large-scale processes. To address these problems, the scale-up capabilities of purification protocols should be investigated and improved as early as possible, preferably when moving from expression studies to small- or pilot-scale protein production [9]. Due to the large investment requirement and the long, complex approval procedure by regulatory agencies, established large-scale purification protocols for biopharmaceuticals often have to be maintained even if new developments would have significant advantages.

14.4
The Individual Steps of a Downstream Process

It is useful to subdivide the downstream processing of recombinant proteins into a few key stages, often referred to as initial processing of the source material and extraction (if necessary), capture, intermediate purification and polishing. These can then be further split into *unit operations*. In each of these stages, predefined goals have to be achieved, so a well-defined purification protocol will sequentially utilize as many different individual properties of the recombinant protein (charge, hydrophobicity, size, affinity, solubility, heat- and pH-stability etc.) as possible to result in the greatest purity. In the case of recombinant proteins, finding the optimal conditions is facilitated by the knowledge of amino acid sequence and associated physicochemical parameters, predicted or verified structural information and, in many cases, the presence of affinity-tags or other fusion partners designed into the expression construct for purification purposes.

14.4.1
Initial Processing and Extraction

The initial processing operations primarily serve to prepare a suitable starting material for the subsequent purification steps. After breaking cells open to release any intracellular product, these operations usually include a liquid-solid separation step to remove cell debris and other particulate matter as well as a conditioning step for the crude cell extract or fermentation broth e.g. diafiltration into a suitable buffer. A concentrating step, using membrane- or hollow fiber-based filtration systems, may also be included here, especially if large volumes of very dilute feed have to be processed. At this stage of the process, the crude extract or broth usually contains a very complex mixture of water-soluble compounds from the plant cell, including substances that may be detrimental to the target protein (e.g. proteinases and oxidants). In particular, when subcellular organelles are destroyed during the extraction process, the liberation of lytic enzymes and reactive secondary metabolites must be anticipated.

This stage of downstream processing is one of the steps in which plant-based systems are unique. While plant cell cultures can be used for high volume production in a contained environment, e.g. for the production of the antineoplastic agent paclitaxel (Taxol) in bioreactors with volumes up to 90,000 l (such cultures are also potentially useful for the production of recombinant proteins, provided expression levels can be boosted significantly [10]), the large-scale cultivation of transgenic plants is carried out in greenhouses or, for reasons of economy and scalability, in the field. As well as considering the influence of varying environmental conditions on product accumulation and quality (e.g. temperature, humidity, UV radiation, soil quality, the presence of fertilizers and the presence of pests, parasites and the chemicals used to treat them), great care must also be taken to minimize the risk of contamination with toxic or noxious soil constituents, chemicals present in the environment and on the harvesting machinery, and chemicals applied to crops and the soil. Leaves and other soft tissue will wilt and begin to undergo degradation after harvest. Therefore, the conditions and duration of storage before initial processing are critical factors. One of the biggest advantages of seed-based production systems is that they facilitate long-term storage, making it possible to separate, in space and time, harvesting and initial processing of the material [11]. Additionally, it is much easier to perform surface cleaning operations on seeds than on vegetative plant material. Looking at the wide range of potential contaminants on field-grown plants, e.g. pesticides, fertilizers, soil bacteria, parasites, animal excreta and other unwanted substances, the importance of this feature is obvious.

Once the material is harvested and introduced into the processing facility, all further operations must be performed according to the regulations of pharmaceutical GMP, including the need for equipment qualification and, at least for commercial production, process validation. Large-scale processing of plant material is facilitated by the fact that machinery, e. g. corn mills, leaf shredders etc., for initial processing on an industrial scale, is readily available from suppliers to the food and feed industries. However, this machinery has to be adapted for the requirements of pharmaceutical process equipment with respect to construction materials, sanitary design, suitability for use in a controlled environment, cleanability etc., and will have to go through the usual procedure of design-, installation-, operational- and performance qualification (DQ, IQ, OQ, PQ). Additionally, it is necessary to evaluate the stability of the target protein against the applied processing conditions, e. g. temperature, pH, shear, foaming etc. Because of the inherent variability in open-field cultivation in terms of environmental conditions and plant health, further critical issues for the validation of processes using field-grown plants will include the definition and specification of batches, and the maintenance of batch-to-batch consistency. This will require acceptance criteria for each batch with respect to the expression levels and activity of the target protein, and the levels of endogenous impurities and adventitious agents. The situation is different for plant cells cultivated in bioreactors, since these represent a controlled environment wherein the cells grow under axenic conditions with a defined supply of oxygen and nutrients and where all relevant cultivation parameters can be continuously monitored and adjusted [12]. The media for plant cell cultivation are generally mineral-based, and are devoid of proteins and other potentially animal-derived products, greatly reducing the danger of contamination with viruses or prions. The link between upstream and downstream is, in the case of plant cells, much more similar to established production systems. The harvesting and initial processing of plant cells that have been cultivated in batch, fed-batch or continuous mode resemble quite closely the procedures for microbial systems. For intracellular products, the cells are separated from the nutrient media by vacuum filtration or centrifugation, washed, and resuspended in extraction buffer. Cell disruption can be achieved in several ways, e. g. by sonication or (probably the most efficient way for larger volumes) by high-pressure homogenization using the standard equipment also used for bacteria and yeast. For secreted products, the fermentation medium is used directly fur further processing after cell removal.

The technical design of the initial processing step for field-grown plants will, to a large extent, be dictated by the source material. Looking at the most widely used plant expression systems for recombinant protein production at the current time (tobacco leaves and maize seeds), the differences are obvious. Leaf material contains a lot of water and can, to some extent, be homogenized and extracted in its own juice through the addition of approximately 1–2 volumes of extraction buffer to control pH, keep the crude extract ready to be pumped and serve as carrier for additives for improved extraction or stabilization of the target protein [13]. Maize seeds, on the other hand, must be subjected to dry milling followed by extraction in a larger proportion (usually 2–5 volumes) of buffer [14].

Given the relatively low expression levels for recombinant pharmaceutical proteins that are currently achieved in plants, at least in comparison to optimized animal and microbial cell lines (usually much less than 100 mg of active recombinant protein per kg of plant material [15]), procedures to extract the target protein quantitatively, preserve its integrity and activity and keep process volumes as low as possible are essential. However, while minimizing the amount of added extraction buffer (which must be prepared according to GMP standards) will help reduce production cost and process time, it will also result in a higher concentration of proteases, oxidants etc. in the extract. The addition of additives may be beneficial for improved extraction or protection of the target protein, but aside from the cost of these substances, their removal during the purification process may be difficult to accomplish and validate.

Before purifying the target protein from the crude plant extract, a clarification step is required to separate particulate material from the liquid phase. Here again, plant material-specific problems have to be addressed. When leaf material is disrupted by shredding, fibers, flakes and other pigmented fine particles are inevitably generated. Even after removal of bulk cell debris e. g. by centrifugation or depth filtration, a proportion of these fine green particles will still be present in the extract and will interfere with subsequent chromatographic purification by clogging the column inlets or interacting with adsorbent beads [13]. Cross-flow microfiltration is a means to remove these fines, and in our experience hollow-fiber modules are superior to membrane cassettes for this particular application. Aside from clarification of the extract for chromatography, filtration also is a means to reduce bioburden in the extract, and depending on the source material it may be important to incorporate a filtration step for this purpose early in a purification strategy. However, this adds another unit operation to the process, resulting in a longer processing time in the early stage of the purification where it is important to keep process time short to minimize exposure of the target protein to proteolytic enzymes, polyphenols and other detrimental components then still present in the extract, in particular as it may not be feasible to work at 4 °C in a large scale process. This dilemma, for the reasons given above, is more pronounced in leaf material than in seeds and needs to be addressed on a case-by-case basis.

In addition to clarification, the extract usually has to be conditioned to match the requirements for the subsequent capture chromatography step, e. g. by reduction of conductivity or adjustment of pH. These manipulations may lead to either immediate or delayed precipitation of extract components, which have to be removed before further processing. Careful design and evaluation of the extraction procedure is therefore an important development task for every individual process.

14.4.2
Chromatographic Purification

Liquid chromatography is the core preparative technique in protein purification, and all supplementary procedures like extraction, centrifugation and filtration, ultimately serve to condition the protein solution for chromatography. A series of chromatographic steps, usually termed capture, intermediate purification and polishing, making use of different intrinsic features of proteins, is usually required to achieve sufficient separation of the target from contaminants [16,17]. Common modes of biochromatography include ion exchange chromatography, affinity chromatography, hydrophobic interaction chromatography, gel filtration and, to a limited extent, reversed phase chromatography. Method development involves selection between these modes, their arrangement in a suitable order and evaluation of their efficacy while taking into consideration the limitations of the target protein, such as incompatibility with organic solvents, susceptibility to precipitation or denaturation and loss of activity outside a certain pH and temperature range. An advantage of working with recombinant proteins is the availability of sequence information that can lead to a prediction of the protein's chromatographic behavior.

The initial chromatographic step generally aims to concentrate the highly dilute starting material and remove bulk impurities, rather than achieving a high degree of purity by high resolution and selectivity. This is reflected by the use of wide columns and adsorbent beads with a large diameter and high binding capacity, which allow the processing of large volumes of liquid at reasonably high linear flow rates and low pressures even with viscous feed or feed containing a certain amount of particulate contamination. Usually, robust techniques like ion exchange chromatography in packed bed columns are used for this step, with an additional requirement that the media are resistant to the harsh cleaning-in-place (CIP)-solutions, e.g. 0.5-1M NaOH, which have to be used to remove tightly adsorbed components and sanitize the column before re-use. Cleaning chromatography columns that have been challenged with a complex feedstream such as leaf extract is a difficult task, and we have frequently observed that pigmented plant extract components accumulate on chromatography media and reduce column lifetime even when recommended CIP procedures have been performed after each purification cycle. This is especially disadvantageous for pharmaceutical production, since all cleaning procedures have to be validated [18].

A particular strategy is usually employed for monoclonal antibody production, which is one of the most promising application areas for molecular farming. Here, highly selective, relatively stable and readily available affinity chromatography ligands (Protein A and Protein G) are routinely used for the capture of immunoglobulins from mammalian cell culture supernatants, giving high recovery and excellent purity even in the first steps of the purification process. These advantages overcompensate for the disadvantages of a protein-based affinity ligand, i.e. high media costs, limited lifetime, restrictions in the choice of CIP reagents and potential for ligand leaching from the column. Protein A affinity chromatography has been successfully adapted for purification of plantibodies on the laboratory scale [19] and preparative scale [13], but its performance with respect to media cleanability and lifetime has not yet been thoroughly evaluated. The problems associated with plant extracts, as a starting material for bioprocessing, will require specific approaches and strategies for the initial purification steps. These will aim to separate the target protein from adverse feed components more rapidly by increasing the specificity of the capture chromatography step and reducing the number of unit operations for shorter processing times. Several recent developments in downstream processing technology may contribute to this effort and are discussed below: fusion protein technologies, affinity tags, synthetic affinity chromatography ligands and expanded bed chromatography.

An impressive example of the fusion protein technique designed specifically for plant biotechnology is the oleosin fusion system, developed by SemBioSys Genetics, which utilizes the unique properties of the oleosin protein of oilseeds to participate in the formation of storage organelles (oil bodies) as the fusion partner for a target protein [20]. A complete platform for production and initial purification of recombinant proteins has been developed around this core technology, and has the potential to eliminate or reduce many of the problems associated with the early steps of downstream processing of plant material. The benefits and practical applications of this technology have been described [21]. Naturally, the technology is limited to oilseeds, narrowing the range of expression hosts, and it also excludes proteins that require post-translational processing in the endoplasmic reticulum (ER). Additionally, for biopharmaceutical production, the possibility that the oleosin fusion partner or the proteolytic cleavage step (which is part of the purification process) might interfere with the folding, solubility, stability, integrity or activity of the target protein will need to be considered. Other fusion protein technologies of potential interest, e.g. using membrane anchors, have been described, but have not yet advanced beyond the research and development level.

Affinity tags are short peptide sequences genetically fused to a recombinant protein. Several of these tags are available, the most widely used being the His$_6$-tag, i.e. six consecutive histidine residues at the N- or C-terminus of the target protein allowing its purification by immobilized metal-ion affinity chromatography (IMAC) [22]. While IMAC is less specific than other affinity methods such as Protein A- or immunoaffinity chromatography, and therefore will result in some co-adsorption of host proteins to the medium, it has the advantage of offering a group-specific affinity capture step at relatively low media cost. In contrast to protein-based affinity ligands, the reactive groups used in IMAC media are small, unaffected by proteases present in the feed and can be subjected to harsh CIP procedures. While originally developed for bacterial expression systems, His$_6$-tags are today also widely used with other expression hosts including plants [23]. For therapeutic applications, affinity tags will likely have to be removed from the final product and, as with the fusion protein technologies described above, efficiency and precision of the cleavage procedure as well as the removal of the cleavage reagent from the final product formulation will have to be demonstrated and verified. Likewise, the potentially negative impact of the tag on product quality, as has been described for some His$_6$-tagged proteins, will have to be evaluated for each individual product. Leakage of metal ions from IMAC-columns is another potential problem with this technology.

The design of synthetic affinity ligands may become an alternative to the fusion technologies described above, since it does not require modification of the expression construct. Instead, it depends on the knowledge of structural and/or functional properties of the unmodified target or on the results of library screening procedures, and the exploitation of this knowledge in biochromatography [24]. The current status of this technology in the field of biopharmaceutical manufacturing has been reviewed recently [25]. Clearly, the availability of affinity chromatography methods based on highly stable, very selective, tailor-made affinity ligands for protein purification, thus circumventing the inherent problems of proteinaceous affinity ligands, would be particularly advantageous for plant-based production, provided such ligands could be supplied in sufficient quantities for large-scale applications, could be readily integrated into the production process (i.e. in terms of binding and elution conditions, throughput, robustness, reproducibility, cleanability and validation) and were economically attractive. The ambiguous results obtained with the use of dye ligands in biopharmaceutical production [25] illustrate both the potential and the problems of this approach.

Finally in this section, we discuss expanded bed adsorption (EBA) [26], a chromatographic technique designed for preparative use in protein purification that addresses the problems of handling large volumes of particulate raw materials in the initial purification step. As stated above, traditional packed bed chromatography inevitably requires a high level of clarification of the column feed, involving laborious, expensive and time-consuming centrifugation and microfiltration steps. The goal of EBA is to allow the application of unclarified feed with a high particle burden directly to the column by innovative design of columns, flow distribution devices and matrix particles. Briefly, EBA is performed in an upward direction through the flow distributor system at the bottom of the column, ensuring plug flow throughout the cross-sectional area, and through the column tube causing the settled adsorbent bed to expand into the headspace of the column to a degree dependent on adsorbent particle density, viscosity and particle load of the feedstream. The high bead density (usually $1.2-1.8 \, \text{g ml}^{-1}$) prevents the beads from being carried out of the column. A correctly expanded bed will appear almost stationary with a height of about 2–3 times the sedimented bed height. The individual adsorbent particles exhibit small, circular movements but no turbulence or channeling when operated at the recommended linear flowrates of about $200-600 \, \text{cm h}^{-1}$. If a particulate feedstream is applied to the stably expanded bed, the particles (cells, debris, aggregates etc.) can pass through the large interstitial space between the adsorbent beads and leave the column through the upper adapter, while the target molecules are adsorbed to the active surface groups of the medium. When the feedstock has passed through the column, a washing step is performed in upward flow until the effluent is particle free. Elution can then be performed either in packed bed mode after reversing the flow direction and lowering the upper adapter to the sedimented bed surface, or in expanded bed mode using upward flow. Thus, EBA has the potential to eliminate unit operations from the downstream process by combining clarification, concentration and capture chromatography into a one-step operation.

With the exception of gel filtration, all modes of biochromatography are, in principle, adaptable to EBA. Separations based on ion exchange [27], Protein A affinity [28] and IMAC [29] have been published. The list of feed includes bacterial fermentation broth [30], cell homogenate [31] and renatured inclusion bodies [32], yeast fermentation broth [29] and cell homogenate [27] as well as mammalian or hybridoma cell culture broth [33]. EBA is also an accepted method in the production of medicinal products. Although, considering the problems in initial processing discussed earlier in this chapter, EBA appears to be an attractive alternative for capture chromatography of plant extracts, it was only recently that the first report of EBA used in the preparative-scale chromatography of plant extracts was published (Protein A-affinity purification of an antibody from several batches between 50 and 200 kg of tobacco leaves [13]). The reasons for this can be deduced from the cited publication: initial clarification steps could not be skipped because the particulate components of the extract, even after removal of bulk cell debris, blocked the column inlet and interacted with the adsorbent particles causing aggregation and destabilization of the expanded bed. These findings agree with our own experience and obviously limit the usefulness of EBA for this particular application. However, the publication also lists positive aspects of the EBA method in comparison with packed-bed chromatography, in particular reduced processing time due to higher linear flow rate and facilitated CIP of the column. Recently, the leading supplier of EBA equipment (Amersham Pharmacia Biotech) introduced a new line of expanded bed columns with a newly designed flow distribution system that may circumvent the inlet-clogging problem. The interaction between feed components and adsorbent beads causing aggregation may vary with the adsorbent matrix material, the properties of the feed needed for the particular application (pH, conductivity) or a combination of both, leaving room for further optimization. For these reasons, and because extracts e.g. from maize seeds or fermenter-grown plant cells may perform differently in EBA, this method should not yet be disregarded for use in molecular farming.

The discussion in the paragraphs above with respect to the chromatographic purification of recombinant therapeutic proteins from plant extracts addresses primarily the early steps of the process due to the influence of the unique properties of the source material. In the later stages of intermediate purification and polishing, the applied technology will not significantly differ from established biotechnological processes and the regulatory requirements for product quality and safety will be the dominant parameters influencing the purification strategy.

14.5
Regulatory Requirements for Downstream Processing of Plant-derived Pharmaceutical Products

Draft documents addressing quality aspects in the production of medicinal products made by transgenic (or "bioengineered") plants were published by both the EMEA and the FDA in 2002 [5,6] and gave a realistic impression of the significant scientific and technological hurdles that will have to be overcome before plants can be considered to be true alternatives for existing biopharmaceutical production systems up to the point of marketed medicinal products. Several of the scenarios that have been outlined in the molecular farming community regarding host plant species, expression strategies, large-scale cultivation etc. may well fail not because of lack of technical feasibility but because of the inability to meet the regulatory guidelines. Therefore, regulations may ultimately decide which plant production systems survive research and development, preclinical studies and early clinical trial phases.

The tenor of the two documents is similar, although certain aspects are emphasized differently or not addressed at all in one or the other. For example, transient expression systems are not addressed in the EMEA document, but they are in the FDA document.. We will concentrate on the EMEA "Points to Consider" and refer to the FDA "Draft Guidance" where appropriate.

It is outside the scope of this chapter to discuss the issues addressed in the section "Development Genetics", although it is obvious that some of these issues will require monitoring and analysis not only in the research and development phase, but throughout normal production. For example, the concern of both agencies about potential immunogenicity or allergenicity of plant-specific carbohydrate structures in glycoproteins will probably lead to a requirement not only to characterize in detail carbohydrate composition and structure during product development, but also to "... routinely control inter- and intra-batch variation in the active substance and finished medicinal product." Similarly, the genetic stability (or lack thereof) of the production system will affect the acceptability of manufacturing batches with regard to lot-to-lot consistency. However, these are potential problems that cannot be addressed and solved by improvement of downstream processing strategies.

The situation is different for some of the topics in the "Cultivation and Harvesting" section, insofar as adventitious contamination by toxins, pesticides, microbes, parasites, viruses etc. is addressed. The detection, control and removal of adventitious contaminants are some of the key tasks for downstream processing and the associated process analytical technology. Its complexity will vary with the degree of containment (fermenter, greenhouse or open field) applied during cultivation and also, as stated earlier in this chapter, on the expression system (leaf *vs.* seed). It will be a major challenge to develop, establish for routine use and validate the analytical methods necessary for comprehensive monitoring of this sort of contamination in field-grown plants.

The risk of viral contamination in plant-based medicinal products, and requirements for strategies to ensure that the product is consistently free of contaminating viruses, is discussed in detail in the EMEA document, while it is not addressed by the FDA. In addition to contamination by insect, bird and animal excreta or carcases, organic fertilizer, production personnel and equipment, the EMEA document lists *plant* virus infection as a source of contamination and claims that "... freedom from contamination with all types of viruses, irrespective of natural tropism, should be demonstrated."

Obviously, the viral safety of plant-derived biopharmaceuticals is a major point of concern for the EMEA, and the requirements for strategies to ensure viral safety, in particular for field-grown plants, are accordingly strict. Such requirements include tests on starting materials, reagents and excipients as well as validated tests (*in vivo* and *in vitro*) on unprocessed and processed bulk and the inclusion of effective, validated virus clearance steps in the purification process. This assessment of viral safety issues in transgenic plant technology, if adopted into the final points to consider document (to be released in 2004), will have great implications not only for downstream processing, but also for the concept of plant-based pharmaceuticals in field-grown plants as a whole, since the absence of human-pathogenic viruses in plants has always been considered and promoted to be an important beneficial feature of plant-based expression systems.

In the "Post-harvest Processing, Formulation, Filling, and Assembly" section of the EMEA document, it is stressed that purification processes should be specified and validated in accordance with the established principles for biotechnological medicinal production and that the manufacturing process, including procedures, equipment and materials should be justified and validated – more or less a paraphrase of the statement that the manufacture has to be carried out in compliance with GMP. The FDA document addresses the initial stages of downstream processing (initial processing and extraction) in more detail, emphasizing the need for early bioburden reduction and filter sterilization.

Both documents then describe in detail the expected analytical test and specification strategies for product characterization in terms of identity, purity, potency and quantity. The EMEA document refers, for further consultation, to established CPMP and *European Pharmacopoeia* precedents and models for similar products. The FDA document refers to the ICH Q6B guideline on test procedures and acceptance criteria for biotechnological/biological products. Many of the applicable test strategies and methods will not differ significantly from the requirements for other biotechnological production systems, but plant-specific analyses include testing for process related impurities such as plant proteases, secondary metabolites (alkaloids, glycosides), adventitious agents (mycotoxins, pesticides, toxic metals) and bioburden as well as for substance-related impurities, including glycoforms. Again, the development and validation of these analytical technologies will be a major effort.

We end this chapter by citing the concluding comment in the EMEA document discussed above: "Transgenic plant technology may provide interesting possibilities for extending the range of recombinant DNA production systems available for consideration by biopharmaceutical manufacturers. The challenge appears to be to emulate the quality attributes of established medicinal products produced in banked microbial and mammalian cell culture systems."

References

[1] E.E. Hood, A. Kusnadi, Z. Nikolov et al., *Adv. Ex. Med. Biol.* **1999**, *464*, 127–147.

[2] H.S. Mason, H. Warzecha, T. Mor et al., *Trends Mol. Med.* **2002**, *8* (7), 324–329.

[3] S. Schillberg, R. Fischer, N. Emans, *Cell. Mol. Life Sci.* **2003**, *60* (3), 433–445.

[4] European Union, *Clinical Trials Directive* (2001/20/EC), Official Journal of the European Communities L121, pp34–44, **2001**.

[5] CPMP, *Points to Consider on Quality Aspects of Medicinal Products Containing Active Substances Produced by Stable Transgene Expression in Higher Plants.* CPMP/BWP/764/02 (Draft), The European Agency for the Evaluation of Medicinal Products (EMEA), **2002**.

[6] FDA, *Guidance for Industry: Drugs, Biologics, and Medical Devices Derived from Bioengineered Plants for Use in Humans and Animals (Draft Guidance).* United States Food and Drug Administration, **2002**.

[7] R.V. DATAR, T. CARTWRIGHT, C.G. ROSEN, *Bio/Technology* **1993**, *11* (3), 349–357.

[8] C. J. A. DAVIS, IN: G. SUBRAMANIAN (ed) *Bioseparation and Bioprocessing Vol I*. Wiley-VCH, Weinheim, pp 125–143, **1998**.

[9] G. SOFER, IN: G. SUBRAMANIAN (ed) *Bioseparation and Bioprocessing Vol I*. Wiley-VCH, Weinheim, pp 497–511, **1998**.

[10] R. FISCHER, N. EMANS, F. SCHUSTER et al., *Biotechnol. Appl. Biochem.* **1999**, *30* (2), 109–112.

[11] E. STOGER, C. VAQUERO, E. TORRES et al., *Plant Mol. Biol.* **2000**, *42* (4), 583–590.

[12] R. A TATICEK, C. W. LEE, SHULER, M. L., *Curr. Opin. Biotechnol.* **1994**, *5* (2), 165–174.

[13] R. VALDES, L. GOMEZ, S. PADILLA et al., *Biochem. Biophys. Res. Commun.* **2003**, *308* (1), 94–100.

[14] M.R. BAILEY, S.L. WOODARD, E. CALLAWAY et al., *Appl. Microbiol. Biotechnol.* **2004**, *63* (4), 390–397.

[15] H. DANIELL, S.J. STREATFIELD, K. WYCOFF et al., *Trends Plant Sci.* **2001**, *6* (5), 219–226.

[16] R. FREITAG, C. HORVATH, *Adv. Biochem. Eng. Biotechnol.* **1996**, *53*, 17–59.

[17] R. L. FAHRNER, H. L. KNUDSEN, C. D. BASEY, et al., *Biotechnol. Genet. Eng. Rev.* **2001**, *18*, 301–327.

[18] G. SOFER, *Dev. Biol.* **2003**, *113*, 61–64.

[19] R. FISCHER, Y.C. LIAO, J. DROSSARD, *J. Immunol. Methods* **1999**, *226* (1–2), 1–10.

[20] G. J. H. VAN ROOIJEN, M. M. MOLONEY, *Bio/Technology* **1995**, *13* (1), 72–77.

[21] C. L. CRAMER, J. G. BOOTHE, K.K. OISHI, *Curr. Top. Microbiol. Immunol.* **1999**, *240*, 95–118.

[22] A. SKERRA, I. PFITZINGER, A. PLUCKTHUN, *Bio/Technology* **1991**, *9* (3), 273–278.

[23] C. VAQUERO, M. SACK, F. SCHUSTER et al., *FASEB J.* **2002**, *16* (3), 408–410.

[24] C. R. LOWE, A. R. LOWE, G. GUPTA, *J. Biochem. Biophys. Methods* **2001**, *49* (1–3), 561–574.

[25] N. E. LABROU, *J. Chromatogr. B* **2003**, *790* (1–2), 67–78.

[26] H. A. CHASE, *Trends Biotechnol.* **1994**, *12* (8), 296–303.

[27] Y. K. CHANG, H. A. CHASE, *Biotechnol. Bioeng.* **1996**, *49* (2), 204–216.

[28] R.L. FAHRNER, G.S. BLANK, G.A. ZAPATA, *J. Biotechnol.* **1999**, *75* (2–3), 273–280.

[29] S. HELLWIG, F. ROBIN, J. DROSSARD et al., *Biotechnol. Appl. Biochem.* **1999**, *30* (3), 267–275.

[30] M. HANSSON, S. STAHL, R. HJORTH et al., *Bio/Technology* **1994**, *12* (3), 285–288.

[31] R. H. CLEMMITT, H. A. CHASE, *Biotechnol. Bioeng.* **2000**, *67* (2), 206–216.

[32] A.-K. B FREJ, *Bioseparation* **1999**, *6* (5), 265–271.

[33] J. THOMMES, A. BADER, M. HALFAR et al., *J. Chromatogr. A* **1996**, *752* (1–2), 111–122.

15
Glycosylation of Plant-made Pharmaceuticals

Véronique Gomord, Anne-Catherine Fitchette, Patrice Lerouge and Loïc Faye

15.1
Introduction

Most therapeutic proteins are glycoproteins, and in this chapter we discuss the advantages and limitations of glycosylation when mammalian proteins and particularly antibodies are produced in plant expression systems.

15.2
Plant Cells can Reproduce the Complexity of Mammalian Proteins

The demand for biopharmaceuticals and particularly therapeutic antibodies is rapidly increasing, and therefore pharmaceutical companies are interested in transgenic production technologies as an alternative to traditional production techniques using cultured mammalian cells. Only animal cells, transgenic animals and plants are able to associate, via disulfide bridges, the light and heavy chains of an antibody (Fig. 15.1). Plant cells can reproduce the complexity of these proteins, as shown in 1989 when a functional antibody was produced for the first time in tobacco plants [1]. Since this pioneering demonstration from Dr. Hiatt's group, many antibodies and antibody fragments have been produced, for therapeutic or diagnostic purposes, in various plant systems (Table 15.1).

Similarities between the protein biosynthesis and maturation machineries of mammals and plants are well illustrated by the ability of plants to produce these various types of recombinant antibodies. Antibodies are complex molecules. Immunoglobulins of the IgG class are tetramers consisting of two identical polypeptides 450 amino acids in length (heavy chains) and two identical polypeptides 250 amino acids in length (light chains). These are linked together by several disulfide bridges. The complexity of a secretory IgA (sIgA) is even greater as these immunoglobulins comprise two IgA molecules linked together by two additional polypeptides (Fig. 15.1). While the assembly of a sIgA requires two different cell types in mammals, such molecules have been produced in a biologically active form in transgenic plants. The

Molecular Farming. Edited by Rainer Fischer, Stefan Schillberg
Copyright © 2004 WILEY-VCH Verlag GmbH & Co. KGaA, Weinheim
ISBN: 3-527-30786-9

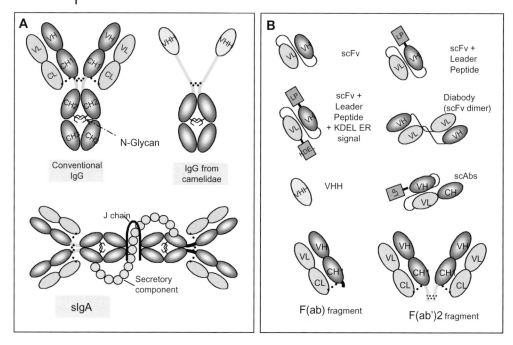

Fig. 15.1 Antibodies and antibody fragments produced in plant systems. **A-** In the conventional IgG, the constant heavy-chain CH1, CH2, CH3 and the variable heavy-chain (VH) domains are shown in dark grey. The constant light-chain (CL) and the variable Light-chain (VL) domains are in light grey. Some camelidae IgG lack a light chain. Their antigen-binding domains consist only of the heavy-chain variable domain (VHH). A secretory IgA (sIgA) is a dimeric IgA made of ten different polypeptides, four light-chains, four heavy-chains, a J-chain that facilitates dimerization and a secretory component. **B-** In the recombinant single-chain Fv fragment (scFv), the variable domains of the light and heavy chains are joined by a peptide linker. scFvs can be complexed into dimers (diabodies), trimers or tetramers (not illustrated). Diabodies have two functional antigen binding domains with similar or distinct antigen specificities. The VHH antigen-binding fragment of IgG from camelidae is highly soluble and stable. The F(ab) fragment contains the light chain and the VH and CH1 domains of the heavy chain. F(ab')$_2$ is divalent and contains two F(ab) fragments linked by a disulfide bond

correct folding and assembly of plant-made antibody molecules within the endoplasmic reticulum (ER), through interactions with a number of chaperones, processing and glycosylation enzymes, demonstrate that co- and post-translational protein maturation events are similar in plants and in mammals. However, further work is still needed to explain why secretory IgA/Gs, which are expected to be secreted to the tobacco cell apoplast, have been found in different subcellular compartments such as the vacuole and the ER [2]. This heterogeneous distribution of a recombinant anti-

Tab. 15.1 Therapeutic antibodies produced in transgenic plants. See Fig. 15.1 for structures of the different antibodies and antibody fragments presented in the second column. Abbreviations: TSP, total soluble protein; FLW, fresh leaf weight; MSP, murine signal peptide; SP, signal peptide; KDEL, ER retention signal. In the column "Expression" + and – means respectively that the antibody was or was not expressed.

Antigens	Type of antibodies	Indications	Transformed plant	Promoter	Targeting signal	Expression	References
Human carcinoembryonic antigen	Mouse/human chimeric IgG1 antibody (cT84.66)	Antibody – mediated cancer therapy (colon cancer, breast cancer and tumor with epithelial origin)	N. tabacum cv petit Havana SR1 (Transient expression)	35S	MSP	+ 1 mg/kg FLW	42
					MSP + KDEL	+	
	scFv T84.66				MSP	+ 5 mg/kg FLW	
	T84.66/G68 diabody		N. tabacum cv petit Havana SR1	35S	Plant codon optimized SP	+ 1–5 mg/kg FLW	43
					SP+ KDEL	+ 4–12 mg/kg FLW	
	scFvT84.66		T. aestivum L. cv bobwhite	Maize ubi1 or 35S	MSP	+ Wheat and rice: 30–100 ng/g FLW	44
			O. sativa L. indica cv M12 and Bengal		MSP+ KDEL	+ Wheat: 50–900 ng/g FLW Rice: 1.5–29µg/g FLW	
Rabies virus protein	Monoclonal antibody (mAb SO57)	Rabies virus neutralization	N. tabacum cv Xanthi	35 S	MSP +KDEL	+ 3 µg/g FLW (0.07% TSP)	33
Human IgG monoclonal antibody	C5 IgG	Anti-human globulin re-agent for phenotyping and crossmatching red blood cells of receivers and donors	M. sativa	35 S	MSP	+ 0.13–1.0% TSP	38

Tab. 15.1 (continued)

Antigens	Type of antibodies	Indications	Transformed plant	Promoter	Targeting signal	Expression	References
Streptococcal surface antigen SAI/II	Guy's 13 IgG IgA/G	Tooth decay	N. tabacum	35S	MSP	+	45
	sIgA/G					+ 200–500 µg/g FLW (5–8% TSP)	46–48
Colon cancer surface antigen	CO-17A IgG	Antibody – mediated cancer therapy	N. benthamiana		MSP / MSP + KDEL		49
HSV-2, protein from herpes simplex virus (HSV)	IgG, IgA, dIgA or sIgA	Immunoprotection against genital herpes and vaginal transmission of HSV	O. sativa			+	50
	IgG1 Fab et F(ab')2		G. max			+	51
					Plant PR1-b SP	+	
Human creatine kinase -MM	MAK33 IgG1	Cardiac disease, mitochondrial disorders, inflammatory myopathies, myasthenia, polymyositis, McArdle's disease, NMJ disorders, muscular dystrophy, ALS, hypo and hyperthyroid disorders, central core disease, acid maltase deficiency, myoglobinuria, rhabdomyolysis, motor neuron diseases,	A. thaliana		A. thaliana 2S2 seed storage protein SP	+ 0.02–0.4% TSP of fresh leaf extract (10–12% TSP of intercellular fluid)	52
	Fab fragment		A. thaliana	35S		+ 1.3% TSP	53
			N. tabacum			0.044% TSP	

Tab. 15.1 (continued)

Antigens	Type of antibodies	Indications	Transformed plant	Promoter	Targeting signal	Expression	References
	MAK33 scFv	rheumatic diseases, and other that create elevated or reduced levels of creatine kinase.	N. tabacum SR1	35S	No SP	+ 6–25 ng scFv/mg TSP	54
					A. thaliana 2S2 seed storage protein SP	+ 8–55 ng scFv/mg TSP	
	MAK33 Fab fragment		A. thaliana	35S	A. thaliana 2S2 seed storage protein SP	+ 6.53% TSP (0.02% TSP in seed)	55
					SP + KDEL	+ 5.9% TSP (0.015% TSP in seed)	
Tumor surface Ig	38C13 scFv	B-cell lymphoma treatment	N. benthamiana	TMV coat protein	Rice α-amylase SP	+ 60 μg/mL of intercellular medium	56
Hepatitis B virus surface antigen (HBsAg)	scFv	Immunoaffinitypurification of recombinan HBsAg	N. tabacum cv petit Havana SR1	35S	No SP	–	57
					Sweet potato Sporamin SP	+ 0.031% TSP	
					Sporamin SP + PP	+ 0.032% TSP	
					Sporamin SP + KDEL	+ 0.22% TSP	
Zearalenone (mycotoxin)	scFv	Passive immunization of animals in their feed	A. thaliana (ecotype Columbia)	Inducible lac	No targeting signal	–	58
					Plant PR1-b SP	+	

body in a plant cell could reflect either improper folding or mis-targeting, and could result in a significant degree of structural heterogeneity in plant-made pharmaceuticals (PMPs) due to different proteolytic maturations occurring on the proteins in the ER, vacuole and apoplast. N-glycan maturation also depends on the localization of a glycoprotein in the secretory pathway [3], and the high level of heterogeneity observed in the glycosylation of IgGs produced in tobacco could also be the result of this heterogeneous distribution [4,5]. In contrast, an IgG produced in alfalfa showed a very homogeneous N-glycan structure, which indicates more efficient secretion and/or folding machinery in this plant expression system (see [6] and chapter 1).

15.3
Plant-made Pharmaceuticals and their Native Mammalian Counterparts Contain Structurally-distinct N-linked Glycans

The ability of plant cells to assemble complex mammalian proteins such as human collagens [7], human growth hormone [8] and antibodies (Table 15.1) clearly illustrates the potential of the plant system for the production of most biopharmaceuticals. However, many therapeutic proteins are glycoproteins and glycosylation is often essential for their stability, solubility, folding and biological activity. When a mammalian glycoprotein is produced in a plant expression system, it is glycosylated on the same Asn residues as it would be in mammals, but its N-glycan structures are different from that of its native counterpart. In plant cells, as in other eukaryotic cells, N-glycosylation begins in the ER through the co-translational addition of an oligosaccharide precursor ($Glc_3Man_9GlcNAc_2$) to specific Asn residues found in the context of potential N-glycosylation-specific sequences (Asn-X-Ser/Thr). Once transferred onto the nascent protein, and while the glycoprotein is transported along the secretory pathway, the N-linked oligosaccharide (N-glycan) undergoes several maturation reactions involving the removal and addition of sugar residues in the ER and the Golgi apparatus (Fig. 15.2). It is only in the late Golgi apparatus that plant and mammalian N-glycan maturation differs, resulting in the addition of core $\alpha(1,6)$-linked fucose and terminal sialic acid residues in mammals, and the addition of bisecting $\beta(1,2)$-xylose and core $\alpha(1,3)$-fucose residues in the plant N-glycans, as shown in Fig. 15.3. These differences are apparent when the glycosylation of antibodies produced in tobacco plants is analyzed. As shown in Fig. 15.4, when the monoclonal antibody Guy's 13 is produced in mammalian cells, it is glycosylated on both of its N-glycosylation sites with oligosaccharides containing core $\alpha(1,6)$-fucose, and about 10% of the glycans contain terminal sialic acid. When produced in transgenic tobacco plants, Guy's13 is also glycosylated on both N-glycosylation sites but the glycan structures are very heterogeneous. A mixture of high-mannose type and complex glycans is present. The high-mannose type glycans contain 5–8 mannose residues, and the complex glycans show the structural characteristics typical of plants, including the presence of bisecting $\beta(1, 2)$-linked xylose and core $\alpha(1,3)$-linked fucose [4]. Similar glycan heterogeneity and structural characteristics have been described for another monoclonal antibody (Mgr48) produced in tobacco [9]. In contrast, when the

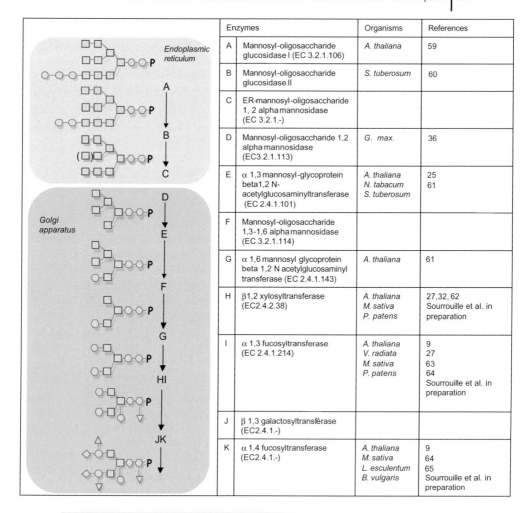

	Enzymes	Organisms	References
A	Mannosyl-oligosaccharide glucosidase I (EC 3.2.1.106)	A. thaliana	59
B	Mannosyl-oligosaccharide glucosidase II	S. tuberosum	60
C	ER-mannosyl-oligosaccharide 1, 2 alpha mannosidase (EC 3.2.1.-)		
D	Mannosyl-oligosaccharide 1,2 alpha mannosidase (EC3.2.1.113)	G. max.	36
E	α 1,3 mannosyl-glycoprotein beta1,2 N-acetylglucosaminyltransferase (EC 2.4.1.101)	A. thaliana N. tabacum S. tuberosum	25 61
F	Mannosyl-oligosaccharide 1,3-1,6 alpha mannosidase (EC 3.2.1.114)		
G	α 1,6 mannosyl glycoprotein beta 1,2 N acetylglucosaminyl transferase (EC 2.4.1.143)	A. thaliana	61
H	β1,2 xylosyltransferase (EC2.4.2.38)	A. thaliana M. sativa P. patens	27,32, 62 Sourrouille et al. in preparation
I	α 1,3 fucosyltransferase (EC 2.4.1.214)	A. thaliana V. radiata M. sativa P. patens	9 27 63 64 Sourrouille et al. in preparation
J	β 1,3 galactosyltransférase (EC2.4.1.-)		
K	α 1,4 fucosyltransferase (EC2.4.1.-)	A. thaliana M. sativa L. esculentum B. vulgaris	9 64 65 Sourrouille et al. in preparation

⬡ GlcNAc ☐ Man ⬡ β1,2Xyl ▽ α1,3Fuc ◇ β1,3 Gal ◯ Glc ▽ α1,4Fuc

Fig. 15.2 Processing of N-glycans in plants. N-glycosylation of plant proteins begins in the endoplasmic reticulum (ER) with the transfer of an oligosaccharide precursor $Glc_3Man_9GlcNAc_2$ to specific Asn residues. This precursor is then modified by glycosidases and glycosyltransferases mainly in the ER and the Golgi apparatus during the transport of the glycoprotein through the secretory pathway. Glycosidases and glycosyltransferases responsible for plant N-glycan maturation are indicated from A to K on the left panel. Most of these enzymes have been recently cloned from different plants as indicated on the right panel.

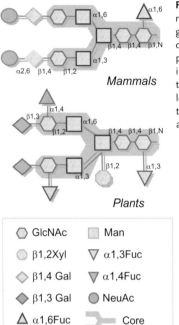

Fig. 15.3 Plant and mammalian N-glycans have different structures. As illustrated here, a core structure (in gray) is common to plant and mammalian biantennary complex N-glycans. However, differences in the glycan processing machineries in plants and in mammals result in the absence of sialic acids in the terminal position of the antennae and the presence of a bisecting β(1,2)-xylose and of an α(1,3)-fucose residue in PMPs instead of the α(1,6)-fucose linked to the proximal N-acetylglucosamine of native mammalian N-glycans.

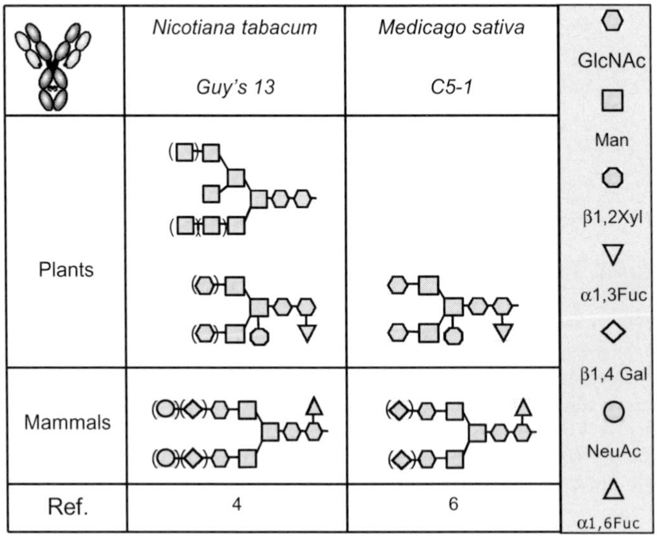

Fig. 15.4 Structure of glycans N-linked to IgG molecules expressed in hybridomas and transgenic plants. Glycans N-linked to plant-derived antibodies are structurally different from their mammalian counterparts. In contrast with antibodies produced in alfalfa, antibodies produced in tobacco plants present a very high glycan heterogeneity.

C5–1 antibody was produced in alfalfa, the glycan component consisted predominantly of a mature oligosaccharide comprising a core $\alpha(1,3)$-fucose residue, a bisecting $\beta(1,2)$-xylose residue and two terminal GlcNAc residues (Fig. 15.4) [6].

15.4
Plant-made Pharmaceuticals Possess Immunogenic N-glycans

While the homogeneity of PMP glycosylation may differ from one plant expression system to the other, all plant species examined thus far for the production of PMPs have the capacity to add the bisecting $\beta(1,2)$-xylose and core $\alpha(1,3)$-fucose residues onto complex N-glycans [4,6,10]. These residues are the constituent glyco-epitopes known to be important as IgE-binding carbohydrate determinants of plant allergens [11–15]. More importantly, it was recently shown that plant N-glycans containing these glyco-epitopes not only show IgE binding activity, but they also cause the release of mediator by human basophils, when at least two of these N-glycans are present on a same protein [16]. Ourselves and others have also reported that the immunization of goats [17] or rabbits [18] with plant-derived glycoproteins elicits the production of antibodies specific for glyco-epitopes containing bisecting $\beta(1,2)$-xylose or core $\alpha(1,3)$-fucose residues. More recently, *in vivo* experiments using BALB/c mice have shown that the administration of antibodies produced in tobacco does not elicit an immunological response against the plant-derived N-linked glycans [19].

The data obtained in laboratory mammals raise the question of the immunogenicity of these glyco-epitopes in the context of a human therapy using PMPs. We recently addressed this issue by re-investigating the immunogenicity of such glyco-epitopes in rodents. We found that immunization with a model glycoprotein, horseradish peroxidase, elicits the production of antibodies specific for $\alpha(1,3)$-fucose and $\beta(1,2)$-xylose-containing glyco-epitopes in C57BL/6 mice and rats, but not in BALB/c mice. Furthermore, we demonstrated that the sera of about 50% of non-allergic human blood donors carry antibodies specific for $\beta(1,2)$-xylose, and that about 25% carry antibodies against core $\alpha(1,3)$-fucose [20]. These antibodies probably result from sensitization to environmental antigens. Although the immunological significance of anti-$\alpha(1,3)$-fucose and anti-$\beta(1,2)$-xylose antibodies is currently a matter of speculation, the presence of such antibodies may at least induce a rapid immune clearance of glycosylated PMPs from the blood stream, which may greatly compromise their effectiveness as *in vivo* therapeutic agents. In addition to accelerated clearance, clinical effects resulting from the immune response caused by the administration of plant-derived therapeutic glycoproteins are also questionable. As a consequence, for a more detailed evaluation of safety concerns relating to the use of plant-derived therapeutic glycoproteins, further experiments have to be carried out in appropriate animal models as well as in humans by administering therapeutic glycoproteins produced in plant and analyzing the immune responses to the plant glyco-epitopes in allergic and non-allergic populations.

Plants are not the only heterogeneous expression system to produce potentially immunogenic N-glycans. When antibodies are produced in non-human mammalian expression systems, they also contain non-human sugar residues such as the N-glycosylneuraminic acid (Neu5Gc) form of sialic acid (antibodies produced in CHO cells and in milk) or terminal α(1,3)-galactose (antibodies produced in murine cells). It has been shown that antibodies containing these sugar residues can also provoke undesirable side effects, including an immune response in humans. The genetic manipulation of CHO cells has been carried out in an attempt to reduce the amount of Neu5Gc in recombinant glycoproteins (see [21] for a recent example). Similarly, the following sections of this chapter describe current efforts to prevent the addition of immunogenic N-glycans to PMPs.

15.5
Current Strategies to Eliminate Immunogenic N-glycans from Plant-made Pharmaceuticals

In order to fully exploit the potential of plants for the production of recombinant therapeutic glycoproteins, it will be necessary to control the maturation of plant-specific N-glycans and thus prevent the addition of immunogenic glyco-epitopes onto PMPs. One of the most drastic approaches is to prevent N-glycosylation all together, by inactivating N-glycosylation sites through the mutation of Asn or Ser/Thr residues. This strategy does not influence the antigen-binding activity of many antibodies used in diagnostic or drug delivery to cancer cells. However, many pharmaceuticals, including antibodies used for their effector functions (such as immune response triggering [22]), require glycosylation for *in vivo* activity and longevity. It was also recently shown that the addition of N-glycans to several recombinant protein or glycoprotein therapeutics increases their *in vivo* activity and half-life. This further illustrates a current tendency in glycoengineering to increase, and not to reduce, the number of glycosylation sites on recombinant pharmaceuticals [23].

Promising results have already been obtained in the production of plant-made glycosylated therapeutic proteins bearing non-immunogenic N-glycans. One of these strategies is based on the inhibition of Golgi glycosyltransferases. The analysis of an *Arabidopsis thaliana* mutant has shown that the inactivation of only one glycosyltransferase, N-acetylglucosaminyltransferase I (GnTI), is sufficient to block the biosynthesis of complex N-glycans in this plant [24]. This glycosyltransferase has been cloned from several other plants (Fig. 15.2), but expression of a GnTI antisense construct failed to inhibit immunogenic N-glycan biosynthesis completely in tobacco and potato [25]. Despite its low efficiency in the prevention of plant glyco-epitope biosynthesis, this pioneer study has stimulated the interest of several laboratories and molecular farming companies, which are now seeking to characterize plant glycosyltransferases responsible for N-glycan maturation. Genes encoding many of these enzymes, especially targets for the inactivation of glyco-epitope biosynthesis (such as β(1,2)-xylosyltransferase and α(1,3)-fucosyltransferase), have been cloned within the past five years in several plant expression systems (Fig. 15.2). In the near future, the development of strategies allowing an efficient inhibition of these glycosyltransferases will prevent the addition of glyco-epitopes to therapeutic proteins produced in higher plants. Already, results obtained in the moss *Physcomitrella patens* have paved the way to these inactivation strategies in higher plants [26]. N-glycosylation is very similar in mosses and higher plants, and *P. patens* genes for α(1,3)-fucosyltransferase and β(1,2)-xylosyltransferase show, 50% and 38% identity respectively, to those from *A. thaliana*. *P. patens* is the only known plant system which shows a high frequency of homologous recombination. This strategy has been used to knock out α(1,3)-fucosyltransferase and β(1,2)-xylosyltransferase genes, thus eliminating the plant-derived glyco-epitopes without any effect on protein secretion [27].

ER-resident proteins bear exclusively high-mannose type N-glycans [28–30]. These oligosaccharide structures are common to plants and mammals, and, for this reason, they are probably not immunogenic. This observation has suggested a second strategy to prevent the addition of immunogenic glycans to PMPs in which recombinant proteins are retained within the ER, i.e. upstream of the Golgi cisternae, therefore preventing the addition of immunogenic glyco-epitopes to maturing plant N-glycans (Fig. 15.5). Addition of the sequence H/KDEL to the C-terminus of a recombinant protein is sufficient for its retention in the plant ER [31]. In this manner, we have shown that a model secretory protein (cell wall invertase) fused to an HDEL retrieval sequence is efficiently retained within the ER. A detailed structural analysis has shown that the invertase-HDEL fusion protein contains predominantly high-mannose type N-glycans but also a detectable level of glycans contain-

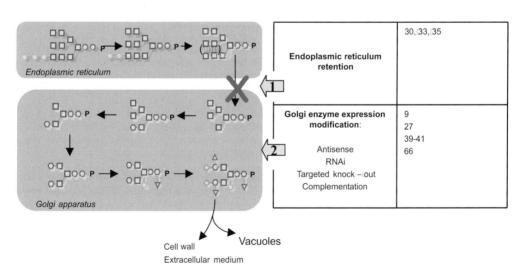

Fig. 15.5 Current strategies to reduce the immunogenicity of and/ or to humanize plant N-glycans. Strategies to reduce structural differences between plant and mammalian N-glycans are based either on retention in the ER, inactivation of endogenous plant glycosyltransferases or expression of mammalian glycosyltransferases in the plant Golgi apparatus. References on the right panel correspond to major publication related to these approaches. Results obtained using these strategies are illustrated in Figs 15.6 and 15.7.

ing the β(1,2)-xylose epitope [30]. This indicates that a small amount of this protein is transported through the secretory pathway at least as far as the medial Golgi (where β(1,2)-xylosyltransferase is located [32]) and then retrieved and transported back to the ER.

Similar results were recently obtained with a human monoclonal antibody, with KDEL sequences fused to the C-termini of both heavy chains, expressed in tobacco [33]. As observed for the invertase-HDEL fusion, about 90 % of the N-linked glycans on this antibody were of the high-mannose type, with 6–9 mannose residues, while a fraction contained the immunogenic β(1,2)-xylose glyco-epitope (Fig. 15.6). However, this antibody was not α(1,3)-fucosylated, a glycan modification occurring in the trans Golgi [34].

The efficiency of the ER-retrieval process is further increased when the KDEL sequence is fused to both the heavy and light chains of the antibody. This was clearly illustrated when a chimeric mouse-human antibody harboring four KDEL retrieval signals on the fully assembled H2L2 form was expressed in tobacco. As shown in Fig. 15.6, this antibody contained exclusively high-mannose type N-glycans with 6–9 mannose residues [35] indicating very efficient recycling based on a N-glycan maturation limited to enzymes located in the ER and cis-Golgi, such as α-mannosidase I [36].

Nicotiana tabacum cv xanthi	*Nicotiana tabacum* cv petit havana SR1
ER retention +KDEL on both heavy chains	**ER retention** +KDEL on the heavy and light chains
90%	100%
33	35

○ GlcNAc   ▦ Man   ◐ β1,2Xyl

Fig. 15.6 Glycosylation of antibodies produced with ER retention signals in tobacco plants. Antibodies fused with two (left panel) or four (middle panel) KDEL-ER retention signals are respectively or exclusively glycosylated with high mannose type, and probably non–immunogenic, N-glycans.

Interestingly, these results demonstrate the possibility that the addition of immunogenic N-glycans to PMPs can be prevented through the use of ER retention signals. However, when compared to their native mammalian counterparts, antibodies produced using this strategy are generally unstable following injection into mice [33]. Such antibodies, with high-mannose type N-glycans, are probably rapidly degraded after binding to the mannose receptor resulting in endocytosis by macrophages as previously observed for antibodies produced in *Lec1* mutant CHO cells [37]. In a similar clearance assay, plant-derived antibodies with complex N-glycans containing bisecting $\beta(1,2)$-xylose or core $\alpha(1,3)$-fucose residues [6] are as stable as their mammalian counterparts in the bloodstream of mice following intramuscular injection [38].

15.6
Towards Humanized N-glycans on PMPs Through the Expression of Mammalian Glycosyltransferases in the Plant Golgi Apparatus

In addition to approaches involving glycosyltransferase inactivation, another attractive strategy to humanize plant N-glycans is to express mammalian glycosyltransferases in plants, which would complete and/or compete with the endogenous machinery of N-glycan maturation in the plant Golgi apparatus. As part of these complementation strategies (summarized in Fig. 15.7) it was hypothesized that the expression of a human $\beta(1,4)$-galactosyltransferase, in the Golgi of plant cells, could lead to a partial humanization of plant N-glycans. Furthermore, it was suggested that the new reaction might compete with the addition of bisecting $\beta(1,2)$-xylose or core $\alpha(1,3)$-fucose. According to this hypothesis, we have shown that the human $\beta(1,4)$-galactosyltransferase, expressed in plant cells, transfers galactose residues onto the terminal N-acetylglucosamine residues of plant N-glycans. Moreover, 30% of N-glycans carried on an antibody, produced in tobacco plants expressing this human galactosyltransferase, bear terminal N-acetyllactosamine sequences identical to those associated to the N-glycans of an antibody produced in mammalian cells (Fig. 15.7) [9]. However, since the N-glycans carried by the tobacco-derived antibodies are very heterogeneous, the action of the human $\beta(1,4)$-galactosyltransferase on this pool of glycans resulted in a highly complex mixture of N-glycans, some partially humanized [9, 39–41]. These strategies developed to glycoengineer plant-made antibodies would be more efficient in plant systems such as alfalfa, where the N-glycosylation of antibodies is restricted to a predominant mature oligosaccharide chain harboring terminal GlcNAc residues. These glycans present the perfect structures for *in vitro* or *in vivo* remodeling to produce human compatible glycan structures. As a proof of this concept, we have shown that *in vitro* galactosylation of an alfalfa-derived antibody, using a $\beta(1,4)$-galactosyltransferase, resulted in the efficient conversion of the plant N-glycan into oligosaccharides having homogeneously galactosylated antennae identical to those of the murine antibody [6].

These results are very promising and several laboratories are currently working to increase the performance of heterologous glycosyltransferases (particularly of hu-

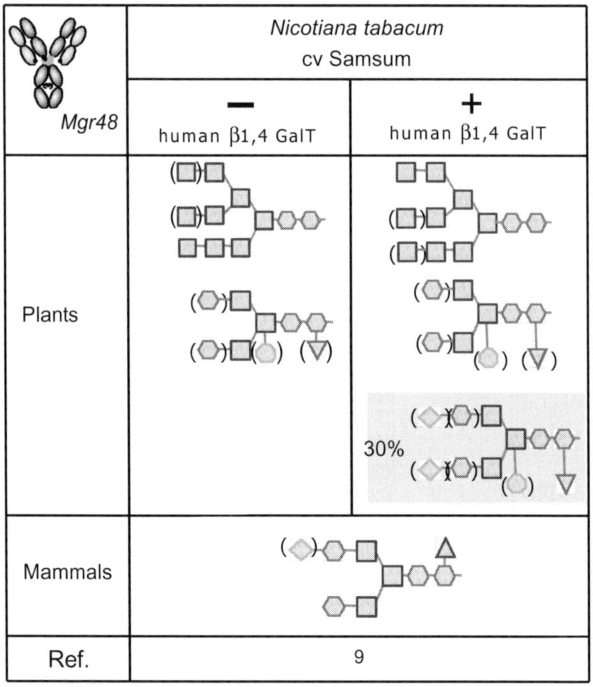

Fig. 15.7 Glycosylation of an antibody produced in tobacco plants expressing a human β(1,4)-galactosyltransferase. As illustrated for Guy's13 in Fig. 15.4, when the monoclonal antibody Mgr48 is produced in wild type tobacco plants (left panel), its glycosylation is structurally different and more heterogeneous than that of its mammalian counterpart (lower panel). When this antibody is produced in tobacco plants expressing the human galactosyltransferase (right panel), 30% of its N-glycans show terminal N-acetyllactosamine sequences identical to those carried by this antibody when it is produced in hybridoma cells.

man galactosyltransferase) through a better control of their targeting in the Golgi cisternae. As is the case for their mammalian counterparts, plant glycosyltransferases are type II membrane proteins. A detailed characterization of β(1,2)-xylosyltransferase from *A. thaliana* at the molecular level has shown that the first 36 amino acids of this glycosyltransferase (i.e. the cytosolic tail plus transmembrane domain) are sufficient for its Golgi retention and also contain sub-Golgi compartment targeting information (Fig. 15.8) [32]. The analysis of several other glycosyltransferases is currently providing a panel of specific signals sufficient for a protein targeting within the different Golgi subcompartments. These signals will help to target exogenous glycosyltransferases in the plant Golgi apparatus for optimal efficiency in the engineering of glycosylation pathway. This has been shown by the expression of a fusion protein containing the first 54 amino acids of *A. thaliana* β(1,2)-xylosyltransferase and the catalytic domain of human β(1,4)-galactosyltransferase [5].

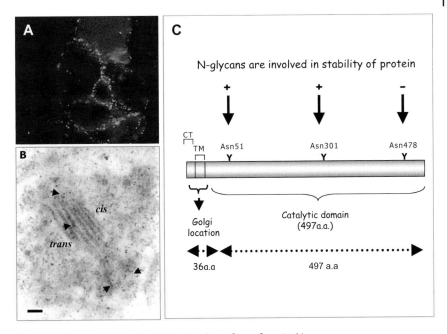

Fig. 15.8 Characterization of a β(1,2)-xylosyltransferase from *Arabidopsis thaliana* (AtXylT). AtXylT was expressed in Sf9 insect cells using a baculovirus vector system. Serial deletions at both the N- and C-termini proved that the integrity of a 497-amino-acid domain located between amino acid 31 and the (luminal) C-terminus is required for AtXylT activity. AtXylT is glycosylated on two of its three potential N-glycosylation sites (Asn51, Asn301, Asn478) and the occupancy of at least one of these two sites (Asn51 and Asn301) is necessary for its stability and activity. The contribution of the N-terminal part of AtXylT, in targeting and intracellular distribution of this protein, was studied by the expression of variably truncated, GFP-tagged AtXylT forms in tobacco cells using confocal (panel A) and electron microscopy (panel B). As illustrated in panel B, the transmembrane domain of AtXylT and its short flanking amino acid sequences are sufficient to localize a reporter protein to the medial Golgi cisternae in tobacco cells.

The presence of sialic acid residues at the termini of N-glycan antennae is very important for the clearance of many mammalian plasma proteins of pharmaceutical interest. The absence of such residues on these circulating proteins results in their rapid elimination from the blood stream, by interactions with galactose-specific receptors on the surface of hepatic cells. Sialic acids are absent from plant cells. The production of sialylated N-glycans in plants, by adapting the maturation machinery of plant N-glycans, would require the transfer of at least five heterologous genes encoding enzymes implicated in sialic acid biosynthesis and transport within the Golgi. The missing enzymes of this metabolic pathway not only need to be expressed in a stable manner, but they also have to be active and correctly targeted in the plant cell.

15.7
Concluding Remarks

The promising results already obtained for the humanization of N-glycans will hopefully permit the creation of plant systems producing PMPs compatible with human therapy in the near future. Current efforts are focused on the association of complementation strategies with strategies allowing the inhibition of plant glyco-epitope biosynthesis. In the near future, our main goal is to produce plant-made antibodies with an N-glycan profile identical to that observed in mammals. In this respect, inactivation of β(1,2)-xylosyltransferase and α(1,3)-fucosyltransferase using strategies more efficient than the antisense approach will be a key step, together with an improvement in the efficiency of β(1,4)-galactosyltransferase activity.

15.8
Acknowledgements

This work was supported by CNRS (UMR 6037 and GDR 2590), University of Rouen, Région Haute Normandie, IFRMP23 and Medicago Inc.

References

[1] A. C. Hiatt, R. Cafferkey, K. Bowdish, *Nature* **1989**, *342* (6245), 76–78.

[2] L. Frigerio, N. D. Vine, E. Pedrazzini et al., *Plant Physiol.* **2000**, *123* (4), 1483–1494.

[3] P. Lerouge, M. Cabanes-Macheteau, C. Rayon et al., *Plant Mol. Biol.* **1998**, *38* (1–2), 31–48.

[4] M. Cabanes-Macheteau, A. C. Fitchette-Laine, C. Loutelier-Bourhis et al., *Glycobiology* **1999**, *9* (4), 365–372.

[5] H. Bakker, D. Florack, H. Bosch et al., WO 03/078637, **2003**.

[6] M. Bardor, C. Loutelier-Bourhis, T. Paccalet et al., *Plant Biotech. J.* **2003**, *1* (6), 451–462.

[7] C. Merle, S. Perret, T. Lacour et al., *FEBS Lett.* **2002**, *515* (1–3), 114–118.

[8] B. Lonnerdal, *J. Am. Coll. Nutr.* **2002**, *21* (3), 218S-221 S.

[9] H. Bakker, M. Bardor, J. W. Molthoff et al., *Proc. Natl Acad. Sci. USA* **2001**, *98* (5), 2899–2904.

[10] B. Samyn-Petit, J. P. Wajda Dubos, F. Chirat et al., *Eur. J. Biochem.* **2003**, *270* (15), 3235–3242.

[11] R. C. Aalberse, V. Koshte, J. G. J. Clemens, *Int. Arch. Allergy Appl. Immunol.* **1981**, *66*, 259–260.

[12] L. Faye, M. J. Chrispeels, *Glycoconj. J.* **1988**, *5* (3), 245–256.

[13] R. Van Ree, R. C. Aalberse, *Int. Arch. Allergy Immunol.* **1995**, *106* (2), 146–148.

[14] G. Garcia-Casado, R. Sanchez-Monge, M. J. Chrispeels et al., *Glycobiology* **1996**, *6* (4), 471–477.

[15] R. Van Ree, M. Cabanes-Macheteau, J. Akkerdaas et al., *J. Biol. Chem.* **2000**, *275* (15), 11451–11458.

[16] S. Westphal, D. Kolarich, K. Foetisch et al., *Eur. J. Biochem.* **2003**, *270* (6), 1327–1337.

[17] A. Kurosaka, A. Yano, N. Itoh et al., *J. Biol. Chem.* **1991**, *266* (7), 4168–4172.

[18] L. Faye, V. Gomord, A. C. Fitchette-Laine et al., *Anal. Biochem.* **1993**, *209* (1), 104–108.

[19] D. Chargelegue, N. D. Vine, C. J. Van Dolleweerd et al., *Transgenic Res.* **2000**, *9* (3), 187–194.

[20] M. Bardor, C. Faveeuw, A. C. Fit-

CHETTE, *Glycobiology* **2003**, *13* (6), 427–434.

[21] S. CHENU, A. GREGOIRE, Y. MALYKH et al., *Biochem. Biophys. Acta*, **2003**, *1622 (2)*, 133–144.

[22] A. WRIGHT, S. L. MORRISON, *J. Exp. Med.* **1994**, *180* (3), 1087–1096.

[23] S. ELLIOTT, T. LORENZINI, S. ASHER et al., *Nature Biotechnol.* **2003**, *21* (4), 414–421.

[24] A. VON SCHAEWEN, A. STURM, J. O'NEILL et al., *Plant Physiol.* **1993**, *102* (4), 1109–1118.

[25] I. WENDEROTH, A. VON SCHAEWEN, *Plant Physiol.* **2000**, *123* (3), 1097–1108.

[26] R. VIETÖR, C. LOUTELIER-BOURHIS, A. C. FITCHETTE et al., *Planta* **2003**, *218 (2)*, 269–275.

[27] A. KOPRINOVOVA, O. LIENHART, E. L. DECKER et al., *Conference on Plant-Made Pharmaceuticals, CPMP 2003*, Quebec City, CA, March 16–19, **2003**.

[28] L. NAVAZIO, B. BALDAN, P. MARIANI et al., *Glycoconj. J.* **1996**, *13* (6), 977–983.

[29] L. NAVAZIO, M. C. NARDI, S. PANCALDI et al., *J. Eukaryot. Microbiol.* **1998**, *45* (3), 307–313.

[30] S. PAGNY, M. CABANES-MACHETEAU, J. W. GILLIKIN et al., *Plant Cell.* **2000**, *12* (5), 739–756.

[31] V. GOMORD, E. WEE, L. FAYE, *Biochimie* **1999**, *81* (6), 607–618.

[32] S. PAGNY, F. BOUISSONNIE, M. SARKAR et al., *Plant J.* **2003**, *33* (1), 189–203.

[33] K. KO, Y. TEKOAH, P. M. RUDD et al., *Proc. Natl Acad. Sci. USA* **2003**, *100* (13), 8013–8018.

[34] A. C. FITCHETTE-LAINÉ, V. GOMORD, A. CHEKKAFI et al., *Plant J.* **1994**, *5* (5), 673–682.

[35] R. SRIRAMAN, M. BARDOR, M. SACK et al., *Plant Biotechnology J.* **2004**, *2* (4), 279–287.

[36] A. NEBENFUHR, L. A. GALLAGHER, T. G. DUNAHAY et al., *Plant Physiol.* **1999**, *121* (4), 1127–1142.

[37] A. WRIGHT, S. L. MORRISON, *J. Immunol.* **1998**, *160* (7), 3393–3402.

[38] H. KHOUDI, S. LABERGE, J. M. FERULLO et al., *Biotech. Bioeng.* **1999**, *64* (2), 135–143.

[39] N. Q. PALACPAC, S. YOSHIDA, H. SAKAI et al., *Proc. Natl. Acad. Sci. USA* **1999**, *96* (8), 4692–4697.

[40] K. FUJIYAMA, N. Q. PALACPAC, H. SAKAI et al., *Biochem. Biophys. Res. Commun.* **2001**, *289* (2), 553–557.

[41] R. MISAKI, Y. KIMURA, N. Q. PALACPAC et al., *Glycobiology* **2003**, *13* (3), 199–205.

[42] C. VAQUERO, M. SACK, J. CHANDLER et al., *Proc. Natl Acad Sci. USA* **1999**, *96* (20), 11128–11133.

[43] C. VAQUERO, M. SACK, F. SCHUSTER et al., *FASEB J.* **2002**, *16* (1), 408–410.

[44] E. STÖGER, C. VAQUERO, E. TORRES et al., *Plant Mol. Biol.* **2000**, *42* (4), 583–590.

[45] J. K. C. MA, T. LEHNER, P. STABILA et al., *Eur. J. Immmunol.* **1994**, *24* (1), 131–138.

[46] J. W. LARRICK, L. YU, C. NAFTZGER et al., *Biomol. Eng.* **2001**, *18* (3), 87–94.

[47] J. K. C. MA, B. Y. HIKMAT, K. WYCOFF et al., *Nature Med.* **1998**, *4* (5), 601–606.

[48] J. K. C. MA, A. HIATT, M. HEIN et al., *Science* **1995**, *268* (5211), 716–719.

[49] T. VERCH, V. YUSIBOV, H. KOPROWSKI, *J. Immunol.* **1998**, *220* (1–2), 69–74.

[50] K. BRIGGS, L. ZEITLIN, F. WANG et al., *Plant Biology* **2000**, July 15–19, 2000, San Diego CA-USA., Abstract 15.

[51] L. ZEITLIN, S. S. OLMSTED, T. C MOENCH et al., *Nature Biotechnol.* **1998**, *16* (13), 1361–1364.

[52] C. DE WILDE, M. DE NEVE, R. DE RYCKE et al., *Plant Sci.* **1996**, *114* (2), 233–241.

[53] M. DE NEVE, M. DE LOOSE, A. JACOBS et al., *Transgenic Res.* **1993**, *2* (4), 227–237.

[54] A. M. BRUYNS, G. DE JAEGER, M. DE NEVE et al., *FEBS Lett.* **1996**, *386* (1), 5–10.

[55] K. PEETERS, C. DE WILDE, A. DEPICKER, *Eur. J. Biochem.* **2001**, *268* (15), 4251–4260.

[56] A. A. McCORMICK, M. H. KUMAGAI, K. HANLEY et al., *Proc. Natl Acad. Sci. USA* **1999**, *96* (2), 703–708.

[57] N. RAMIREZ, M. AYALA, D. LORENZO et al., *Transgenic Res.* **2002**, *11* (1), 61–64.

[58] Q. YUAN, W. HU, J. J. PESTKA et al., *Appl. Env. Microb.* **2000**, *66* (8), 3499–3505.

[59] M. BOISSON, V. GOMORD, C. AUDRAN et al., *EMBO J.* **2001**, *20* (5), 1010–1019.

[60] M. A. TAYLOR, H. A. ROSS, D. McRAE et al., *Plant J.* **2000**, *24* (3), 305–316.

[61] R. STRASSER, J. MUCHA, H. SCHWIHLA et al., *Glycobiology* **1999**, *9* (8), 779–785.

[62] R. STRASSER, J. MUCHA, L. MACH et al., *FEBS Lett.* **2000**, *472* (1), 105–108.

[63] H. LEITER, J. MUCHA, E. STAUDACHER et al., *J. Biol. Chem.* **1999**, *274* (31), 21830–21839.

[64] I. B. WILSON, D. RENDIC, A. FREILIN-GER et al., *Biochim. Biophys. Acta.* **2001**, *1527* (1–2), 88–96.

[65] R. LEONARD, G. COSTA, E. DARRAMBIDE et al., *Glycobiology* **2002**, *12* (5), 299–306.

[66] E. G. WEE, D. J. SHERRIER, T. A. PRIME et al., *Plant Cell* **1998**, *10* (10), 1759–1768.

16
Biosafety Aspects of Molecular Farming in Plants
ULRICH COMMANDEUR and RICHARD M. TWYMAN

16.1
Introduction

Plants can be used to synthesize a wide range of industrial and pharmaceutical proteins, providing new commercial opportunities in the agriculture and biotechnology industries. Crops that were once used solely for the production of food, feed or raw materials can now produce recombinant proteins on an agricultural scale [1–3]. Although plants are relative newcomers in the molecular farming marketplace, they have numerous advantages over the more traditional production systems, particularly in terms of cost, convenience, scalability and product safety [4,5]. In a commercial setting, the cost of production decreases with increasing scale, and field-grown transgenic plants therefore represent the most lucrative of all the plant-based production platforms. However, controversy surrounds the biosafety of molecular farming in field plants, particularly their potential impact on human health and the environment [6–8].

Specific biosafety risks fall into two major categories, which we describe as the risk of *transgene spread* and the risk of *unintended exposure* [6]. The risk of transgene spread can be defined as the potential for transgene DNA sequences to spread outside the intended host plants and production site. This can result in the growth of transgenic crops in fields reserved for non-transgenics, the growth of transgenic crops in non-cultivated areas, the spread of foreign DNA to other plants (and possibly to microbes and animals) and the uncontrolled production of recombinant proteins in natural settings. Mechanisms of transgene spread include the dispersal of transgenic plants or seeds by human and animal activities or the weather, outcrossing via transgenic pollen, and horizontal gene transfer from plants to other organisms. The risk of unintended exposure can be defined as the potential for any non-target organism (including humans) to come into contact with the recombinant protein produced by a transgenic plant. Many different mechanisms can be involved, including herbivory and parasitism, the exposure of pollinating insects to transgenic pollen, the exposure of microbes in the rhizosphere to root exudates, the exposure of non-target microbes and animals to proteins secreted in the leaf guttation fluid, and the release of recombinant proteins by dead and decaying transgenic plant material, and

Molecular Farming. Edited by Rainer Fischer, Stefan Schillberg
Copyright © 2004 WILEY-VCH Verlag GmbH & Co. KGaA, Weinheim
ISBN: 3-527-30786-9

the contamination of food or feed crops during harvesting, transport, processing and/or waste disposal. In many cases, transgene spread can lead to unintended exposure because the naturalization of transgenic plants outside the intended production site results in the wider exposure of non-target organisms. While these risks apply to all field-grown transgenic crops regardless of their use, those used for molecular farming deserve special attention because of the pharmacological or toxic properties of many of the recombinant proteins they produce. A final reason for concern, at least to the biotechnology industry, is that the spread of proprietary transgenes into wild species places intellectual property in the public domain. In this chapter, we discuss the biosafety issues associated with molecular farming and some of the emerging strategies that are being used to address them.

16.2
Transgene Spread

16.2.1
Classes of Foreign DNA Sequences in Transgenic Plants

Three different classes of DNA sequence need to be considered when addressing the biosafety aspects of molecular farming. The first class can be described as the *primary transgenes*, i.e. the genes and surrounding elements required to express the desired recombinant product. Note that the term *transgene* has a much broader meaning than the word *gene*, from which it is derived, and generally refers to a DNA cassette that may include one or more actual genes plus any regulatory elements and other sequences needed for proper expression. Primary transgenes are absolutely required in molecular farming since without them there would be no production of the desired protein. The impact of such sequences on the survival and fecundity of wild species is difficult to predict but it is certainly undesirable for proteins that have potent pharmacological or immunological effects when administered to humans or animals to be expressed in natural populations of plants and microorganisms, or in crops intended for the human and domestic animal consumption. The second class of sequences can be described as the *secondary transgenes*, i.e. the genes and surrounding elements that are needed during transformation and regeneration but which are not essential for continued production of the target recombinant protein. This group includes selectable marker genes, reporter genes and genes encoding other accessory proteins that are used to manipulate primary transgenes or their expression (e.g. recombinases), and the regulatory elements required for their expression. These sequences need to be introduced during the gene transfer process but can be discarded when stable plant lines are available. The impact of secondary transgenes on the survival and fecundity of wild species is also difficult to evaluate but there is a great deal of concern that certain markers could have negative effects if they spread outside the intended transgenic plants. In particular, there is concern that herbicide-resistance markers could spread to weedy plants, producing a new generation of 'superweeds' [9], and that antibiotic resistance markers could spread to pathogenic bacteria, severely compromising the use of antibiotics in human health-care. The final category of sequences can be described as *superfluous DNA*, and comprises those sequences that are required neither for transformation nor for recombinant protein synthesis, but which tend to be introduced during the transformation process [10]. Essentially, this means vector backbone sequences from plasmid vectors, which are linked to the primary and secondary transgenes.

16.2.2
Mechanisms of Transgene Pollution – Vertical Gene Transfer

Vertical gene transfer is the movement of DNA between plants that are at least partially sexually compatible. This is the most prevalent mechanism of transgene spread and occurs predominantly via the dispersal of transgenic pollen, resulting in the formation of hybrid seeds with a transgenic male parent [11]. Gene flow from transgenic to non-transgenic populations of the same crop occurs by this method if the two populations are close enough for wind- or insect-mediated pollen transfer. Very high rates of gene flow from crops to related wild species have also been documented along this route. For example, Kling [12] noted that 50% of wild strawberries growing near a field of cultivated transgenic strawberries contained marker genes from the transgenic population. Similarly, herbicide resistance genes have introgressed from transgenic oilseed rape (*Brassica napus*) into its weedy cousin *B. campestris* by hybridization [13]. As discussed below, a number of potential solutions to the problem of transgene pollution have been based on preventing the spread of transgenic pollen, either by physical or genetic containment. However, hybrid seeds can also be generated with the transgenic plant as the female parent if the transgenic crops are fertilized by wild type pollen. In this case, transgene pollution would occur via seed dispersal, either during growth, harvesting or during transport. Seed dispersal from fully transgenic plants can also result in the colonization of natural ecosystems and is more prevalent if seeds can lie dormant for extended periods.

16.2.3
Mechanisms of Transgene Pollution – Horizontal Gene Transfer

Horizontal gene transfer is the movement of genes between species that are not sexually compatible and may belong to very different taxonomic groups. The process is common in bacteria, resulting in the transfer of plasmid-borne antibiotic resistance traits from harmless species or strains to pathogenic ones, but there are few examples of natural gene transfer between bacteria and higher eukaryotes. *Agrobacterium* spp. represent a special case where gene transfer occurs naturally from bacteria to plants if the bacterium contains an appropriate virulence plasmid. There is a perceived risk that horizontal gene transfer from transgenic plants to bacteria in the soil or in the digestive systems of animals could yield new bacterial strains expressing primary and/or secondary transgenes. These traits could have unpredictable effects on relationships between different organisms, e.g. they could render harmless bacteria pathogenic, or could be passed on to pathogenic species making them more difficult to control. There is a specific concern that antibiotic resistance markers and transgenes encoding pharmaceutical proteins could be acquired by human pathogens.

The risks of horizontal transgene transfer from plants to microbes are considered to be extremely small because of the lack of evidence, over millions of years of evolution, that natural plant genes have followed this route [14,15]. For example, Kay *et al.* [16] demonstrated horizontal transfer of marker genes from the chloroplasts of transplastomic tobacco plants to opportunistic strains of *Acinetobacter* spp., but transfer was achieved only under highly idealized conditions in which the bacteria were modified to contain a sequence homologous to the plant's transgene. No gene flow was demonstrated to wild type strains of the bacterium. Even if gene transfer from plants to bacteria did occur in nature, it would be necessary for the transgene to be maintained in the recipient bacterial population. In the case of antibiotic resistance markers there might be strong selective pressure for transgene maintenance due to the widespread use of antibiotics. However, since all natural plants are already liberally covered with antibiotic-resistant bacteria, these would appear to be a much more likely source of resistance genes that could jump to human pathogens [17]. DNA can be taken up from saliva by oral bacteria, and cells lining the gastrointestinal tract can take up and incorporate DNA from the gut [18,19]. Again however, there is a conspicuous lack of evidence that such mechanisms have resulted in the stable incorporation of a plant gene into a bacterial population. Studies with glyphosate-resistant transgenic plants showed that the DNA was completely digested in the gastric environment within a few minutes. Antibiotic resistance genes are the focus of attention because of their potentially strong and general selective advantage in human pathogens. Other transgenes, with much more specific therapeutic applications, would not provide the same benefits as antibiotic resistance and would likely be eliminated even if transfer from plants to bacteria were inevitable. These seemingly insurmountable barriers indicate that horizontal gene transfer is unlikely to represent a significant hazard, and biosafety research has therefore focused on ways to prevent transgene spread by vertical gene transfer.

16.3
Combating the Vertical Spread of Transgenes

16.3.1
Choosing an Appropriate Host

An appropriate choice of host species can go a long way to prevent or minimize transgene spread by dispersal or vertical gene transfer. In general terms, plants that produce large amounts of pollen or large numbers of seeds should be avoided, especially if the seeds are small and easily dispersed. Plants that are often grown as open-pollinated varieties or those that cross spontaneously with wild relatives are also to be avoided, while self-fertilizing plants would be a better choice. Certain plants have been singled out as inappropriate hosts by regulatory organizations such as APHIS.

For example, alfalfa and canola have been highlighted as unsuitable because they are bee-pollinated, sexually compatible with abundant and local weed species and the seeds can lie dormant for several years, making volunteer plants difficult to isolate and destroy [20]. In the end, however, the search for the ideal crop in terms of biosafety will often frustrate the very principles upon which molecular farming in plants is based, i.e. large-scale production, rapid scale-up due to prolific seed production, and the use of existing agricultural and processing infrastructure. There is no single field crop that meets all biosafety demands, and further steps in addition to the selection of a host species must therefore be taken to limit outcrossing and other forms of vertical gene transfer.

16.3.2
Using Only Essential Genetic Information

One way in which the risk of transgene spread can be minimized is to limit the amount of new genetic material incorporated into the production crop. As discussed above, while only the transgene encoding the recombinant protein is required for protein production, transformation usually involves a host of other sequences including superfluous backbone elements and selectable markers. The standard method for producing a transgenic plant line is to introduce the primary transgene along with a selectable marker, which allows the propagation of transformed plant material at the expense of non-transformed material. The use of selectable markers is perhaps one of the major issues in biosafety because traditional markers, which exploit herbicide or antibiotic resistance as selectable traits, are each thought to represent significant environmental or health threats. It is also standard practice to transform plants with plasmid vectors containing the expression cassette. This results in the integration of vector backbone sequences along with the functional primary and secondary transgenes. Not only are such sequences superfluous to requirements, but they also have numerous undesirable effects in transgenic plants, acting as triggers for de novo methylation and promoting extensive rearrangement of the foreign DNA sequences prior to integration [21]. They may also carry additional functional DNA sequences such as selectable markers, promoters and origins of replication used in bacteria, which could become active after gene transfer to non-target organisms.

Ideally, it would be possible to produce transgenic plants carrying just the primary transgene, without recourse to marker genes and other superfluous sequences. The negative impact of these sequences has been established only in the last few years, and only recently have efforts been made to dispense with them. In the case of *Agrobacterium*-mediated transformation, it has been realized that inefficiency in the T-DNA processing step results in the co-transfer of vector sequences in 30–60% of transformation events depending on plant species, *Agrobacterium* strain and transformation method [22]. Since plasmids are pre-requisite for this mode of gene transfer, the only way to guarantee clean transformation (transformation without vector sequences) is to flank the T-DNA with counterselectable marker genes that kill any plant cells containing them [23,24]. With direct DNA transfer methods (such as

PEG-mediated protoplast transformation, electroporation and particle bombardment), vector sequences are generally present in all transformants because whole
plasmids are used in the transformation procedure. An efficient and practical alternative is to carry out transformation using minimal cassettes, i.e. linear constructs
containing just the promoter, open reading frame and polyadenylation signal
[25–28]. Not only does this avoid vector backbone integration but it appears to circumvent another problem specific to direct DNA transfer methods, which is the formation of large, highly complex, multicopy transgene loci containing many rearran-
gements [10, 29, 30]. Such loci are undesirable because they tend to be unstable, and
in many cases contain inverted repeats or truncated transgenes that have the potential to form DNA secondary structures or to express hairpin RNAs, both of which
can trigger transgene silencing [31,32]. In contrast, transformation with minimal
cassettes leads to the generation of very simple integration patterns with the majority
of transgenic loci represented by a single transgene copy [25–28].

Dispensing with selectable markers is more difficult because stable transformation is a rare event and markers are required to identify the very few transformed
plant cells in a large background of nontransformed ones. It is possible, although
quite laborious, to screen plant cells for the incorporation of a primary transgene
using the polymerase chain reaction, without relying on any type of marker. However, most 'marker-free' transformation strategies involve removal of selectable markers *after* transformation has been achieved [33,34] (Sect. 16.3.3). An alternative approach is to use an innocuous scorable marker gene such as *gus*A (encoding the bacterial enzyme β-glucuronidase) or *gfp* (encoding the jellyfish green fluorescent pro-

Tab. 16.1 Novel marker genes that avoid the use of potentially toxic antibiotics, herbicides and
drugs for the selection of transgenic plants. *ESR1*, enhancer of shoot regeneration 1; *CKI1*, cyto-
kinin-independent 1.

Marker gene	Product/phenotype	Sources	Selective agent	Refs
Innocuous selectable marker genes				
*xyl*A	Xylose isomerase	*Streptomyces rubignosus* *Thermoanaerobacterium* *sulfurogenes*	D-Xylose	62, 63
*man*A	Phosphomannose isomerase	*Escherichia coli*	D-Mannose	64
*gus*A	β-Glucuronidase	*Escherichia coli*	Benzyladenine-*N*-3- glucuronide	65, 66
Growth regulator genes				
ipt	Isopentyl transferase	*Agrobacterium tumefaciens*	None	67
pga 22	Isopentyl transferase	*Arabidopsis thaliana*	None	35
rol	Hairy root phenotype	*Agrobacterium rhizogenes*	None	68
ESR1	Transcription factor	*Arabidopsis thaliana*	None	69
CKI1	Histidine kinase	*Arabidopsis thaliana*	None	35

tein) [34]. Even better, a bacterial gene or preferably a plant gene can be used as an innocuous selectable marker, i. e. a gene that would have no conceivable negative effects in wild populations. Examples of such markers include growth regulators (e. g. *ipt* or *CKI1*) and metabolic markers (e. g. *manA* or *BADH*) under inducible control [35]. Such markers could be used to restrict the growth of plants under non-permissive conditions but would not affect the growth or reproduction of wild plants [35,36]. Table 16.1 lists some of the new innocuous markers that can be used in transgenic plants.

16.3.3
Elimination of Markers After Transformation

Where the use of conventional markers is inescapable, an acceptable strategy is the elimination of these genes after transformation, leaving transgenic plants containing the primary transgene alone (Table 16.2). This can be achieved either by segregation or recombination, the former requiring independent cointegration of the marker and primary transgene and the latter requiring the use of site-specific recombination systems such as Cre-*lox*P or FLP-*FRP*.

Tab. 16.2. Strategies for the elimination of marker genes and superfluous DNA

Strategy	Advantages	Disadvantages	References
Pre-transformation strategies			
Flanking counter-selectable markers	Eliminates vector backbone	Reduces transformation efficiency	23, 24
Minimal cassettes	Eliminates vector backbone. Simpler transgenic loci. Higher expression levels. Reduced silencing	Particle bombardment only	25–28
Marker free transformation	No markers needed	Laborious detection by PCR	70
Post-transformation strategies			
Segregation	Simple crossing procedure	Requires independent cointegration of primary and marker transgenes	37
Transposon-mediated repositioning and segregation	Simple crossing procedure. Independent co-integration not required	Depends on transposons to separate transgenes and marker genes. Generates transposon footprint	38
Marker excision by site-specific recombination	Very clean excision, small footprint	Complex cloning procedure. Requires additional transgene encoding Cre recombinase	39
Marker excision using the λ *att*B system	Very clean excision, small footprint. Spontaneous excision	Efficiency?	40

It is surprisingly difficult to persuade separate transgenes to integrate at different loci allowing segregation in later generations. Where two separate plasmids are used to coat microprojectiles, cointegration at the same locus is the predominant outcome (usually as a highly complex concatemer). The introduction of separate binary vectors into *Agrobacterium tumefaciens*, and even the use of different *A. tumefaciens* strains for co-infection, also generally results in co-integration, although this depends on the strain. For example, Komari *et al.* [37] were able to achieve marker gene segregation in a small number of R1 transgenic plants following a transformation strategy involving co-infection with two different *A. tumefaciens* strains. More recently, it has been shown that particle bombardment with minimal cassettes can yield a large number of independent cointegration events, resulting in efficient marker gene segregation in later generations [28]. An alternative and rather elegant way to achieve the same goal is to clone the primary transgene and marker gene in a single construct, but enclose the marker gene within the active elements of a transposon such as *Activator*. Integration is followed by transposition, resulting in the relocation of the marker gene to a different genomic site. As discussed above, the marker can then be eliminated by crossing [38].

The need for crossing can be avoided by building a marker excision strategy into the transformation construct. In most cases, this involves the use of a two-component site-specific recombination system such as Cre-*lox*P [40]. Cre is a recombinase that recognizes short sequences known as *lox*P. If two *lox*P sites are in the same orientation, Cre recombinase activity will excise any DNA between them, so marker genes flanked by *lox*P sites can be efficiently excised from transgenic plants if Cre is present. Cre can be expressed transiently [40] or crosses can be carried out between primary transgenic lines and Cre-transgenic lines to generate hybrids containing both *cre* and the *lox*P-flanked marker, allowing the marker gene to be removed. Where this strategy is used, further crossing may be required to remove the *cre* transgene, unless a 'self-excising' *cre* transgene is integrated [41]. More recently, the *att*B system from bacteriophage λ has been developed for use in transgenic plants because spontaneous recombination occurs at a high frequency, leading to marker removal [42].

Site-specific recombination has also been used to reduce the complexity of multicopy transgenic loci generated by particle bombardment. As discussed above, such loci are prone to transgene silencing and structural instability, and are unsatisfactory from a biosafety perspective because the complex organization means that uncharacterized transcripts and proteins could be produced with unpredictable effects. Simplification is possible either by inserting the transgene at a predefined locus or by streamlining the locus structure after transformation. Both these processes can be achieved using a site-specific recombination system such as Cre-*lox*P. Site-specific integration of transgenes can occur if the genome contains a recombinase recognition site such as *lox*P that has been introduced in a previous round of transformation. Transgene integration occurs at a low efficiency if an unmodified recombination system is used because the equilibrium of the reaction favors excision. However, high-efficiency Cre-mediated integration has been achieved in tobacco using mutated *lox*P sites [43]. Post-integration locus simplification in transgenic wheat has been achieved by incorporating a single *lox*P site within the transgene. Cre expression then drove recombination between the tandemly-arranged *lox*P sites until only one site remained, reducing the transgenic locus to a single copy. This resulted in increased transgene expression accompanied by reduced methylation at the transgenic locus [44].

16.3.4
Containment of Essential Transgenes

For indispensable primary transgenes, the only way to avoid transgene spread from field plants to compatible crops and wild species is by containment. The aim of containment is to prevent seed and pollen dispersal, prevent the survival of dispersed seeds and pollen, or prevent gene flow from viable pollen. The containment may be physical and based on habitat barriers. For example, transgenic plants can be maintained in greenhouses, in artificially-irrigated desert plots miles from any other plants, or in underground caverns and caves [45]. Alternatively, the physical containment may be focused on individual plants. For example, flowers can be emasculated before viable pollen has developed, or the flowers/fruits may be concealed in plastic bags. Isolation zones are often placed around transgenic crops. These can be barren, but a more suitable alternative for insect-pollinated crops is to provide a zone of non-insect-pollinated plants which would discourage the insects from leaving the transgenic zone. Barrier crops, i.e. a border of non-GM plants of the same species as the transgenic crop, are also useful as these can absorb much of the pollen released by transgenic plants and can then be destroyed after flowering.

Biological containment measures provide additional barriers to gene flow and many different strategies have been tested. In some cases, natural genetic barriers have been exploited. For example, molecular farming in self-pollinating species (e.g. rice, wheat, pea) or crops with no sexually compatible wild relatives near the site of production provide a first level of defense against gene flow. Similarly, crops with asynchronous flowering times or atypical growing seasons are useful. Cleistogamy (self-fertilization before flower opening) is an extension of the above, and could be engineered into crops used for molecular farming by modifying the architecture of flower development. In practice, however, there is always a residual risk of outcrossing. Another potential strategy, yet to be fully explored, is the exploitation of apomixis (embryo development in the absence of fertilization). Transformation strategies can also be adapted to take advantage of natural barriers. An example of this approach is genomic incompatibility, which is suitable for polyploid species such as wheat. Many cultivated crops are polyploid but have distinct genomes, only a subset of which are compatible with related wild species for interspecific hybridization. In the case of wheat, only the D genome is compatible with wild *Aegilops* species. Therefore, wheat plants used for molecular farming should carry the transgene(s) on the A or B genomes, a fact that can be established by fluorescence in situ hybridization (FISH) before the plants are transferred to the field.

These natural mechanisms may be replaced or augmented with artificial genetic strategies which are themselves controlled by transgenesis. Such strategies include male sterility, chloroplast transformation, conditional transgene excision and transgene mitigation. Male sterility is achieved by interfering with flower development, or more specifically pollen development, often through the expression of a ribonuclease that prevents the differentiation of the male reproductive organs (e.g. [46]). For example, Bayer Crop Sciences have developed and commercialized a male sterile variety of oilseed rape expressing barnase. The barnase inhibitor (barstar) is also expressed, but it is controlled by an inducible promoter allowing propagation of the transgenic line under laboratory conditions but not in the field. This strategy prevents outcrossing by pollen dispersal but not by pollen immigration, so there remains the possibility of transgene pollution by seed dispersal.

An alternative to male sterility is chloroplast transformation, i.e. the introduction of foreign DNA into the chloroplast genome rather than the nuclear genome. This limits gene flow because the pollen of many crop species does not contain chloroplasts, and where chloroplasts are present, functional DNA is either not transferred to the egg during fertilization, or is degraded during generative and sperm cell development. There are several advantages to molecular farming by chloroplast expression in addition to the biosafety benefits, including the high transgene copy numbers in photosynthetic cells, the absence of position effects and transgene silencing phenomena which can result in low yields in nuclear transgenic plants, and the opportunity to carry out multigene engineering using operons [47]. Thus far, the technology is only applicable to three field species used for molecular farming: tobacco, tomato and potato. However, transformed chloroplasts in these species have been used successfully for the production of diverse products, including biopolymers, vaccines and human growth hormone (see Chapter 8). One possible disadvantage is that proteins produced in chloroplasts are not glycosylated so this system cannot be used for the production of complex glycoproteins [47]. It is also notable that chloroplast inheritance is not strictly maternal in some species, so while gene flow by pollen dispersal may be limited, it may not be eliminated. As with male sterility, chloroplast transformation does not prevent transgene spread by volunteer seed dispersal.

Another genetic barrier to transgene flow is seed sterility, which is achieved by using suicide genes to destroy the developing plant embryo. This mechanism is employed in Monsanto's notorious 'terminator technology' in which a ribosome inhibitor protein is expressed under the control of an embryonic promoter, but in a manner that is regulated by tetracycline. Among several variations of the technique originally discussed in a patent application assigned to Pine Land Corporation, one involved the tetracycline-depended expression of Cre recombinase, which would lead to the excision of the suicide gene if it was flanked by *lox*P sites [48] (Figure 16.1). A 'recoverable block of function' system, based on the constitutive expression of barnase and the inducible expression of barstar, has also been developed [49].

Transgenic mitigation can also be used to prevent the spread of transgenes to wild plants. This involves the inclusion of a tightly linked transgene that confers a trait that is selectively neutral to the crop but disadvantageous to wild plants. Examples might include dwarfing genes or genes that control seed dormancy or shattering. A more sophisticated strategy is conditional transgene excision. In this strategy, plants are created with the transgene flanked by *lox*P sites. A *cre* transgene is also

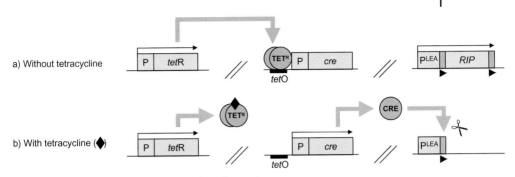

Fig. 16.1   The 'terminator technology' which can be used to prevent the growth of volunteer plants from dispersed transgenic seed. P = constitutive promoter, *tetR* = TET repressor gene, TET^R = TET repressor protein, *tetO* = TET operator sequence, ◆ = tetracycline, *cre*/CRE = Cre recombinase gene/protein, P^{LEA} = late embryogenesis abundant promoter, *RIP* = gene for ribosome inactivating protein, shaded blocks are *lox*P sites in orientation shown by solid triangle. After [48]

present, and this is expressed under the control of a cell-specific or inducible promoter, such that the transgene is physically removed before flowering. If the *cre* transgene is also present within the *lox*P sites, then this transgene will be removed from the plant at the same time, but only when its 'clean-up' task is complete. One potential drawback of this approach is that incomplete transgene excision will leave a residual population of transformed cells from which transgenic gametes could arise.

16.4
Unintended Exposure to Recombinant Proteins

16.4.1
Environmental Risks of Unintended Exposure

The recombinant proteins produced by transgenic plants constitute another risk to the environment. One immediate concern is the possible negative effect of recombinant proteins on non-target organisms, particularly insects and microorganisms that interact directly with the plant and herbivores that may eat transgenic plant material laced with industrial enzymes or protein drugs. Such proteins might have direct toxicity effects, or they might accumulate in the food chain and therefore affect animals that do not interact with the transgenic plants at all. Toxicity may result from direct consumption (e.g. the ingestion of toxins by aphids, and knock-on effects to ladybugs and birds further up the food chain), by simple exposure to the plant (e.g. the effects of pollen on butterflies and moths), from the exudation of recombinant protein into the rhizosphere or leaf guttation fluid (most likely to affect microorganisms) and by the consumption of dead and decaying plant material by saprophytes. Many recombinant proteins expressed in plants are directed to the secretory pathway in order to fold or assemble properly. Such proteins accumulate in the apoplast, the space beneath the cell wall, but there is some leakage into the guttation fluid and root exudate that may change the biochemical environment on the exposed leaf surface or in the rhizosphere. The long term effects of recombinant pharmaceuticals accumulating in the soil and in drainage water have not been investigated and provide scope for all manner of unseen hazards. Finally, the processing of transgenic plants will produce waste containing residual recombinant proteins. An important biosafety issue, particularly for large-scale molecular farming enterprises, is what to do with this waste plant material. A pertinent danger is that such material will be allowed to decay in the environment, providing further opportunities for both protein pollution and transgene escape.

Like transgene spread, the risk of unintended exposure to recombinant proteins can be addressed to a certain extent by physical containment, since this restricts the impact of protein toxicity to a very localized environment. In other words, although microbes in the rhizosphere of contained transgenic plants are exposed to the same extent as those associated with uncontained transgenic plants, the microbes themselves are contained, thus limiting knock-on effects to non-target organisms. However, further barriers to protein toxicity can be put in place by controlling transgene expression or protein structure, therefore limiting the availability of the protein even to closely interacting organisms.

16.4.2
Addressing the Risks of Unintended Exposure

16.4.2.1 Controlling Transgene Expression

The exposure of non-target organisms to recombinant proteins can be minimized by restricting expression to particular tissues. For example, a number of promoters have been identified that restrict gene expression to seeds, tubers or fruit. This prevents the consumption of recombinant proteins by insects and other animals feeding on green plant tissue, and likewise prevents other forms of contact, such as the exposure of pollinating insects to recombinant proteins expressed in pollen grains. By avoiding transgene expression in roots, leaching of the recombinant protein into the soil (and consequent disruption of the rhizosphere) is also prevented. If restricted expression strategies are used in combination with effective management (e. g. specific harvesting times) then vegetative transgenic material can decay safely in the environment with little risk of protein contamination in the environment or unintended exposure. In monocots, where seed expression is the normal strategy, various seed-specific promoters usually derived from seed storage protein genes have been employed to control transgene expression. Examples include promoters from maize zein, rice glutelin and pea legumin genes [50–52]. Care must be taken, however, because although these promoters are described as seed-specific, a low level of activity is present in other tissues. One pertinent example is the bean USP (unknown seed protein) promoter, which has been used in transgenic peas. Although predominantly seed-specific, this promoter also drives low level transgene expression in pollen grains, which could pose a risk to pollinating insects if it were to be used in insect-pollinated plants (Chapter 12). An alternative strategy is to bring the transgene under inducible control, such that the recombinant protein would be expressed only when the plant was exposed to a certain chemical inducer [53]. One of the most promising developments in this area is the use of inducible expression systems to prevent recombinant protein expression until after the crop has been harvested, as has been shown for recombinant glucocerebrosidase using a tomato promoter induced by mechanical stress [54]. A more recent example is the peroxidase gene promoter from sweet potato (*Ipomoea batatas*), which is induced by hydrogen peroxide, wounding or ultraviolet light [55]. In all cases, an effective waste-management policy is necessary to dispose of the waste generated by product processing and extraction.

16.4.2.2 Controlling Protein Accumulation and Activity

In addition to the control of transgene expression, the protein can also be targeted to a specific intracellular compartment. This would not necessarily protect herbivores from exposure to the protein, but it might limit adventitious contact. For example, by adding a KDEL tetrapeptide tag to the C-terminus of a recombinant protein which has been targeted to the secretory pathway with a suitable N-terminal signal sequence, there is efficient retrieval from the Golgi apparatus to the endoplasmic reticulum (ER) [56]. This helps to prevent proteins being secreted to the apoplast, phloem or xylem, where contact with the plant's environment, including microbes and insects, becomes more likely. This is also the case for recombinant proteins containing a heterologous transmembrane domain, which are anchored in the plasma membrane or in the vacuolar membrane depending on other targeting information [57]. The chloroplast or vacuole are alternative destinations for recombinant proteins produced in plants, and help to protect the plant from toxicity effects as well as preventing unintended exposure [58]. Recombinant proteins can also be produced as inactive precursors that have to be processed by proteolytic cleavage before they attain full biological activity. This strategy has been used by Prodigene Inc. for the production of proteases such as trypsin [59,60] and was also used for the expression leech hirudin [61]. As is the case for targeting to the chloroplast and vacuole, the expression of inactive precursors not only limits the extent of protein toxicity in the environment, but also protects the host plant from any negative effects the recombinant protein might have on growth or development.

16.4.2.3 Contamination of the Food Chain During Processing

We have discussed ways in which transgenic plant material, or products derived therefrom, could enter the food chain of humans or domestic animals. These include transgene spread to food and feed crops, contact between transgenic plants and non-target organisms, adventitious herbivory of transgenic plants and leaching of recombinant proteins into the environment through poor waste management. Another major source of contamination is the unintentional mixing of transgenic and non-transgenic crops during harvesting, transport, refining and processing, which has resulted in some highly-publicized incidents including the discovery of recombinant DNA in Linda McCartney food products and the discovery of unregistered Starlink corn in maize products. A more pertinent example in molecular farming is the recent ProdiGene incident in which stray maize plants expressing pharmaceutical proteins were found growing among a soybean crop.

The problem of contamination is compounded by the use of existing facilities to process both food/feed crops and crops used for molecular farming. Ideally, there should be a clear distinction between transgenic plant material used for molecular farming and any normal plant material being processed in the same facility, which is intended for human or domestic animal consumption. A rigorous series of regulatory practices should be in place from the farm to the factory, ensuring complete isolation of transgenic material during growth, harvesting, transport, storage, processing, extraction and waste disposal, and this should supported by validated procedures for cleaning shared equipment. The accidental mixing of transgenic and non-transgenic harvest products is more likely when those products appear visually identical. Therefore, an important step towards identity preservation is the use of non-commercial crop varieties, visually striking varieties (e.g. white tomatoes) or non-food/feed crops that could not possibly be introduced into food or feed processing by misidentification (e.g. tobacco).

16.5
Conclusions

Molecular farming in field plants provides an opportunity for the economical and large-scale production of pharmaceuticals, industrial enzymes and technical proteins that are currently produced at great expense and in small quantities. However, this opportunity is not risk free, and measures for environmental protection must be put into place to make sure that the benefits of molecular farming are not outweighed by risks to human health and the environment. In this chapter, we have summarized the available strategies that can be used to limit the amount of unnecessary foreign DNA incorporated into transgenic plants, prevent transgene spread to non-production plants and other organisms, and limit the exposure of non-target organisms, including humans, to the recombinant products synthesized in plants. The production of well-characterized transgenic plants will allow more effective risk assessment and transgene tracking, and combinations of management and containment strategies will help to prevent transgene spread, protein toxicity and contamination of the food and feed chains. Whatever precautions are taken, it is unlikely that these undesirable occurrences will be completely eliminated so it is possible that the benefits of plant-based protein synthesis will be exploited less controversially in highly contained bioreactors using aquatic plants, single celled plants, or plant cell suspension cultures, albeit with the loss of many of the economical and scalability advantages of field crops. These systems are described in more detail in Chapters 2, 3, 7 and 13.

References

[1] G. Giddings, G. Allison, D. Brooks et al., *Nature Biotechnol.* **2000**, *18* (11), 1151–1155.

[2] R. Fischer, N. Emans, *Transgenic Res.* **2000**, *9* (3–4), 279–299.

[3] E.E. Hood, *Enzyme & Microbial Technol.* **2002**, *30* (3), 279–283.

[4] J.K.-C. Ma, P.M.W. Drake, P. Christou, *Nature Rev. Genet.* **2003**, *4* (10), 794–805.

[5] R.M. Twyman, E. Stoger, S. Schillberg et al., *Trends Biotechnol.* **2003**, *21* (12), 570–578.

[6] U. Commandeur, R.M. Twyman, R. Fischer, *AgBiotechNet* **2003**, *5* (ABN 110), 1–9.

[7] H, Daniell, *Trends Plant Sci.* **1999**, *4* (12), 467–469.

[8] P. N. Mascia, R. B. Flavell, *Curr. Opin. Plant Biol.* **2004**, *7* (2), 189–195.

[9] J. Gressel, *Transgenic Res.* **2000**, *9* (4–5), 355–382.

[10] N. Smith, J.B. Kilpatrick, G.C. Whitelam, *Crit. Rev. Plant Sci.* **2001**, *20* (3), 215–249.

[11] K. Eastham, J. Sweet, *European Environment Agency, Environment Issue Report No: 28*, **2002**.

[12] J. Kling, *Science* **1996**, *274* (6559), 180–181.

[13] T.R. Mikkelsen, J. Jensen, R.B. Jorgensen, *Theor. Appl. Genet.* **1996**, *92* (3–4), 492–497.

[14] KM Nielsen, AM Bones, K Smalla et al., *FEMS Microbiol. Rev.* **1998**, *22* (2), 79–103.

[15] K. Schluter, J. Futterer, I. Potrykus, *Bio/Technology* **1995**, *13* (10), 1094–1098.

[16] E. Kay, T.M. Vogel, F. Bertolla et al., *Appl. Env. Microbiol.* **2002**, *68* (7), 3345–3351.

[17] K. Smalla, S. Borin, H. Heuer et al., in: *Proceedings of the 6th International Symposium on The Biosafety of Genetically Modified Organisms.* University Extension Press, University of Saskatchewan, Canada, pp 146–154, **2000**.

[18] D.K. Mercer, K.P. Scott, C.M. Melville et al., *FEMS Microbiol. Lett.* **2001**, *200* (2), 163–167.

[19] D.K. Mercer, K.P. Scott, W.A. Bruce-Johnson et al., *Appl. Environ. Microbiol.* **1999**, *65* (1), 6–10.

[20] Anon, Docket No. 02D-0324, CBER 200134, Federal Register **2002**, *67*, 57828–57829.

[21] A. Kohli, S. Griffiths, N. Palacios et al., *Plant J.* **1999**, *17* (6), 591–601.

[22] A. Wenck, M. Czako, I. Kanevski et al., *Plant Mol. Biol.* **1997**, *34* (6), 913–922.

[23] V. Ramanathan, K. Veluthambi, *Plant Mol. Biol.* **1995**, *28* (6), 1149–1154.

[24] M.E. Kononov, B. Bassuner, S.B. Gelvin, *Plant J.* **1997**, *11* (5), 945–957.

[25] X.D. Fu, L.T. Duc, S. Fontana et al., *Transgenic Res.* **2000**, *9* (1), 11–19.

[26] N.T. Loc, P. Tinjuangjun, A.M.R. Gatehouse et al., *Mol. Breeding* **2002**, *9* (4), 231–244.

[27] JC Breitler, A Labeyrie, D Meynard et al., *Theor. Appl. Genet.* **2002**, *104* (4), 709–719.

[28] P.K. Agrawal, A. Kohli, R. M Twyman et al., *Plant Physiol.* (in press).

[29] S.A. Jackson, P. Zhang, W.P. Chen et al., *Theor. Appl. Genet.* **2001**, *103* (1), 56–62.

[30] R.M. Twyman, A. Kohli, E. Stoger et al., in: J.K. Setlow (ed) *Genetic Engineering: Principles and Practice (Volume 24)*, Kluwer/Plenum Press, NY, pp 1–18, **2002**.

[31] R.H.A. Plasterk, R.F. Ketting, *Curr. Opin. Genet. Dev.* **2000**, *10* (5), 562–567.

[32] S.M. Hammond, A.A. Caudy, G.J. Hannon, *Nature Rev. Genet.* **2001**, *2* (2), 110–119.

[33] P.D. Hare, N.H. Chua, *Nature Biotechol.* **2002**, *20* (6), 575–580.

[34] B. Miki, S. McHugh, *J. Biotechnol.* **2004**, *107* (3), 193–232.

[35] J.R. Zuo, Q.W. Niu, Y. Ikeda et al., *Curr. Opin. Biotechnol.* **2002**, *13* (2), 173–180.

[36] H. Puchta, *Plant Cell Tiss. Org. Cult.* **2003**, *74* (2), 123–134.

[37] T. Komari, Y. Hiei, Y. Saito et al., *Plant J.* **1996**, *10* (1), 165–174.

[38] A.P. Goldsbrough, C.N. Lastrella, J.I. Yoder, *Bio/Technology* **1993**, *11* (11), 1286–1292.

[39] E.C. DALE, D.W. OW, *Proc. Natl Acad. Sci. USA* **1991**, *88* (23), 10558–10562.

[40] A.P. GLEAVE, D.S. MITRA, S.R. MUDGE et al., *Plant Mol. Biol.* **1999**, *40* (2), 223–235.

[41] L. MLYNAROVA, J.P. NAP, *Transgenic Res.* **2003**, *12* (1), 45–57.

[42] E. ZUBKO, C. SCUTT, P. MEYER, *Nature Biotechnol.* **2000**, *18* (4), 442–445.

[43] H. ALBERT, E.C. DALE, E. LEE et al., *Plant J.* **1995**, *7* (4), 649–659.

[44] V SRIVASTAVA, OD ANDERSON, DW OW, *Proc. Natl Acad. Sci. USA* **1999**, *96* (20), 11117–11121.

[45] E. S. TACKABERRY, F. PRIOR, M. BELL et al., *Genome* **2003**, *46* (3), 521–526.

[46] C. MARIANI, M. DEBEUCKELEER, J. TRUETTNER et al., *Nature* **1990**, *347* (6295), 737–741.

[47] H DANIELL, MS KHAN, L ALLISON, *Trends Plant Sci.* **2002**, *7* (2), 84–91.

[48] M.J. OLIVER, J.E. QUISENBERRY, N.L. TROLINDER et al., *US Patent 5,723,765*, **1998**.

[49] V. KUVSHINOV, K. KOIVU, A. KANERVA et al., *Plant Sci.* **2001**, *160* (3), 517–522.

[50] E. STOGER, M. SACK, Y. PERRIN et al., *Mol. Breeding* **2002**, *9* (3), 149–158.

[51] E. STOGER, C. VAQUERO, E. TORRES et al., *Plant Mol. Biol.* **2000**, *42* (4), 583–590.

[52] Y. PERRIN, C. VAQUERO, I GERRAND et al., *Mol. Breeding* **2000**, *6* (4), 345–352.

[53] M. PADIDAM, *Curr. Opin. Plant Biol.* **2003**, *6* (2), 169–177.

[54] C.L. CRAMER, J.G. BOOTHE, K.K. OISHI, *Curr. Top. Microbiol. Immunol.* **1999**, *240*, 95–118.

[55] K.Y. KIM, S.Y. KWON, H.S. LEE et al., *Plant Mol. Biol.* **2003**, *51* (6), 831–838.

[56] S. MUNRO, H.R.B. PELHAM, *Cell* **1987**, *48* (5), 899–907.

[57] S. SCHILLBERG, N. EMANS, R. FISCHER, *Phytochem. Rev.* **2002**, *1* (1), 45–54.

[58] L.W. JIANG, S.S.M. SUN, *Trends Biotechnol.* **2002**, *20* (3), 99–102.

[59] S.L. WOODARD, J.M. MAYOR, M.R. BAILEY et al., *Biotechnol. Appl. Biochem.* **2003**, *38* (2), 123–130.

[60] J.A. HOWARD, E.E. HOOD, *US Patent 6,087,558*, **1998**.

[61] D.L. PARMENTER, J.G. BOOTHE, G.J. VAN ROOIJEN et al., *Plant Mol. Biol.* **1995**, *29* (6), 1167–1180.

[62] A. HALDRUP, S.G. PETERSEN, F.T. OKKELS, *Plant Cell Rep.* **1998**, *18* (1–2), 76–81.

[63] A. HALDRUP, S.G. PETERSEN, F.T. OKKELS, *Plant Mol. Biol.* **1998**, *37* (2), 287–296.

[64] M. JOERSBO, I. DONALDSON, J. KREIBERG et al., *Mol. Breeding* **1998**, *4* (2), 111–117.

[65] M. JOERSBO, F.T. OKKELS, *Plant Cell. Rep.* **1996**, *16* (3–4), 219–221.

[66] F.T. OKKELS, J.L. WARD, M. JOERSBO, *Phytochemistry* **1997**, *46* (5), 801–804.

[67] S. ENDO, T. KASAHARA, K. SUGITA et al., *Plant Cell Rep.* **2001**, *20* (1), 60–66.

[68] H. EBINUMA, K. SUGITA, E. MATSUNAGA et al., *Plant Cell Rep.* **2001**, *20* (5), 383–392.

[69] H. BANNO, N.-H. CHUA, *US Patent 6,407,312*, **2002**.

[70] S, KOMARNYTSKY, A. GAUME, A. GARVEY et al., *Plant Cell Rep.* **2004** (in press, published online 10 Feb 2004).

17

A Top-down View of Molecular Farming from the Pharmaceutical Industry: Requirements and Expectations

FRIEDRICH BISCHOFF

17.1
Introduction

Plant biologists are technology-driven, and many complain about ignorance and hesitative conservatism when they encounter a top-down view of molecular farming in discussions with representatives of the pharmaceutical industry. On the other hand, business people are used to looking at the wider picture. They find essential issues such as downstream processing and product quality addressed insufficiently if at all in molecular farming research. Although knowledge and experience of industrial requirements is accumulating in the green biotechnology field, it is often communicated inadequately. Therefore, the first part of this chapter highlights the requirements and expectations of industrial manufacturers, and the second part considers solutions and answers to the issues that have been raised.

17.2
Industrial Production: The Current Situation

Although proteins can be expressed in many heterologous production systems, including bacteria such as *Proteus mirabilis* [1], fungi such as *Pichia pastoris* [2, 3] and *Aspergillus awamori* [4] and insect cells [5, 6], the pharmaceutical industry has narrowed down process development to a small number of platform technologies:

1. Mammalian cell suspension cultures are the preferred choice for large-scale recombinant protein production in stirred-tank bioreactors. The most widely used systems are Chinese hamster ovary (CHO) cells and the murine myeloma lines NS0 and SP2/0. In half of the biological license approvals from 1996–2000, CHO cells were used for the production of monoclonal antibodies and other recombinant glycosylated proteins, including tPA (tissue plasminogen activator) and an IgG1 fusion with the tumor necrosis factor (TNF) receptor, the latter marketed as Enbrel [7].
2. The bacterium *Escherichia coli* is preferred for the production of small, aglycosylated proteins like Insulin, Proleukin (interleukin-2), Kineret (interleukin-1 recep-

Molecular Farming. Edited by Rainer Fischer, Stefan Schillberg
Copyright © 2004 WILEY-VCH Verlag GmbH & Co. KGaA, Weinheim
ISBN: 3-527-30786-9

tor antagonist), Neupogen (granulocyte colony stimulating factor, G-CSF) and Interferon-beta.

3. The yeast *Hansenula polymorpha* serves as the production system for a recombinant hepatitis B vaccine (Berna Biotech AG; [8] and yeast is also used to produce granulocyte-macrophage colony stimulating factor (GM-CSF), marketed as Leukine by Schering AG.

The restriction to a small number of platform production systems is driven by regulatory affairs, risk-benefit evaluation and time constraints that disfavor new production systems. Sauer *et al.* [9] have demonstrated that choosing a current platform system can significantly decrease process development timelines. The situation is different for vaccines and tissue-replacement products, which are produced in many different cell lines using specialized media and cultivation systems. However, it is unlikely that these specialized technologies will be used for bulk production. Consequently, molecular farming has a chance to build up a new platform technology for bulk products if it offers strong advantages, meets requirements and solves current drawbacks in existing production systems.

After the approval of the first product, recombinant insulin, in 1982, progress in the development of new recombinant protein pharmaceuticals was slow ([10], Fig. 17.1). The number of biotechnology-derived drugs and vaccines approved by the US Food and Drug Administration (FDA) has increased significantly only since 1995. More recently, sales of biologics have skyrocketed, e.g. from $900 million in 1999 to an estimated $3.5 billion in 2001 for monoclonal antibodies [11]. The annual global market for biopharmaceuticals is estimated to have increased from 12 billion US$ to 30 billion US$ in 2003 [12]. 500 candidate biopharmaceuticals are undergoing clinical evaluation and over one hundred protein-based therapeutics are in the

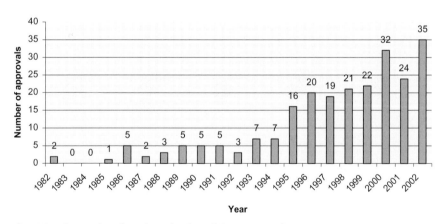

Fig. 17.1. The number of new biotechnological drugs approved each year from 1982–2002. Source: Biotechnology Industry Organization [10].

development pipeline for the next few years. Forty of these are likely to reach the market by 2005 [13]. Also, many monoclonal antibody products are currently in clinical trials, with the probability of 5–10 new antibodies being approved each year [14]. Das and Morrow (2002) [15] have predicted that sales of recombinant antibodies will reach $8 billion by 2004.

The major issue facing current production technologies is the need to increase capacities and the related investments sharply. A typical manufacturing facility costs € 150–400 million and takes four years to build. Due to regulatory guidelines, equipment and certain materials are dedicated to one specific product. For instance, chromatographic material used for the purification of one protein cannot be used for a different one. The production facility generally needs to be separated from other processes, so only a few companies have multi-purpose equipment. Consequently, it is very difficult to switch from the production of one protein to another, and even more difficult to produce two or more proteins in parallel. This means that the economy of scales plays an important role, i.e. it would be better to produce the most successful protein in the largest amounts. Unfortunately, one cannot predict which of several drugs will be the most successful. In phase III of clinical testing, the decision to build up a production line is difficult to make, since half of the projects fail at this stage. In addition, the time and effort required to obtain approval by the FDA slows down the entire process and demands further resources. As a consequence, many companies hesitate to invest in new production facilities and then suddenly run out of production capacity when therapeutics are ready to be produced in larger volumes.

Manufacturing capacity was estimated to be 575,000 liters in 2002 and was predicted to increase to 1.1–1.4 million liters by the end of 2005 [13,16]. However, demand is expected to increase quicker than production capacity. The urgent bottlenecks are highlighted by two examples: the $1 million monthly reservation fee Abgenix is paying to the contract manufacturer Lonza according to Arthur D. Little experts [11] and the problems faced by Immunex when the company was unable to meet the increasing demand for Enbrel, a rheumatoid arthritis drug. Immunex share prices plummeted by nearly 75% between August 2000 and August 2001, and the company had to purchase unused capacity at a contract manufacturer from Medimmune. As a consequence, Immunex decided to invest $400–500 million in November 2001 to build a new facility in Rhode Island and also gained access to a Wyeth plant in Ireland with expected completion in 2005. As similar manufacturing bottlenecks are not unknown to other companies, several have decided to add to their capacities ([16], Fig. 17.2). One might speculate whether low capacities at IDEC might have favored a fusion between IDEC and Biogen.

In addition to the urgent problem of capacity, manufacturers have to cope with the operating costs of production, which are increased by the need for skilled personnel and expensive media components. Another cost driver is the inherent contamination risk when using mammalian cell culture systems. All materials must be checked closely for bacterial and viral contamination, and the presence of prions and endotoxins. This affects not only the manufacturing process, but also downstream materials and even human serum albumin (HSA) used for formulations. In the end, production costs add up to $100–1000 per gram of therapeutic protein.

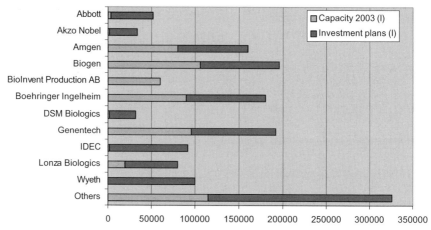

Fig. 17.2. Major increases in fermentation capacities up to 2006 [from Ref. 16]. Note that IDEC and Biogen merged.

Because of the financial problems faced by public health insurance companies in Europe, there is general pressure to cut the prices of expensive biologics. Political systems support biotechnology start-ups that promise to produce cheaper generic biologics once the patents run out. In Europe, the generics company Sandoz (part of Novartis) has already filed for approval of a generic version of the expensive human growth hormone called Omnitrop (somatropin). The company was able to submit its application after the European Commission adopted a regulatory route for biogenerics in mid-2003. And one can imagine that politicians would similarly pave the way for generic therapeutics produced by Molecular Farming. Additionally, decreased prices would enable developing countries to afford these drugs. To defend biggest-selling biological drugs against generic versions, the established biotech companies attempt to extend molecule patents by filing patents on production methods. Consequently, there is a great interest in circumventing patents by using new production technologies provided they also offer cost incentives. The expectation is that Molecular Farming might be such a new production technology.

17.3
Expectations

As a consequence of the above, interest in molecular farming stems from anticipated time and cost savings once the technology is established. The time aspect has three components. The first time constraint occurs in development. Although the development phase is beyond the scope of this chapter, it should be mentioned briefly that choosing the correct protein to produce is a matter of *time to production*. In the case of antibodies, one would like to produce small amounts of several antibodies, eventually selecting those with the highest stability, affinity and optimal performance according to any other relevant criteria. In the case of vaccines, several protein fragments must be produced to test immunogenic efficacy. Using mammalian cell culture, it generally takes 6–12 months to produce a sufficient amount of recombinant protein for toxicity and other crucial tests. Molecular farming might offer an advantage in this respect, since small but sufficient amounts of purified recombinant protein can be obtained quickly by multiple routes.

One option is to use viral vectors that produce high levels of protein in leaves, albeit transiently [17]. This has been demonstrated in the case of hepatitis B virus surface antigen and single chain antibodies using a tobacco mosaic virus vector [18]. In order to produce a full-size monoclonal antibody, *Nicotiana benthamiana* plants were co-infected with two recombinant tobacco mosaic virus constructs encoding the heavy and light chains, respectively [19]. The overall yield can be higher than with other methods because the viral infection spreads to all cells, resulting in a large number of transgene copies through virus replication. This speedy production platform enabled Large Scale Biology Corp. (LSBC) to develop a personalized vaccine against B-cell non-Hodgkin's lymphoma (NHL), a disease with immunologically unique tumors for each patient. The expectation is that each patient can be vaccinated with a plant-expressed anti-NHL scFv manufactured specifically for that individual from the patient's tumor-associated antigens.

Another option, transient transformation, is more familiar to plant biologists. Leaf tissue is infiltrated by *Agrobacterium tumefaciens* carrying a T-DNA vector encoding the protein of interest. The transgene is placed under the control of a plant promoter, which is activated after the T-DNA is transferred into the host plant cell. This method has been shown to be very efficient in the case of tobacco [20]. Agroinfiltration can generate milligram amounts of a recombinant protein within a week [21]. An advantage is that stable plants can be generated in parallel. For plants that can be propagated vegetatively, large amounts of biomass can be produced rapidly for the initial extraction and validation of the heterologous protein. In conclusion, small amounts of the therapeutic protein for pre-clinical tests could be delivered more quickly than is the case with CHO cells, provided that transformation of the host plant species is rapid and efficient and the protein is produced in leaves.

The second time constraint is linked to upscaling. Certainly, scaling up production in plants that produce hundreds of seeds each is one of the greatest advantages of molecular farming. Even for plants propagated by cuttings or tubers (e.g. potato), the agrobiotechnology industry has shown that scale-up is still rapid and economical, since it simply involves increasing the crop acreage and storage capacity. In contrast, at least four years are required to build a new production line for mammalian cell culture and this is a risky venture.

The third time constraint depends on whether the product can be extracted from seeds or fruits. This uncouples protein expression and purification. Large batches of seeds containing the recombinant protein can be produced and stored at low costs. Provided the protein remains stable in the stored seeds, purification can be carried out on demand or shifted according to free capacities. The advantage of one large harvest, with seeds mixed to uniformity, is that this allows production on demand. In contrast, mammalian cell culture is prone to minor batch-to-batch variations in yield and quality. Each fermentation batch has to be processed immediately through down-stream purification. Due to its biologically limited stability storage times had to be minimized.

Probably the most exciting aspect of molecular farming from the industrial perspective is the low initial capital investment compared with mammalian cell culture production. The reduced capital input reflects the lower costs of laboratory equipment and materials for plant molecular biology. For example, the costs involved in establishing small, non-sterile greenhouse facilities are dwarfed by those required for a sealed pilot fermentation plant.

The ultimate decision for big capital investment comes when therapeutic biologics enter phase III of clinical testing. The protein used for phase III testing has to be produced with a process that is identical to that envisaged for future routine production. The process and the equipment cannot easily be changed at a later stage due to regulatory constraints. For instance, it is not possible to switch simply to a fermenter of doubled volume. Alternatively, one could duplicate the process facilities, but this would take years. The supply of a product that has passed phase III might become difficult if the drug sells well and if only minor capacities were implemented during production process design due to financial constraints. However, it is critical to exploit successful products to the maximum extent in an industry in which only 30% of launched drugs return a positive net present value (NPV). This above-average income offsets other products in company portfolios that never provide a return on capital employed (ROCE). The above-mentioned case of supply shortage for Enbrel is such an example. It is evident that molecular farming allows managers in pharmaceutical companies to scale up production more easily if sales are higher than expected. Doubling production is just a matter of sowing or planting twice as many plants.

The production of edible vaccines in fruits and vegetables provides another large cost saving. Given the pre-requisite of guaranteed immunogenicity and adequate vaccine level in the fruit, the expensive purification stage could be omitted and would reduce costs considerably. In the case of β-glucuronidase produced in corn, 88% of operating cost has been attributed to protein extraction and downstream processing [22]. A big share which could be saved in case of edible vaccines.

Improved safety is often highlighted by researchers as one of the advantages of molecular farming. Indeed, contamination of mammalian cell cultures remains a significant issue since antibiotics are used in a number of large-scale industrial processes [7]. However, microbial contamination is also an issue in the case of plants. This concern was already addressed for example in regulatory guidelines (summarized in [23] and is a subject of current research in the area of medicinal plants [24]. The food industry has already developed some cost-effective solutions. For example, lye peelers, which use 5–15% sodium hydroxide solutions and elevated temperatures, effectively remove potato skin and would inactivate soil-borne pathogens on the tuber surface.

In addition, the regulatory authorities are concerned about contamination with viruses and prions, such as the causative agent of bovine spongiform encephalopathy (BSE), which could be present in mammalian cell cultures. It is necessary to check all material used for protein production and purification to eliminate such contaminants. The regulatory authorities have recommended the use of serum-free, animal protein-free media as well as concomitant changes in cell line development and process design [7], but a small number of processes still require calf serum. Instead of human serum albumin (HSA) purified from human blood, recombinant HSA can be used. Furthermore, one has to show that chromatographic filtration steps reduce viral titers by several orders of magnitude. It is argued that molecular farming would not be subject to such controls because plant viruses have been eaten by humans since prehistoric times with no adverse effects, excluding the existence of potentially harmful pathogens. However, draft guidelines from the EMEA show that concerns about viral contamination persist although not scientifically justified [25] (www.emea.org). The final outline of the regulatory guidelines remains to be seen.

In summary, alternative production systems are attractive if they provide solutions for the above-mentioned problems. Molecular farming is one such system, with compelling advantages in terms of time, money and safety. In addition, however, it must fulfill the requirements discussed in the following section.

17.4
Requirements

Consistent efficacy and biological equivalence are the conditions *sine qua non* for regulatory authorities and consequently they are the major concern of executives in the pharmaceutical industry. Miele, an FDA official, stated in 1997 that "Recombinant macromolecules produced in plants should be biochemically, pharmacologically and clinically comparable to their counterparts produced in traditional cell substrates or animals or purified from human sources, if such products are available." Unfortunately, this paper [26] was largely ignored by most molecular farming researches, and very few studies have been carried out to assess the biological equivalence of proteins produced in plants and mammalian cells. A belief that the approval of plant-derived pharmaceuticals is outside of the scope of research might jeopardize the whole field of molecular farming. Managers will turn down any molecular farming project where there is a high risk that the product will not enter the market. Answering the following key issues will minimize the risk of failure.

17.4.1
Equivalence of the Recombinant Product to the Original Protein

A panoply of physical, chemical and biological or immunological tests should be instituted for the purpose of establishing comparability between products derived from plants and those from animals or humans. A combination of SDS-PAGE and mass spectrometry to determine the molecular weight, protein sequencing, isoelectric focusing (IEF), HPLC, peptide mapping (e.g. tryptic mapping) and carbohydrate mapping should be applied. *In vitro* potency assays can often be used to screen for possible differences in bioactivity. In relevant animal models, the equivalence of plant-derived products can be shown in terms of pharmacokinetic studies and biological activity. In addition to equivalence, batch-to-batch consistency needs to be shown. The recent case of Eprex versus Procrit (both erythropoeitin) had alarmed authorities as serious side effects (red cell aplasia) occurred in patients treated with Procrit made at a J&J factory in Puerto Rico while none occurred with batches made at Amgen's facility (sold as "Eprex" in the USA). This has lead to speculations that differences in the manufacturing process had altered the final product.

17.4.2
Processing in the Endoplasmic Reticulum (ER)

Many proteins have to pass through the ER and Golgi apparatus to be processed correctly, to fold efficiently or to form disulfide bridges. These proteins need be expressed as pre-proteins, which contain an N-terminal signal peptide that must be cleaved off during translocation into the ER. The efficacy of plant signal peptides in targeting fusion proteins to the ER has been demonstrated by several methods, e.g. electron microscopy [27]. The proper cleavage of signal peptides from heterologous pre-proteins has been reviewed [28]. Most recombinant proteins are cleaved correctly, thus preventing the accumulation of pre-proteins instead of mature proteins, although some exceptions have been reported (e.g. [29]). Interestingly, expression of the entire prepro-HSA protein in tobacco and potato resulted only in partial processing of the precursor, and the secretion of pro-HSA into the medium. In contrast, fusion of the tobacco PR-S protein signal sequence directly to the HSA gene resulted in the production of mature HSA, indistinguishable from the original human protein [30]. The tobacco Pr1b signal peptide was fused to the cholera toxin B subunit [31], and correct processing was demonstrated by SDS-PAGE. Furthermore, the antigenicity of the CTB protein was identical to its native counterpart as judged by immunodiffusion and immunoelectrophoresis. Only in a few cases has proper cleavage of the plant signal peptide been demonstrated by N-terminal amino acid sequencing: (i) the proteinase inhibitor II signal sequence fused to HSA produced in potato [32], (ii) the rice α-amylase signal peptide attached to a single-chain antibody fragment expressed in *N. benthamiana* [18], (iii) the barley α-amylase signal peptide attached to avidin expressed in maize, and (iv) the rice RAmy3D signal peptide attached to human lysozyme expressed in rice cell culture [33]. In this last report, the authors demonstrated that the recombinant enzyme had the same molecular mass and isoelectric point as native human lysozyme [33]. The integrity of purified aprotinin, commercially produced in transgenic maize seeds, was thoroughly analyzed and compared to native bovine aprotinin with respect to molecular weight, the presence of disulfide bonds, pI, N-terminal amino acid sequence and trypsin inhibition activity [34]. The proper formation of disulfide bridges was shown for recombinant trout growth hormone expressed as a fusion with the PR1b signal peptide in tobacco [35].

Some reports indicate that ER-targeting, using a C-terminal KDEL sequence, is necessary to achieve high protein levels [36–39]. It has been suggested that proteins accumulating in the cytosol suffer proteolytic degradation and therefore do not reach very high levels. Interestingly, adding a KDEL signal to several scFv antibodies resulted in considerably improved expression levels (0.2%) even though the antibodies remained exclusively in the cytosol [36, 40]. This indicates that the KDEL signal may generally stabilize proteins. Even so, the level of scFv accumulation with ER-targeting and the KDEL signal was fivefold higher (1%). Therefore, many research teams have used the C-terminal addition of KDEL to further improve the yield of ER-targeted proteins (e.g. [39]. Some exceptions of the rule are known [29, 33, 41]. For instance, the recombinant heat-labile toxin of enterotoxigenic *E. coli* joined to a barley α-amylase signal sequence yielded higher protein levels if targeted to the maize cell surface than if retained in the ER [42]. Although many researchers favor the addition of the KDEL or KDEI retention signal (e.g. [40]), most pharmaceutical companies would need the effect of the KDEL tag on immunogenicity to be assessed thoroughly. For instance, an ER-retained recombinant hepatitis B virus surface antigen (HBsAg) containing a KDEL tag was favored as a vaccine candidate because mice immunized with this protein showed a stronger immune response than those immunized with unmodified plant-derived HBsAg. However, the immunoreactivity of mice against uncleaved signal peptides was not assessed [43, 44]. The fact that a recombinant protein with a KDEL-tag is not equivalent in the strict sense to the native protein is often overlooked. One should be aware that it is necessary to clarify whether the addition of KDEL might influence biological activity when used in therapy. Most pharmaceutical companies would prefer to have smaller amounts of the correct protein.

17.4.3
Glycosylation in the Golgi

Targeting recombinant proteins to the secretory pathway is essential for correct folding, the formation of disulfide bridges and glycosylation [45]. Glycosylation influences many properties of recombinant proteins including biological activity, folding, solubility, stability and blood clearance (reviewed in Ref. [46]). Glycosylation of the Fab regions has been shown to affect the binding activity of some monoclonal antibodies, in contrast to glycosylation in the Fc region, which rarely has an impact on immunoreactivity. However, the correct glycosylation of the C_H2 domain is necessary for the efficient binding of IgGs to the Fc receptor of immune cells [47]. Therefore, in general, recombinant proteins produced for pharmaceutical purposes should have the same glycosylation profile as the native human protein.

CHO cells have been favored because the glycans they attach to proteins are very similar to those of human cells. If other platform technologies result in differing glycosylation patterns, new regulatory approval for the recombinant proteins must be obtained, certainly in the case of generic biologics. However, one has to bear in mind that glycosylation patterns vary in any heterologous system. The glycosylation of tPA was shown to depend on CHO cell growth rate and temperature [48]. Similarly, growth conditions have been shown to influence the glycosylation of an IgG antibody expressed in tobacco leaves [49]. Interestingly, the galactosylation of serum IgG glycans increases also in healthy pregnant women [50]. Reportedly, MedImmune's

Synagis, a humanized monoclonal antibody, consists of several glycoforms. This indicates that a certain degree of microheterogeneity would be acceptable by the FDA.

While the above reflect rather minor variations, the major differences in glycosylation between plants and animals might not be seen in the same light [26]. Indeed, plants (like insects and yeast) do not incorporate sialic acid into their glycan chains, and the fucose linkage is different to that found in mammals. Furthermore, plant glycans contain immunogenic xylose residues, which are never found in human glycoproteins. For instance, recombinant avidin obtained from transgenic maize was glycosylated at Asn-17 with a carbohydrate chain smaller than that of chicken egg avidin [51]. Similarly, the N-linked oligosaccharides attached to erythropoeitin are smaller in tobacco than in mammals [52]. The N-glycan composition of a murine monoclonal IgG1 antibody (Guy's 13) was studied detailing and compared with the glycan structures of the same monoclonal antibody expressed in tobacco. Although the N-glycosylation sites were the same in both cases, different glycans were detected, and the tobacco glycans contained terminal xylose and fucose residues [53]. The unusual sugars could be detected easily using the lectin concanavalin A or anti-$\beta(1,2)$-xylose or anti-$\alpha(1,3)$-fucose antibodies [54]. Notably, the analysis of another IgG antibody produced in tobacco revealed that a proportion lacked terminal xylose and fucose residues in contrast to endogenous proteins [49]. An ER-retained anti-HBsAg antibody had predominantly oligomannoside type N-glycans attached to the same N-glycosylation sites as seen in the mouse antibody. However, complex type N-glycans were also detected, containing $\alpha(1,3)$-fucose attached to the core GlcNAc residue [55]. Due to the presence of non-mammalian xylose/fucose residues, plant-derived drugs might be recognized as foreign antigens in mammals.

Fortunately, this seems not to be the case: no immunogenic reaction was seen in mice when they were immunized subcutaneously with the Guy's 13 antibody expressed in tobacco [56]. Furthermore, in human clinical trials, a modified version of the tobacco-derived Guy's 13 antibody was administered on six occasions, and did not elicit an immunological reaction against the plant-specific N-linked glycans [57]. Similarly, no adverse reactions were reported when a soybean-derived antibody directed against herpes simplex virus 2 was delivered to the mouse vagina [58]. Similar results were also reported by EPIcyte: a mouse monoclonal antibody produced in maize did not elicit an immune response in mice [59]. In the few clinical trails conducted so far (see below) problems with allergic reactions towards plant-derived drugs have not been reported. It was suggested that plant glycans are known to the immune system as components of our daily food intake. The rationale is that food ingredients cross the intestine barrier into the blood stream, and therefore cannot stimulate a constant immune response. More recent results oppose this view: antibodies specific for core-fucose and core-xylose-epitopes were found in the blood of non-allergic donors, probably resulting from food or environmental sensitization [60].

17.4.4
Differential Glycosylation – Implications on Immunogenicity of vaccines

Whereas immunogenic reactions against the therapeutic protein should be avoided in the case of plant-derived antibodies, an immune reaction is needed for vaccines expressed in plants. However, the modified glycosylation of plant-derived vaccines compared to the original protein might change the immunogenicity of the vaccine and might even cause allergies. Some indications for altered immunogenicity were found when horseradish peroxidase was injected into mice and rats. They raised not only antibodies against the protein moeity, but also against core-fucose and core-xylose-epitopes [60].

On the other hand, for the hepatitis B surface antigen, the glycan was shown not to be required for immunogenicity, and in contrast to the soybean-derived vaccine the yeast-derived vaccine has no glycosylation [61]. Interestingly, the glycan chains are lost with storage of the partially purified extract from vaccine-producing soybean cell cultures [44]. Furthermore, engineered mutations that increased the amount of N-linked glycosylation on the coat protein of human T-lymphotropic virus (HTLV-I) did not alter the immunogenicity of the virus, but enhanced immune recognition [62].

Previous studies had shown that the measles virus hemagglutinin (H) protein is highly susceptible to altered glycosylation. Glycosylation is required at two or more of the four sites usually glycosylated in the native protein, otherwise protein folding, stability and protease susceptibility are adversely affected [63]. The H protein produced in transgenic tobacco and carrot had a molecular weight lower than the native protein, probably due to differences in the glycosylation pattern, but it retained its antigenic and immunogenic properties [33, 64]. This indicates that plant-type glycan structures did not affect the protein's conformation. This is further corroborated by the recognition of the plant-derived protein using monoclonal antibodies or human serum antibodies produced in response to a wild-type measles infection [33]. Oral as well as intraperitoneal immunization of mice induced high titers of IgG antibodies that neutralized the virus in vitro [33, 64]. Interestingly, antibodies induced by injecting transgenic carrot leaves extract were of both the IgG1 and IgG2a subclasses, whereas the immune response in mice injected with measles virus hemagglutinin protein produced in mammalian cells was essentially restricted to IgG1. However, overall antibody levels were comparable [64]. Taking all the data together, it seems that differences in glycosylation have little impact on the immunogenicity of plant-derived vaccines (reviewed in Ref. [65]. Their efficacy has also been demonstrated in clinical tests (see below). Nevertheless, the pharmaceutical industry would prefer unchanged glycosylation patterns in order to avoid possible side effects caused by non-human glycan structures.

17.4.5
Glycosylation and Stability

Glycosylation affects protein stability as well as immunogenicity. A prominent example is tobacco-derived erythropoeitin (EPO), which possesses smaller N-linked oligosaccharides compared to commercial EPO and lacks sialic acid residues. The plant-derived protein was shown to be more active in vitro i.e. inducing the differentiation and proliferation of erythroid cells stronger than commercial EPO. Strikingly, this difference was eliminated if the commercial EPO was desialylated. However, neither the tobacco-derived protein nor the desialylated recombinant EPO showed activity in vivo [52]. The authors suggested that EPO with glycans lacking terminal sialic acids is probably trapped by the asialoglycoprotein receptor on hepatocytes and excluded from the circulation before it can exert biological activity in rats. Intriguingly, the contrary effect i.e. prolonging half-life of an EPO analog was achieved by increasing sialic acid content. This modification was considered as an advantage by the FDA. Consequently, the EPO analog Aranesp produced by Amgen Inc. was approved in 2001.

Glycosylation also affects the stability of antibodies expressed in plants. For example, an ER-retained antiviral monoclonal antibody was barely detectable 10 days after injection into mice, whereas the same antibody obtained from murine/human hybridoma cell lines was still abundant in blood samples [66]. This finding was surprising since there was no difference between the antibodies in terms of their rabies virus neutralizing activity, and no differences in efficacy measured in terms of post-exposure prophylaxis for hamsters injected with rabies virus. The higher stability of the mammalian antibody might again be dependent on terminal sialic acids in complex N-glycans [67]. In contrast to mammalian antibodies, the ER-retained anti-viral antibody from tobacco displayed predominantly oligomannose-type N-glycans (90%), although no $\alpha(1,3)$-fucose residues were detected. Oligomannose-type N-glycans are likely to be recognized by mannose receptors on e.g. liver macrophages, resulting in increased blood clearance of the plant-derived antibody. However, the shorter half-life of the tobacco-derived antibody may offer also an advantage for therapeutic use because there will be less interference between passive and active immunity compared to the current commercial antibody vaccine [66]. Studying the degradation of antibodies expressed in plants, Stevens et al. [68] discovered that the tobacco-derived antibody was less stable in plant extracts than the control antibody produced by hybridoma cells. The examples given emphasize the importance of glycosylation to ensure stability of plant-derived biological drugs.

For other plant-derived antibodies, stability was shown to be similar to mammalian counterparts. For instance, a humanized anti-herpes simplex virus monoclonal antibody (IgG1) was expressed in soybean and showed stability in human semen and cervical mucus over 24 h similar to the antibody obtained from mammalian cell culture. In addition, the plant-derived and mammalian antibodies were tested in a standard neutralization assay with no apparent differences in their ability to neutralize HSV-2. As glycans may play a role in immune exclusion mechanisms in mucus, the diffusion of these monoclonal antibodies in human cerival mucus was tested. No differences were found in terms of the prevention of vaginal HSV-2 transmission in a mouse model, i.e. the plant-derived antibody provided efficient protection against a vaginal inoculum of HSV-2 [58]. This shows that glycosylation differences do not necessarily affect efficacy.

Some research groups prefer IgAs because they are more stable than IgGs and more resistant to proteolysis. Plants can assemble secretory IgAs, which consist of four chains (heavy and light chains, J chain and secretory component). Additionally, secretory IgAs are more efficient at binding antigens. These characteristics allow IgAs to treat mucosal sites such as the gastrointestinal tract against infections [69]. As repeated large doses of antibody are required for topical passive immunotherapy, transgenic plants could be the only cost-effective production system.

In conclusion, glycosylation and processing should result in plant-derived pharmaceutical proteins mainly identical to wild-type protein even though different forms may have exactly the same characteristics as the original protein. However, differences in glycosylation might be acceptable by the FDA if the isoforms offers advantages e. g. Aranesp (darbepoetin alpha) and glucocerebrosidase.

17.4.6
Equivalence of Enzymes

Drugs used in replacement therapies are often enzymes, as exemplified by glucocerebrosidase and iduronidase. The treatment of Gaucher patients is based on the external supply of the recombinant enzyme glucocerebrosidase and costs $100,000–400,000 per annum [70] despite the recent approval of CHO-derived cerebrosidase (Ceredase, Genzyme Corp.). Enzymatically active glucocerebrosidase has also been obtained from transgenic tobacco and purified using an epitope tag. The plant-derived enzyme is comparable to the human or CHO-derived glucocerebrosidase based on kinetic studies and affinity binding characteristics. Glucocerebrosidase from tobacco comigrates with the human placental-derived enzyme in SDS-PAGE, and is glycosylated [70,71]. Other examples for equivalence of plant-derived enzymes are gastric lipase (produced by Meristem Therapeutics) and neomycin phosphotransferase II (NPTII). The equivalence of latter enzyme produced in *E. coli*, cotton seed, potato tuber and tomato fruit has also been assessed. Microbial and plant-produced NPTII proteins have comparable molecular weights, immunoreactivity, epitope structures, N-terminal amino acid sequences and biological activity. Interestingly, all NPTII proteins remained unglycosylated irrespectively of the production organism [72].

17.4.7
Degradation

Reliable extraction protocols are needed to prevent the degradation of plant-derived antibodies. Many researchers use protease inhibitors, such as phenylmethanesulfonylfluoride (PMSF) [30, 49, 67, 73] leupeptin [56, 74] or mixtures thereof [66, 75]. Following detailed analysis, however, the degradation appears to occur in planta rather than during extraction. This hypothesis is corroborated by the finding that antibody degradation fragment patterns are not changed following the addition of protease inhibitors or protein protective agents [74]. Some researchers have noted a protein band of 120–125 kD often seen in preparative SDS-PAGE gels for plant-derived antibodies. Van Engelen *et al.* [76] and Stevens *et al.* [68] interpreted the 125 kD fragment as F(ab')$_2$-like fragment whereas Sharp and Doran [74] found cross-reaction of this band with an anti-Fc antibody. But all authors agree that this fragment and others stem from proteolytic activity inside the plant cells, most probably starting with the proteolytic removal of (part of) the constant region (Fc). Interestingly, this break-down results in relatively stable products [68]. In general, it seems that degradation is lower for ER-retained proteins in storage organs. For example, an ER-retained scFv purified from potato tubers in total protein extract was found to be stable for at least 16 days at 4 °C [75]. In summary, the role of degradation needs to be studied more extensively for plantibodies. Avoiding degradation might also improve yields in a more efficient way than simply increasing expression.

For orally delivered plant vaccines some crop processing might be necessary to en-sure consistent antigen dosage. The subsequent loss in compartmentalization could expose the antigen to proteases, polyphenol oxidases and plant phenolics. Conse-quently, immunogenic epitopes could be destroyed or changed. Serum-derived hepa-titis B surface antigen (HBsAg) proved to be remarkably protease resistant, which was attributed to its extensive disulfide cross-linking yielding dimers and higher multimers [77]. Unfortunately, this extensive cross-linking does not occur in the plant-derived HBsAg resulting in degradation of the HBsAg dimers in plant extracts [44]. Different proteinase inhibitors, including leupeptin, aprotinin, E-64, pefabloc and pepstatin had no effect on HBsAg stability in potato extracts. Only the combina-tion of leupeptin and ß-mercaptoethanol protected the antigen [78]. Interestingly, under optimized detergent conditions, protein stability was extended to 1 month in tomato, but not potato extracts [44]. The N-terminal addition of a signal peptide re-sulted in a more stable, but uncleaved HBsAg fusion protein that formed multimers and was more potent than unmodified plant-derived HBsAg [29]. Despite the prote-ase-sensitive nature of potato-derived HBsAg, uncooked transgenic potatoes elicited an increased immune response when compared with a similar oral dose of commer-cial yeast-derived rHBsAg [79]. Similarly, rHBsAg partially purified from transgenic tobacco leaves generated a qualitatively similar immune response in mice when compared to yeast-derived rHBsAg [80]. This indicates that despite some degrada-tion, both the B- and T-cell epitopes of HBsAg are preserved when the antigen is ex-pressed in transgenic plants. Other stability studies were carried out on clover plants expressing the *Mannheimia haemolytica* A1 leukotoxin 50 fusion protein [81]. After harvest, clover plants were allowed to dry at room temperature for 1–4 days. No de-gradation of the fusion protein was observed and the protein induced an immune re-sponse in injected rabbits.

17.4.8
Efficacy in Clinical Trials

As highlighted by the erythropoeitin example discussed above, efficacy and bioequi-valence shown *in vitro* do not guarantee efficacy in vivo. But, the latter is a key re-quirement for the success of protein therapeutics expressed in plants. One of the concerns regarding the use of plant-derived edible vaccines is that humans ingesting transgenic plants might not respond to immunization because the same plant spe-cies is part of their regular diet. To date, a number of clinical trials using edible vac-cines have been carried out or are under way. The first human trial was carried out in 1997. Fourteen healthy adult volunteers were given either wild type or transgenic potato, the latter containing a recombinant B subunit of the heat-labile enterotoxin of *E. coli* (Lt-B) as antigen. The raw potatoes were generally well tolerated. Antibody-secreting B cells were detected seven days after eating one dose of transgenic potatoes. Ten of eleven volunteers developed IgGs against Lt-B protein, but neutralizing titers were only achieved in eight volunteers [82]. Similar efficacy was shown in mice ingesting an ER-retained form of Lt-B, although the expression level was quite low (<0.01% of TSP). The mice developed both serum and gut mucosal antibodies specific for Lt-B [83]. In different sets, mice were shown to be partially protected against challenge with the holotoxin after oral delivery of Lt-B-producing potato [84] or maize expressing a codon-optimized Lt-B [85]. In contrast, Lauterslager *et al.* [86] found that oral immunization elicited systemic and local IgA responses only in parenterally primed, but not naive mice.

Many studies have been carried out on plant-derived Hepatitis B surface antigen (HBsAg) showing its equivalence to the commercially available HBsAg from yeast (e.g. [80]). HBsAg was successfully expressed in tobacco, potato, tomato, soybean, lupin and lettuce [43, 44, 87, 88], but only transgenic lettuce was used for two human trials with five and twelve volunteers, respectively. After two or three immunizations with the transgenic lettue, all test participants showed low levels of HBsAg-specific antibodies. Unfortunately, protective levels were only transiently reached in two volunteers, and this might have been due to the low expression level of the recombinant HBsAg (<0.01%). No apparent side effects occurred in volunteers within 20 weeks after first ingestion of transgenic lettuce [88, 89]. In pre-clinical animal studies, uncooked transgenic potatoes containing as little as 0.0001% HBsAg had to be administered together with an adjuvant (cholera toxin) due to the limited amount mice could ingest in 24 h. Subsequent parenteral boosting was necessary to reach protective antibody levels [43, 79]. In conclusion, the efficacy of the commercial yeast-derived vaccine (also containing an adjuvant) has not yet been matched by molecular farming in plants.

The Norwalk virus capsid protein (NVCP) has been expressed in transgenic tobacco and potato at levels of 0.02–0.07% [90], but in potato only 50% of NVCP correctly assembled in virus-like particles (VLP). Having ingested 2–3 doses of raw transgenic potato, 19 out of 20 volunteers showed a significant increase in the number of specific IgA antibody-secreting cells. But only four and six people developed serum anti-NCVP IgGs and IgMs, respectively [91]. None of the volunteers showed changes in serum anti-patatin IgG after ingestion of transgenic potatoes, which contain patatin as major storage protein [92]. In 2002, another clinical trial was carried out using transgenic spinach expressing epitopes from the rabies virus glycoprotein and nucleoprotein fused to the coat protein of alfalfa mosaic virus. After oral immunization, five of nine naive volunteers and three of five volunteers primed with the conventional vaccine showed significant elevation in rabies-specific antibodies [93]. As it is risky to assess the protective level against rabies in humans, a mouse model was used to test the plant-derived rabies vaccine. After parenteral immunization, mice were protected against challenge infection [93]. As far as animal edible vaccines are concerned, Prodigene Inc. were able to demonstrate the efficacy of a corn-derived vaccine against challenge with swine-transmissible gastroenteritis coronavirus (TGEV) [85, 94]. Protection against challenge was also reported for a potato-derived rabbit hemorrhagic disease virus vaccine in rabbits [95] and for a potato-derived VP1-vaccine protecting against food and mouth disease virus [96]. In summary, most studies indicate that efficacy could be still improved, probably through higher expression levels in plants.

In contrast to edible vaccines, plant-derived antibodies and therapeutic proteins are mainly prepared for clinical testing by biotech companies, and fewer details have been published. According to the company's press release, EPIcyte is slated to become the first company to enter Phase I clinical trials with a human herpes antibody (called HX8) produced in plants. Trials in 2003 should show the efficacy of HX8 to prevent the transmission of herpes simplex virus 1 and 2. A phase II SBIR grant has been awarded to EPIcyte, which is allied with Dow. The efficacy of such an antibody from soybean for prevention of vaginal HSV-2 infection in mice has been shown [58]. Tobacco-derived secretory IgAs, marketed as CaroRx^{TM}, were designed to prevent oral bacterial infections contributing to dental carries. In a clinical trial, the plant-derived IgA gave specific protection against colonization by oral streptococci for over four months [57]. No adverse effects or immunological response against the IgA have been observed in more than 40 patients receiving topical oral application of the IgA [57]. According to another company's homepage, Planet Biotechnology is currently undergoing Phase II US clinical trials under a US FDA-approved Investigational New drug (IND) application. The companies NeoRx and Monsanto Protein Technologies (formerly Agracetus or NSC Technologies) joined to produce a humanized antibody in corn to be used for Avicidine cancer treatment. Unfortunately, all that has been reported about the clinical trails is that the maize-derived antibody behaved similarly to the murine antibody [97]. Some months later, the project was discontinued by the partner Janssen because of a high incidence of severe side-effects in Phase II trials with the murine antibody. No serious adverse side-effects were reported by Large Scale Biology Corp. (LSBC) after completion of phase I clinical trials with their personalized cancer vaccine, by August 2002. Each patient in the study received a tobacco-derived scFv manufactured specifically for that individual from the patient's tumor-associated antigens (non-Hodgkin's lymphoma). 15 of the 16 individuals showed no immune reaction to the plant-derived single-chain antibody and ten mounted a humoral or cellular response to the treatment. Researchers at LSBC are working together with the FDA to design PhaseII/III trials [98]. Even more promising is LSBC's human α-galactosidase for which the FDA granted orphan drug status in January 2003. The enzyme is also produced in tobacco infected with a re-engineered virus and showed positive results in pre-clinical studies using an animal model of Fabry disease [98]. Positive responses to a treatment with corn-derived gastric lipase given to 15 cystic fibrosis patients resulted from a phase II multicenter trial within 2002. Meristem Therapeutics and its exclusive partner Solvay S.A. reported about reduction of fecal lipid content during these clinical trails. According to Giddings *et al.* [98], Ventria Bioscience (formerly Applied Phytologics) started trials with α-1-antitrypsin from transgenic rice and hopes to get product approval by 2004. It will be interesting to see in years to come, if plant-derived antibodies and enzymes will show full efficacy.

17.4.9
The Optimal Production System

Stability and degradation are also issues when it comes to choosing the production system. Industry would ask for a plant production system that offers the possibility of storage and transport from the field to the downstream processing facilities. As discussed above, this can be difficult in the case of leaves and leaf extracts due to protein degradation. Losses of enzymatic activity in leaf extracts were observed in a feasibility study [99]. There is one exceptional report concerning an ER-retained scFv antibody expressed in tobacco leaves that was extracted after drying and storage for one week, and was still active [100]. Khoudi *et al.* [101] reported antibody stability in dried alfalfa hay as well as in extracts made in pure water. Seeds as natural storage organs seem to be the better alternative for several reasons: (i) Seeds contain a less complex mixture of proteins and lipids. This is an advantage for purification (see below). (ii) The stability of recombinant proteins is likely to be higher in seeds. Seeds have low protease activity and contain fewer phenolics than leaves. The advanced state of dehydration confers enhanced stability, allowing seeds to be stored for periods of several years without any notable degradation of proteins or loss of activity, as shown for phytase [102]. In rapeseed, a β-glucuronidase-oleosin fusion protein was stable for long periods [103]. In *Arabidopsis*, an scFv antibody expressed under the control of the β-phaseolin and arcelin-5 promoters accumulated to 36% TSP in homozygous seeds and was stable for at least one year [104], while a level of 2% TSP was achieved using same constructs in *Phaseolus acutifolius*. In tobacco seeds, an active scFv accumulated to 0.7% TSP, and the seeds could be stored for one year at room temperature without protein degradation or loss of antigen-binding activity [100, 105]. Similar stability was observed for another scFv expressed in dry wheat and rice grains [106]. Furthermore, *E. coli* Lt-B and the S antigen of TGEV were stable for longer than a year when expressed in maize seeds and extracted in the germ meal fraction [94]. Avidin was stable in whole maize kernels, but not in flaked material, and could withstand temperatures up to 50 °C for at least 7 days without loss of activity [51]. The potato tuber is another natural storage organ, but unfortunately it only contains 2% protein. Artsaenko *et al.* [75] reported a 50% loss of functional antibody after 18 months in storage, starting from a pre-harvest yield of 2% TSP in potato tubers. To fulfill the industrial standard of current logistics in medicinal plant drug production, germplasm with low protease activity is needed to ensure stability for at least one year in storage. Despite the advantage of seed-based production, however, it might not be possible to express all proteins in seeds [35].

Stable expression of the recombinant protein is another pre-requisite. For production in CHO cell culture, narrow specifications of expression level are requested by the FDA. Consequently, plant-derived proteins need to be produced according to a defined scheme and within defined specifications in order to generate material suitable for clinical trials. Therefore, the expression level must be maintained within a defined range. For example, Prodigene Inc. assessed whether the *E. coli* Lt-B protein was uniformly distributed throughout the defatted maize germ by analyzing samples taken randomly from the bulk material [42]. The amount of an IgG expressed as percentage of total soluble protein was analyzed in tobacco leaves from the top (0.15–0.24%), middle (0.13–0.19%) and base (0.14–0.21%) of the plants, indicating no dependence on leaf age [68]. However, it will be always difficult to control expression levels in the field. Therefore, expression in a controlled environment is more appropriate to ensure specified expression levels. Post-harvest induced expression (see below) would fit to this requirement.

The ideal crop should be amenable to transformation and regeneration, and should also facilitate rapid scaling up of production when required. Tobacco and rapeseed are prolific seed producers (up to one million seeds produced per plant in the case of tobacco). However, it would be necessary to establish a homogenous line after several generations, since even in the third generation, large differences between progenitor lines can occur [34] and selection and back-crossing with elite germplasm over several generations might be necessary to achieve economical protein yields [14]. Consequently, seed propagation makes it necessary to monitor the stability of transgene expression after re-amplification from the master seed bank [26]. Therefore, in terms of speed and fidelity, crops reproduced by vegetative propagation (e. g. cuttings or sprouts) might be preferred. Additionally, the guideline draft by EMEA [25] (www.emea.org) favors vegetative propagation. For example, no apparent loss in the ability to synthesize and accumulate a fully functional recombinant antibody was reported in alfalfa plants produced through stem propagation [101].

Another valuable feature of the ideal production system would be a high biomass yield in combination with established, efficient harvesting and processing technologies and infrastructure. Recovery of the therapeutic protein often involves grinding the plant tissue and immediate cold buffer extraction and removal of solids. The availability of harvest and recovery technologies (e. g. milling or oil-extraction) would reduce costs and investments at the beginning. A crop would offer a considerable benefit if the recombinant protein was expressed at high levels in a defined part of the plant, e. g. the wheat germ or oil-fraction, which can be easily obtained using established technologies. For example, the fractionation of transgenic seeds by dry and wet-milling removed seed components (fiber, starch, oil) that did not contain the recombinant protein [22]. The maize embryo fraction is rich in soluble protein and can easily be separated from other seed tissue to increase the concentration of the recombinant protein [65]. This enriched fraction would decrease storage volumes, transport costs and the extract volume to be handled in the subsequent steps. Additionally, fractionation would decrease the complexity of the extract and reduce the level of plant metabolites that interfere with downstream processing thus facilitating purification (see below). Furthermore, fractionation procedures and milling will result in a defined particle size distribution which is helpful to design first filtration or cleaning steps prior to protein purification. For established crops, a large selection of pesticides is available that would facilitate the selection of chemicals appropriate for the production of plant pharmaceuticals. Finally, the ability to grow the crop of choice on both hemispheres could ensure constant supply and reduce storage capacities and the risk of supply problems due to harvest losses in one region.

Importantly, the optimal production crop must offer a high standard of safety, also in view of liabilities [107, 108]. Since biosafety is discussed in more detail in chapter 16, I will only briefly summarize the expectations of pharmaceutical industry:

(i) Transgene containment: Out-crossing to wild relatives or non-transgenic plots of the same crop should be prevented by appropriate choice of the crop plant and the location of the transgenic plot, by physical containment, or by genetic containment mechanisms.

(ii) Identity preservation: A crop with a visible marker, such as a defined seed color, would help to prevent mixing with consignments of the same crop during harvesting, transport and processing. Examples could include white tomatoes, black barley, red carrots and pink potatoes.

(iii) Product containment: A plant variety that is non-toxic, but has an unpalatable taste, would help to prevent involuntary ingestion of the recombinant protein. Strategies that direct expression to inedible organs or causes the protein to be expressed after harvest (see below) would provide a similar measure of safety.

17.4.10
Post-harvest expression

Post-harvesting technologies have already been established. CropTech Corp. is using the inducible MeGA^{TM} promoter to produce glucocerebrosidase in tobacco leaf tissues. Although expression levels vary depending on the plant line and the precise induction protocol, over 1 mg of the enzyme per gram fresh weight has been achieved in crude extracts [71]. Malting of rice or barley seeds is a technical process analogous to natural seed germination. Under defined malting conditions, many parameters like protein content and protein quality can be controlled. The companies Maltagen and Ventria Bioscience took advantage of the classical malting technology and developed protein production systems using transgenic grains. Under the control of germination-specific promoters, coordinated expression of several transgenes (e.g. light and heavy chain genes to produce a full size antibody) can be achieved. The germination-specific promoters allow exact regulation of the expression level during malting. The influence of the environment on the integrity, stability, glycosylation and expression level of the heterologous protein [49, 68] is excluded since the promoters (e.g. α-amylase promoter) are inactive during the growth period in the field, and active only under controlled malting conditions. One expects that this would avoid transgene silencing problems especially arising when (multiple copies of) strong promoters are used to achieve high-level expression [109].

17.4.11
Purification

In contrast to the protein recovery methods discussed above, protein purification is still based predominantly on laboratory-developed procedures that are often not directly scalable because of the high costs of the chemicals employed, the difficulties in controlling foaming, and pumping problems related to non-homogeneity and viscosity. For example, Fischer *et al.* [110] developed a three-step protocol for the recovery of antibodies from extracts of tobacco suspension cells, starting with cross-flow filtration followed by protein A affinity chromatography and gel filtration. More than 80% of the expressed recombinant antibody was recovered. Some studies have been carried out concerning the choice of defatting solvents [111] and the recovery of recombinant proteins from canola by cation exchange chromatography [112]. However, highly efficient purification protocols for molecular farming are still lacking, even though such protocols are absolutely necessary with regard to economic and regulatory affairs [26]. Protein recovery and purification costs are estimated to account for 88% of operating costs [22]. Similarly, for the commercial production of insulin in *E. coli*, chromatography accounts for 30% of operating expenses and 70% of equipment costs. Therefore, the low production costs of molecular farming would not counterbalance the higher downstream processing costs associated with plant tissue if the efficiency of purification from plants could not be increased dramatically.

Affinity tags such as His_6 [21, 113], Myc [75] and FLAG [70] have been used to facilitate purification by affinity chromatography. However, therapeutic proteins containing additional affinity tags will certainly not obtain approval. Therefore, economical and viable methods to cleave off the affinity tag are required. A slightly different, and very elegant approach is the use of an oleosin tag [114] in combination with simple flotation-centrifugation technology. SemBioSys Inc. has developed an oleosin-fusion platform that includes cleavage of the oleosin-fusion protein with proteases like factor Xa in a cost-efficient manner.

An intriguing problem in recombinant protein purification is the presence of co-purifying proteins. An excellent study has been carried out on the removal of corn trypsin inhibitor (CTI) which co-purified with aprotinin on a trypsin-agarose affinity column [115]. After grinding and milling, protein extraction at pH 3 reduced the amount of CTI in the extract and increased the aprotinin content in the mass fraction. After subsequent filtration and affinity adsorption, the remaining CTI was captured on an agarose-IDA-Cu^{2+} column while the recombinant aprotinin was collected in the flow-through with a purity of at least 79% [115].

The most important issue in downstream processing is often neglected: quality management. Similar to purification schemes for CHO-derived biologics, many analytical parameters also have to be registered for the purification of plant-derived products. Despite the fact that molecular farming involves plants that are part of our normal diet, the injection of unwanted co-purifying plant constituents into humans is likely to provoke totally different and even fatal responses. Therefore, good manufacturing practice (GMP) would require the quantification of endotoxins and pesticide residuals, the removal of any co-purifying plant proteins and metabolites, determination of the extent of product aggregation, stability in final formulations, inactivation of eventual viral contaminations, and particle load. The costs for these measures can easily double production costs and should be included in the cost calculations of molecular farming when compared with classical fermentation.

17.5
Conclusions

The pharmaceutical industry anticipates that molecular farming will save time and money compared to traditional production systems. Because of bottlenecks and production costs, many biologics will never reach the market and the intended patients, or will do so only with great delays, if molecular farming fails. However, a number of points in the production of plant-derived proteins have yet to be addressed appropriately. In order to fulfill all requirements and obtain regulatory approval, the questions outlined above have to be answered for each recombinant protein. Last but not least, economical factors will decide whether molecular farming in plants will increase the number of available products.

References

[1] J. F. RIPPMANN, M. KLEIN, C. HOISCHEN et al., *Appl. Environ. Microbiol.* 1998, *64* (12), 4862–4869.

[2] R. FISCHER, J. DROSSARD, N. EMANS et al., *Biotechnol. Appl. Biochem.* 1999, *30* (2), 117–120.

[3] P. HOLLINGER, *Methods Mol. Biol.* 2002, *178*, 349–357.

[4] A. SOTIRIADIS, T. KESHAVARZ, T. KESHAVARZ-MOORE, *Biotechnol. Prog.* 2001, *17* (4), 618–23.

[5] R. A. TATICEK, C. W. T. LEE, M. L. SHUKLER, *Curr. Opin. Biotechnol.* 1994, *5* (2), 165–174.

[6] M. F. GOOSEN, *Bioprocess Technol.* 1993, *17*, 1–17.

[7] L. CHU, D. K. ROBINSON, *Curr. Opin. Biotechnol.* 2001, *12* (2), 180–187.

[8] G. GELLISSEN, *Appl. Microbiol. Biotechnol.* 2000, *54* (6), 741–750.

[9] P. W. SAUER, J. E. BURKY, M. C. WESSON et al., *Biotechnol. Bioeng.* 2000, *67* (5), 585–597.

[10] www.bio.org

[11] R. ANDERSSON, R. MYNAHAN, *In vivo: The Business and Medicine Report*, Windhover Information Publication, 2001.

[12] G. WALSH, *Nature Biotechnol.* 2003, *21* (8), 865–870.

[13] B. GOODMAN, Shalom Equity Fund Newsletter 2002, *11* (16), 1–3.

[14] E. E. HOOD, S. L. WOODARD, M. E. HORN, *Curr. Opin. Biotechnol.* 2002 *13* (6), 630–635.

[15] R. C. DAS, K. J. MORROW, *Antibody Therapeutics: Production, Clinical trials, and Strategic issues* 2002,

[16] FROST, SULLIVAN, *Biopharmaceuticals Industry Analysis – Quantification of Supply and Demand of Manufacturing Capacities*, Frost & Sullivan, London, 2003.

[17] M. H. KUMAGAI, J. DONSON, G. DELLA-CIOPPA et al., *Gene* 2000, *245* (1), 169–174.

[18] A. A. McCORMICK, M. H. KUMAGAI, K. HANLEY et al., *Proc. Natl Acad. Sci.* 1999, *96* (2), 703–708.

[19] T. VERCH, V. YUSIBOV, H. KOPROWSKI, *J. Immunol. Methods* 1998, *220* (1–2), 69–75.

[20] J. KAPILA, R. DE RYCKE, M. VAN MONTAGU et al., *Plant Sci.* 1996, *122* (1), 101–108.

[21] C. VAQUERO, M. SACK, J. CHANDLER et al., *Proc. Natl Acad. Sci. USA* 1999, *96* (20), 11128–11133.

[22] R. L. EVANGELISTA, A. R. KUSNADI, J. A. HOWARD et al., *Biotechnol. Prog.* 1998, *14* (4), 607–614.

[23] K. E. STEIN, K. O. WEBBER, *Curr. Opin. Biotechnol.* 2001, *12* (3), 308–311.

[24] W. KNEIFEL, E. CZECH, B. KOPP, *Planta Med.* 2002, *68* (2), 5–15.

[25] www.emea.org

[26] L. MIELE, *Trends Biotechnol.* 1997, *15* (2), 45–50.

[27] K. DÜRING, S. HIPPE, F. KREUZALER, J. SCHELL, *Plant Mol. Biol.* 1990, *15* (2), 281–293.

[28] A. R. KUSNADI, Z. L. NIKOLOV, J. A. HOWARD, *Biotechnol. Bioeng.* **1997**, *56* (5), 473–484.

[29] P. SOJIKUL, N. BUEHNER, H. S. MASON, *Proc. Natl Acad. Sci. USA* **2003**, *100* (5), 2209–2214.

[30] P. C. SIJMONS, B. M. M. DEKKER, B. SCHRAMMEIJER et al., *Bio/Technology* **1990**, *8* (3), 217–221.

[31] X. G. WANG, G. H. ZHANG, C. X. LIU et al., *Biotechnol. Bioeng.* **2001**, *72* (4), 490–494.

[32] L. FARRAN, J. J. SANCHEZ-SERRANO, J. F. MEDINA et al., *Transgenic Res.* **2002**, *11* (4), 337–346.

[33] Z. HUANG, I. DRY, D. WEBSTER et al., *Vaccine* **2001**, *19* (15–16), 2163–2171.

[34] G. Y. ZHONG, D. PETERSON, D. E. DELANEY et al., *Mol. Breeding* **1999**, 5, 345–356.

[35] D. BOSCH, J. SMAL, E. KREBBERS, *Transgenic Res.* **1994**, 3, 304–310.

[36] A. SCHOUTEN, J. ROOSIEN, F. A. VAN ENGELEN et al., *Plant Mol. Biol.* **1996**, *30* (4), 781–793.

[37] U. CONRAD, U. FIEDLER, *Plant Mol. Biol.* **1998**, *38* (1–2), 101–109.

[38] E. TORRES, C. VAQUERO, L. NICHOLSON, *Transgenic Res.* **1999**, *8* (6), 441–449.

[39] H. XU, F. U. MONTOYA, Z. WANG et al., *Protein Expr. Purif.* **2002**, *24* (3), 384–394.

[40] A. SCHOUTEN, J. ROOSIEN, J.M. DE BOER et al., *FEBS Lett.* **1997**, *415* (2), 235–241.

[41] A. M. BRUYNS, G. DE JAEGER, C. DE NEVE et al., *FEBS. Lett.* **1996**, *386* (1), 5–10.

[42] S. J. STREATFIELD, J. R. LANE, C. A. BROOKS et al., *Vaccine* **2003**, *21* (7–8), 812–815.

[43] L. J. RICHTER, Y. THANAVALA, C. J. ARNTZEN et al., *Nature Biotechnol.* **2000**, *18* (11), 1167–1171.

[44] M. L. SMITH, H. S. MASON, M. L. SHULER, *Biotechnol. Bioeng.* **2002**, *80* (7), 812–822.

[45] V. GOMORD, E. WEE, L. FAYE, *Biochimie.* **1999**, *81* (6), 607–618.

[46] A. WRIGHT, S. L. MORRISON, *Trends Biotechnol.* **1997**, *15* (1), 26–32.

[47] P. M. RUDD, T. ELLIOTT, P. CRESSWELL et al., *Science* **2001**, *291* (5512), 2370–2376.

[48] D. C. ANDERSEN, T. BRIDGES, M. GAWLITZEK et al., *Biotechnol. Bioeng.* **2000**, *70* (1), 25–31.

[49] I. J. W. ELBERS, G. M. STOOPEN, H. BAKKER et al., *Plant Physiol.* **2001**, *126* (3), 1314–1322.

[50] J. M. PEKELHARING, E. HEPP, J. P. KAMERLING et al., *Ann. Rheum. Dis.* **1988**, *47* (2), 91–95.

[51] E. E. HOOD, D. R. WITCHER, S. MADDOCK et al., *Mol. Breeding* **1997**, *3* (4), 291–306.

[52] S. MATSUMOTO, K. IKURA, M. UEDA et al., *Plant Mol. Biol.* **1995**, *27* (6), 1163–1172.

[53] M CABANES-MACHETEAU, A. C. FICHETTE-LAINE, C. LOUTELIER-BOURHIS et al., *Glycobiol.* **1999**, 9, 365–372.

[54] M. BARDOR, L. FAYE, P. LEROUGE, *Trends Plant Sci.* **1999**, *4* (9), 376–380.

[55] N. RAMINEZ, M. RODRIGUEZ, M. AYALA et al., *Biotechnol. Appl. Biochem.* **2003**, *38* (Pt3), 223–230.

[56] D. CHARGELEGUE, N. D. VINE, C. J. VAN DOLLEWEERD et al., *Transgenic Res.* **2000**, *9* (3), 187–194.

[57] J. MA, B. HIKMAT, K. WYCOFF et al., *Nature Med.* **1998**, *4* (5), 601–606.

[58] L. ZEITLIN, S. S. OLMSTED, T. R. MOENCH et al., *Nature Biotechnol.* **1998**, *16* (13), 1361–1364.

[59] J. MORROW, *Genet Eng. News* **2002**, *12* (1), 54–59.

[60] M. BARDOR, C. FAVEEUW, A.-C. FITCHETTE et al., *Glycobiology* **2003**, *13* (6), 427–434.

[61] M. L. SMITH, M. E. KEEGAN, H. S. MASON et al., *Biotechnol. Prog.* **2002**, *18* (3), 538–550.

[62] S. F. CONRAD, I. J. BYEON, A. M. DIGEORGE et al., *Biomed. Pept. Proteins Nucleic Acids* **1995**, *1* (2), 83–92.

[63] A. HU, R. CATTANEO, S. SCHWARTZ et al., *J. Gen. Virol.* **1994**, *75* (5), 2173–2181.

[64] E. MARQUET-BLOUIN, F. B. BOUCHE, A. STEINMETZ et al., *Plant Mol. Biol.* **2003**, *51* (4), 459–469.

[65] H. DANIELL, S. J. STREATFIELD, K. WYCOFF, *Trends Plant Sci.* **2001**, *6* (5), 219–226.

[66] K. KO, Y. TEKOAH, P. M. RUDD et al., *Proc. Natl Acad. Sci. USA* **2003**, *100* (13), 8013–8018.

[67] B. W. GRINNELL, J. D. WALLS, B. GER-
LITZ, *J. Biol. Chem.* **1991**, *266* (15),
9778–9785.

[68] L. H. STEVENS, G. M. STOOPEN, I. J. W.
ELBERS et al., *Plant Physiol.* **2000**, *124*
(1), 173–182.

[69] J. W. LARRICK, L. YU, C. NAFTZGER et al.,
Biomol. Eng. **2001**, *18* (3), 87–94.

[70] C. L. CRAMER, J. G. BOOTHE, K. K. OISHI,
Curr. Topics Microbiol. Immunol. **1999**,
240, 95–118.

[71] C. L. CRAMER, D. L. WEISSENBORN,
K. K. OISHI, D. N. RADIN in: M. R. L.
Owen, J. Pen (Eds.), Transgenic plants:
a production system for industrial and
pharmaceutical proteins, WILEY, Chi-
chester, **1996**, 299–310.

[72] R. L. FUCHS, R. A. HEEREN, M. E. GUS-
TAFSON et al., *Bio/Technology* **1993**, *11*
(13), 1537–1542.

[73] A. HIATT, R. CAFFERKEY, K. BOWDISH,
Nature **1989**, *344* (6265), 469–470.

[74] J. M. SHARP, P. M. DORAN, *Biotechnol.
Bioeng.* **2001**, *73* (5), 338–346.

[75] O. ARTSAENKO, B. KETTIG, U. FIEDLER
et al., *Mol. Breeding* **1998**, *4* (4), 313–
319

[76] F. A. VAN ENGELEN, A. SCHOUTEN,
J. W. MOLTHOFF, *Plant Mol. Biol.* **1994**,
26 (6), 1701–1710.

[77] D. L. PETERSON, *J. Biol. Chem.* **1981**,
256, 6975–6983.

[78] B. DOGAN, H. S. MASON, L. RICHTER
et al., *Biotechnol. Prog.* **2000**, *16* (3),
435–441.

[79] Q. KONG, L. RICHTER, Y. F. YANG et al.,
Proc. Natl Acad. Sci. USA **2001**, *98* (20),
11539–11544.

[80] Y. THANAVALA, Y.-F. YANG, P. LYONS
et al., *Proc. Natl Acad. Sci. USA* **1995**,
92 (8), 3358–3361.

[81] R. W. LEE, J. STROMMER, D. HODGINS
et al., *Infect. Immun.* **2001**, *69* (9),
5786–5793.

[82] C. O. TACKET, H. S. MASON, G. LO-
SONSKY et al., *Nature Med.* **1998**, *4* (5),
607–609.

[83] T. A. HAQ, H. S. MASON, J. D. CLEMENTS
et al., *Science* **1995**, *268* (5211), 714–716.

[84] H. S. MASON, T. A. HAQ, J. D. CLEMENTS,
C. J. ARNTZEN, *Vaccine* **1998**, *16* (13),
1336–1343.

[85] S. J. STREATFIELD, J. M. JILKA, E. E. HOOD

et al., *Vaccine* **2001**, *19* (17–19), 2742–
2748.

[86] T. G. LAUTERSLAGER, D. E. FLORACK,
T. J. VAN DER WAL et al., *Vaccine* **2001**, *19*
(17–19), 2749–2755.

[87] H. S. MASON, D. M.-K. LAM, C. J. ARNT-
ZEN, *Proc. Natl Acad. Sci. USA* **1992**, *89*
(24), 11745–11749.

[88] J. KAPUSTA, A. MODELSKA, M. FIGLERO-
WICZ et al., *FASEB J.* **1999**, *13*, 1796–
1799.

[89] J. KAPUSTA, A. MODELSKA, T. PNIEWSKI
et al. in: *Progress in Basic and Clinical
Immunology* (Ed. by Mackiewicz),
Kluwer Acad., New York, **2001**,
299–303.

[90] H. S. MASON, J. M. BALL, J. J. SHI et al.,
Proc. Natl Acad. Sci. USA **1996**, *93* (11),
5335–5340.

[91] C. O. TACKET, H. S. MASON, G. LO-
SONSKY et al., *J. Infect. Dis.* **2000**, *182*
(1), 302–305.

[92] H. J. HIRSCHBERG, J. W. SIMONS, N.
DEKKER et al., *Eur. J. Biochem.* **2001**,
268 (19), 5037–5044.

[93] V. YUSIBOV, D. C. HOOPER, S. V. SPITSIN
et al., *Vaccine* **2002**, *20* (25–26), 3155–
3164.

[94] B. J LAMPHEAR, S. J. STREATFIELD,
J. M. JILKA et al., *J. Cont. Release* **2002**,
85 (1–3), 169–180.

[95] S. CASTANON, J. M. MARTIN-ALONSO, M.
S. MARIN et al., *Plant Sci.* **2002**, *162* (1),
87–95.

[96] C. CARRILLO, A. WIGDOROVITZ,
K. TRONO, et al., *Viral Immunol.* **2001**, *14*
(1), 49–57.

[97] *Scrip* **1998**, *2321*, 21.

[98] G. GIDDINGS, G. ALLISON, D. BROOKS
et al., *Nature Biotechnol.* **2000**, *18* (11),
1151–1155.

[99] S. AUSTIN, E. T. BINGHAM, R. G. KOE-
GEL et al., *Ann. NY Acad. Sci.* **1994**, *721*,
235–244.

[100] U. FIEDLER, J. PHILLIPS, O. ARTSAENKO
et al., *Immunotechnology* **1997**, *3* (10),
205–216.

[101] H. KHOUDI, S. LABERGE, J.-M. FERULLO
et al., *Biotechnol. Bioeng.* **1999**, *64* (2),
135–143.

[102] J. PEN, T. C. VERWOERD, P. A. VAN PARI-
DON et al., *Bio/Technology* **1993**, *11* (7),
811–814.

[103] G. J. VAN ROOIJEN, M. M. MOLONEY, *Bio/Technology* **1995**, *13* (1), 72–77.

[104] G. DE JAEGER, S. SCHEFFER, A. JACOBS et al., *Nature Biotechnol.* **2002**, *20* (12), 1265–1268.

[105] U. FIEDLER, U. CONRAD, *Bio/Technology* **1995**, *13* (10), 1090–1093.

[106] E. STÖGER, C. VAQUERO, E. TORRES et al., *Plant Mol. Biol.* **2000**, *42* (4), 583–590.

[107] S. SMYTH, G. G. KHACHATOURIANS, P. W. B. PHILLIPS, *Nature Biotechnol.* **2002**, *20* (6), 537–541.

[108] J. L. FOX, *Nature Biotechnol.* **2003**, *21* (1), 3–4.

[109] B. LECHTENBERG, D. SCHUBERT, A. FORSBACH et al., *Plant J.* **2003**, *34* (4), 507–517.

[110] R. FISCHER, Y. C. LIAO, J. DROSSARD, *J. Immunol. Methods* **1999**, *226* (1–2), 1–10.

[111] Y. BAI, Z. L. NIKOLOV, *Biotechnol. Prog.* **2001**, *17* (1), 168–174.

[112] C. ZHANG, C. E. GLATZ, *Biotechnol. Prog.* **1999**, *15* (1), 12–18.

[113] C. VAQUERO, M. SACK, F. SCHUSTER et al., *FASEB. J.* **2002**, *16* (3), 408–410.

[114] D. L. PARMENTER, J. G. BOOTHE, G. J. VAN ROOIJEN et al., *Plant Mol. Biol.* **1995**, *29* (6), 1167–1180.

[115] A. R. AZZONI, A. R. KUSNADI, E. A. MIRANDA, Z. L. NIKOLOV, *Biotechnol. Bioeng.* **2002**, *80* (3), 268–276.

18

The Role of Science and Discourse in the Application of the Precautionary Approach

KLAUS AMMANN

18.1
Introduction

The precautionary approach (PA) is an important element of environmental law that is used to address a potential risk whether or not that risk can be demonstrated or its consequences identified. The static use of a sole, generally accepted definition of the PA is extremely difficult, since this cannot meet the multitude of needs in important legislative tools introduced in many conventions designed to protect biodiversity. The way out will be a more discursive model, a model that allows for adaptation to specific conditions and which enforces solution-oriented procedures.

Before discussing the application of the PA in molecular farming, it is necessary to make some preliminary remarks on the transatlantic divide concerning the regulation of genetically modified (GM) crops. Economic globalization has not, thus far, led to a convergence in the regulation of agricultural biotechnology in the European Union and the United States. While the EU has taken a precautionary approach to the regulation of biotech products, the US has decided that such products are not significantly different from those made using more traditional methods. Consequently, the US government has yet to implement any novel legislation or risk assessment procedures to regulate them. These varying regulatory responses provide an interesting contrast and background for debate concerning the PA. It is particularly interesting to note that although agricultural biotech products were developed for highly competitive and globally integrated agri-business markets, biotechnology regulation has nevertheless followed very different paths within the EU and the US.

Recently, US biotechnology policy has shown signs of gravitating towards the EU model, with noticeable changes in the regulatory climate hopefully also occurring in the EU [1]. It will be challenging to initiate a more fruitful dialog on the PA since we cannot afford hesitation in the light of increasingly difficult agricultural production combined with urgent needs to feed a rapidly growing population.

In other fields of the biotech debate, the contrasts are much sharper. Some non-governmental organizations (NGOs) such as Greenpeace and Friends of the Earth are clearly abusing the PA and employing it as a weapon in their uncompromising fight against GM crops. Patrick Moore, a Greenpeace founder, has stated that many

Molecular Farming. Edited by Rainer Fischer, Stefan Schillberg
Copyright © 2004 WILEY-VCH Verlag GmbH & Co. KGaA, Weinheim
ISBN: 3-527-30786-9

environmentalists reject consensus politics and sustainable development in favor of continued confrontation, ever-increasing extremism and left-wing politics [2]. On the other side, biotech companies have built up enormous activities to cope with risk assessment, making it difficult for smaller companies to follow suit [3].

Discussions about the PA usually focus on definitions. Such definitions are plentiful, they depend on the scientific and social background of their authors, and they all contain elements of truth and error. One of the basic problems with the PA is that there is no such thing as an overall definition. The application of the PA is always heavily context-dependent. It is no use solving problems associated with applying the PA by means of a generally accepted definition, since it is difficult to define a principle sharply where uncertainty is the main element. The definition of terms and concepts like uncertainty always depend on the scientific, social, cultural and economic background of individuals employing them.

18.2
Other Roots to Problems with the Precautionary Approach

Problems with the application of the PA also have other roots, two of which are discussed in more detail below:

- The lack of knowledge about the origin of the PA and how it was first defined.
- The problem that the PA is too closely based on factual knowledge alone.

18.2.1
The Roots of the Precautionary Approach and Environmental Debate

Although the idea of precaution in environmental matters has been around since the 1970s, the term was first introduced under Principle 15 in the Convention on Biological Diversity (CBD), 1992 [4]:

> "In order to protect the environment, the *precautionary approach* shall be widely applied by the States according to their capabilities. Where there are threats of serious or irreversible damage, lack of full scientific certainty shall not be used as a reason for postponing cost-effective measures to prevent environmental degradation."

Such statements are written in a spirit that accepts the deteriorating environment as a proven fact. The environment obviously suffers from human activity of all kinds, e.g. air and soil pollution, including pollution from heavy metals and dioxins, which most agree is killing the forests. In the beginning, however, environmentalists often exaggerated the risks involved. Although this helped to bring such issues onto the table, we now face a credibility gap in Europe which manifests as the dying forest syndrome: despite the claims, the forests just refuse to die... There was a time in the 1970s when environmental debates in Europe became derailed, when activists started to mix deontic knowledge (knowledge of how things ought to be) with factual knowledge (see Sect. 18.4). However, while the CBD was under development, there was no doubt that factual knowledge had to predominate in order to trigger some decisions. In the CBD text, one can also clearly see a timeline: decisions had to be taken early, but on the baseline of growing hazards.

Other kinds of knowledge were debated in the Cartagena Protocol [5], which included bracketed text allowing decisions taken by Importing Parties to be reviewed if there was reasonable evidence that such decisions had not been based on scientific, socio-economic, cultural or precautionary principles. Australia suggested that Exporting Parties should be able to request reviews under similar circumstances. Later, the adoption of cultural and social issues was deleted from the Protocol, but Article 26 has been established based on all these considerations during the negotiations [6]. The draft article on socio-economic considerations proposed by the African group included taking into account the length of time before any impacts might be seen, and proposed a seven-year notification period prior to export. The African group proposal contained an extensive list of socio-economic considerations to be included in risk assessment, including anticipated changes in the existing social and economic patterns, and possible threats to biological diversity, traditional crops or other products.

Finally, Article 26 was included:

1. The Parties, in reaching a decision on import under this Protocol or under its domestic measures implementing the Protocol, may take into account, consistent with their international obligations, socio-economic considerations arising from the impact of living modified organisms on the conservation and sustainable use of biological diversity, especially with regard to the value of biological diversity to indigenous and local communities.
2. The Parties are encouraged to cooperate on research and information exchange on any socio-economic impacts of living modified organisms, especially on indigenous and local communities.

Environmentalists soon used deontic and instrumental knowledge to formulate strategies for the solution of targeted environmental problems. It was a peaceful debate, where everybody was optimistic and keen to solve the problems within a few years or, at the most, decades. Then a lady whose name is well known today changed all this: Rachel Carson's Silent Spring demonstrated that the long-term effects of dichlorodiphenyltrichloroethane (DDT) could seriously harm bird life. The DDT issue even made everyone forget the good points of this particular pesticide, namely that it saved hundreds of millions of lives by killing mosquitoes [7,8]. Gradually, environmentalists started to realize that ecological problems are so-called 'wicked problems', which are not readily solved because solutions are so difficult to find.

In the past, there were difficult days filled with endless debates about flux, modeling, circulation ecology, and interdisciplinary or even transdisciplinary collaboration as the best way to solve research problems and to find swift solutions for environmental problems. In the end it was accepted that what was called interdisciplinary or even transdisciplinary research too soon degenerated into multidisciplinary structures, structures that were unavoidable, since research money was limited and had to be divided up equitably.

Interdisciplinary work requires at least some mutual understanding and eventual reaction to what the research partner does. Transdisciplinary work should include a planning phase in order to fix a common research goal, and to try and get the necessary disciplinary groups to work together. In the end, the goal is to produce an amalgamate of all research activities. It is beginning to become clear why this worthy ideal could be so difficult to achieve [9–17].

In 2004, a report was published concerning the impact of agricultural biotechnology on biodiversity [18]. The report was 100 pages long, with 300 citations, and served to demonstrate how hugely complex the interaction between modern and traditional agriculture and biodiversity really is. There was no room for simple slogans and each scientific field experiment had to be evaluated individually and with care. Note that the report [18] was restricted to biotechnology and biodiversity, and it did not deal with any other factors, desirable as this might have been.

Matters are complicated further when we try to expand inter(trans)disciplinary work beyond natural sciences, including social sciences such as sociology, history, philosophy etc. This inevitably springs the trap of statistical debates and the 'factualization' of the research work of all the included disciplines. This is of course a dead end and will never ever lead to solutions with a broad consensus, which will also become politically important.

18.2.2
Discussion About the PA is Too Closely Related to Factual Knowledge Alone

This might seem a paradoxical heading given the discussion above, but once factual knowledge is placed in the correct proportion to all other kinds of knowledge and a true systems approach is utilized, we will cut through the Gordian knot easily. We must realize that problems encountered in discussions about the PA are 'wicked problems', problems with a social and cultural context. This automatically means that linear planning will resolve nothing, and this is why it is virtually impossible tackle the problem of scientific lacunae directly, a problem intricately linked to PA discussions.

The challenge of wicked problems is exacerbated by social complexity – the number and diversity of stakeholders in the problem-solving process. Social complexity means that the environment of a project team is populated by individuals, other project teams, and other organizations that have the power to undermine the project if their stake is not considered – or if they are not at least included in the thinking and decision-making process [19].

Unfortunately, even the planning of green biotechnology has now evolved into a wicked problem with complex structures and no obvious causal chains. This applies also to the PA. These problems cannot be determined completely in a quantitative and scientific manner, and there are no existing solutions in the sense of definitive and objective answers alone. Wicked problems have been addressed mainly through formalized (linear) methods that are suitable only for the solution of tame problems.

Often solutions have been found empirically, by trial and error. Acceptable solutions can be found, and gifted planners or regulators often develop good intuitive abilities, also taking into account socio-economic factors. However, the linear approach that often works properly for tame problems usually ends in a fiasco when tackling wicked problems.

18.3
The First and Second Generation Systems Approaches

18.3.1
First Generation Systems Approach

Much hope has been placed in the first generation systems approach, which certainly had its merits (NASA missions, toll bridges, defense systems, supercrops etc). Planning goals were clearly defined and all decisions were oriented towards these goals. In general it can be said that the first generation systems approach has been followed by an era of disappointment, since it has not produced what was expected. A number of large and complex projects such as urban renewal, improving the environment, tackling the nutrition problems of mankind etc. can only be considered as failures, or partial failures such as the Green revolution.

The main reason is that the classic paradigm of (rational) science and technology is not applicable to the problems of open ecological and/or societal systems. It is very important to realize that problems in biotechnology are not solely problems of science, but also problems of society. This does not mean that risk assessment should not be science-based; on the contrary. It would be a big mistake to assume that the involvement of open structures in ecology and human society would give excuses to deviate from the path of science when it comes to questions of safety and regulation, or even worse, to abuse scientific language in order to achieve an ideologically-stamped agenda as certain members of the newly grown (protest or biotech) industry are doing.

18.3.2
Second Generation Systems Approach

Professional management tools that are based on a second-generation systems approach should not be mixed up with so-called future workshops, with their frequent and inconsiderate use of pin walls when activist groups start their planning. Rarely have those actions led to sustainable results. Too often, future workshops (German: Zukunftswerkstätten) start with fulminate brainstorming and lots of enthusiasm, but later the participants go home to live their normal lives, tending to forget about the big decisions taken earlier. If the workshops would be properly carried through after Jungk and Müllert [20], their results would be certainly better.

There should also be a distinction between the second-generation systems approach and collaborative learning workshops, which can seem delightful and thus also successful (as can be their subsequent decisions), but which rarely achieve sustainable results either, because such events actually lack a collaborative decision making process. It is important to avoid the following misunderstanding: Decision-making is not in its fundamental structure a democratic process; it is a process where people genuinely involved are participants. To be even more explicit, partners in the decision making process should have their own and genuine interest in the cause. This avoids the danger of manipulation through clever public relations, and through the use of populist and, even worse, fundamentalist arguments.

Consensus conferences and citizens' conferences are extremely helpful in cases of conflicts with the public, but here again it is difficult for any criticized processes to be changed for the better, and negative trends are rarely turned around. The difficulty is that a citizens group cannot be expected to learn about the complexity of necessary solutions after only a few days of intensive briefing.

Another kind of internal consensus conference is designed by the promoters of the syntegrity approach, which brings together corporate people in order to analyze internal dynamics and processes, and to discern negative effects. Despite the effort now involved in the design of new planning and management methods, negative results still predominate and are in fact part of a planning crisis, stemming from the 1970s and still continuing today.

It is primarily the paradox of rationality that has been severely underestimated in the first generation systems approach. The more questions we ask, the more answers are possible and *vice versa*. Limitations of technological solutions are always hidden in the open ecological and social systems. Just compare the infamous case of DDT spraying in the past. Constraints in possible secondary effects in ecology should be examined carefully. This is well demonstrated in the case of Bt pollen and its effect on the monarch butterfly larvae, the result of a highly sophisticated laboratory study where press interpretation was way out of proportion – even though the author himself warned about this. If the farmers had been asked, they would have been able to say that feeding and pollination times rarely overlap, and that the plants fed on by the monarch caterpillars are actually weeds which they attack and attempt to eliminate with herbicides.

In order to tackle wicked problems it is necessary to go through an extensive process of argumentation, also called objectification, not to be mixed up with an objective approach to the problem. There is rational planning, but there is no way to start to be rational. One should always start a step earlier, since there are important trends and facts that will make straightforward rational thinking and action useless in solving wicked problems. It is not the theory component, but rather the political component of the knowledge, which determines the vector of the action. This is the zero step, so important in the publications of Horst Rittel [13–15]. This is also the basis of the understanding of the term *symmetry of ignorance* [13–15, 21]. As an example, consider the fact that experts can be wrong and farmers know better in certain agricultural situations because they are better observers out in the field. Agriculture is especially well suited to the second-generation systems approach.

The knowledge needed to address wicked problems is not concentrated in a single source. It is absolutely essential to involve all partners in the problem solution process, including the general population (mainly farmers' organizations and consumer organizations), governmental regulators, non-governmental organizations, life science companies and the scientists. There is no monopoly over knowledge; no one can decide alone on the PA. Having illustrated the difficulties in solving wicked problems, we need a new approach in problem solving, in order to avoid the pitfalls of ignoring bottom up feedback. As Adam Kahane stated in early 2004, one should only let people participate if they are part of the problem.

However, it is only possible to keep to this rule if another important rule is also followed: All partners in the planning process have to avoid hidden agendas, which can be achieved if a minimum amount of respect is paid to each other partner. Nobody should be criticized for speaking up in his own interest. It is wrong to perpetuate reciprocal accusations of 'abuse of the PA for the purposes of conducting a trade war' or denigrating the PA for reasons of global unhindered trade or self-advantage.

It is obvious in these times of growing difficulties in communicating about biotech products, especially in agriculture, that all partners still have a lot of homework to do.

The biotech companies are populated with people who are convinced about their own products (in most cases rightly so), since they know precisely about safety standards and regulatory processes. So far so good, but these people live in a world of euphemisms and perfection, and they develop over time a lack of understanding of criticism from outside.

The scientists often are naïve enough to stick to factual, instrumental and explanatory knowledge alone. Many miss a very important point, as Hannah Arendt put it: "One of the noblest tasks of scientists is to make out of facts public opinion". The regulators should find ways and means to cope up with the growing speed of new developments. One of the main reasons why events in Europe turn sour is the fact that European regulation is way behind regulation in the United States (although picking up in the last few years). On the other hand, this is an excellent time to see more clearly the geographical differences in regulation.

Some of the big NGOs have developed into powerful protest industries and are not interested in a thorough scientific analysis, since this could blur populist arguments that they need to keep up in order to get more donors, which are in fact their 'shareholders'. The public is often lost between the two camps and, surprisingly enough, only a minority feels the need for better education, whatever this would mean according to the two camps described above. And what about the press? Journalists like to write stories, stories that are there to enhance the number of printed copies of their own newspaper. Consequently, they often write what the public wants to hear – and the professional science journalists, who dare to swim against mainstream, are very few, since this needs a profound knowledge, a talent for foresight and, last but not least, some courage. We should have more investigative journalism in this field.

18.4
How to Solve Wicked Problems in Biotechnology and the Environment

What we need to solve these wicked problems is an action-oriented approach. Risk assessment and management must be seen as a second-generation planning strategy in developing a professional framework for decision-making. Strategies have to be developed which recognize the consequences of our actions on one side, yet specify our knowledge on the other. This knowledge has to be gained step by step and case by case: If we want clearly to distinguish our present state of knowledge (or indeed our ignorance) from appropriate decisions to be made, which are not based on our views and opinions, we need to go through the following steps.

What is the problem?
What do we want?
What are the alternatives?
How do we compare them?
How can we reach a solution?

All participants need to keep in mind that there are various types of planning knowledge (arranged according to the five questions asked above). Examples given below are grouped together as simple keyword illustrations, taken out of their context in real planning examples. They cannot be regarded as examples of realistic situations. This would be exactly the task of a second-generation planning process.

Factual knowledge is the knowledge of what actually happens (quantitative data or empirical, observed data). Examples: gene flow species by species or region by region; facts about insect resistance in agriculture.

Deontic knowledge is the very important knowledge of what ought to be. Examples: knowledge about new crops that enhance agricultural production; new agricultural techniques to avoid erosion; new biological approaches to fight insect pests; the benefits of segregating imports for Europe.

Explanatory knowledge is information that explains why things are so or why certain effects will happen. Here is where it is possible to determine the direction of the solution. Examples: the way Bt proteins affect specific pest and beneficial insects; what are the main reasons for unwelcome erosion effects; mechanisms of vertical gene flow; mechanisms of resistance development.

Instrumental knowledge is information about how to steer certain processes, on how to achieve certain goals, i.e. knowledge that needs to be balanced against regulation and safety. Examples: how to build Bt and other genes into crops and how to stabilize them; how to avoid vertical gene flow; how to avoid unwelcome soil erosion; how to avoid early pest resistance.

Conceptual knowledge is knowledge that allows conflicts to be avoided before they occur. This is knowledge about complex situations, taking into account all previous kinds of knowledge and also weighting them against arguments coming from open ecological and societal systems. Example: concepts about transgenic crops compatible with the idea of sustainable agriculture. It is a matter of developing conceptual knowledge of precision agriculture based on the best practices and the introduction of new techniques.

18.5
How to Achieve Such Demanding Planning Goals

This is essentially the process or argumentation or objectification as discussed above. The hopes of this process are to:

- *forget less, to raise the right issue*
- *look at the planning process as a sequence of events*
- *stimulate doubt by raising questions, to avoid short-sighted explicitness*
- *control the delegation of judgment: experts have no absolute power, and scientific knowledge is always limited.*

18.6
There is no Scientific Planning

Solving practical problems such as the development of sustainable transgenic crops cannot be dealt with by making the planning process more science-based. Dealing with wicked problems is always political because of its deontic premises. Science only generates factual, instrumental and in the best cases, explanatory knowledge.

The planner (here the regulator who must take decisions using the PA) is not primarily an expert, but a midwife of problem solving, a teacher more than a doctor. Moderate optimism and careful, seasoned disrespect allows doubt to be raised, and is a virtue, not a disadvantage of an action plan manager.

The planning process for wicked problems has to be understood as an argumentative process. It should be seen as a venture (or even an *ad*venture) among conspirators, where one cannot anticipate all the consequences of the plans. Second-generation systems methods try to make this deliberation explicit, to support it, to find the means to make the process more powerful, and to get it under better control for all participants. A caveat is certainly justified here, since we are dealing not only with human beings in a discourse, but also with the environment, which has basically no voice [22]. Finally, it is necessary to mention the abuse of the PA, so clearly visible in many aspects of the GM debate. A blatant example is the case of US aid to states of southern Africa, which suffered in 2000 and 2001 from severe food shortages resulting in an estimated 14 million people facing starvation. The delivery of thousands of tons of transgenic maize, initially without additional information (this information gap was later filled), pushed many of those states into a dilemma that was worsened by NGOs with a vested interest in campaigning against GM crops despite the humanitarian costs in this case. Whereas Zambia flatly refused to accept the GM maize, Malawi accepted the desperately needed staple food due to the severe public health challenges, but is also calling for a better dialogue between the donors and the developing countries [23].

18.7
Outlook

It is beyond logic and our present day knowledge to predict the many surprising outcomes of debates about genetic engineering, such as those discussed above. It is clear that many new generations of GM plants will be produced in the near future. The pipeline is filled with fascinating new products, and three are discussed here. First, there will soon be a new generation of GM plants in which the transgenes derive from near relatives. There are many very useful non-alien transgenes, which will enhance resistance, adaptation to special ecological requirements etc. Second, we will see lots of pharming applications with relevance in medicine. Complex molecules cannot easily be synthesized, so they need to be built using natural pathways in higher organisms, controlled by transgenes or even completely novel genes that remain to be developed. We will soon be able to steer the molecular evolution of complex organisms in a much more targeted way. Third, we will see a massive development of renewable energy sources in agriculture, which will be of utmost importance given the present day situation in oil politics. For all these new developments, it is imperative to let planning methods grow up to new horizons as well, and this is for several reasons: New development horizons will confront us with more knowledge gaps than ever, and they need to be taken care of in a highly professional way.

Such new approaches should lead to *precision biotechnology* for better crop design in the future. Using a simple example, precision biotechnology means that on one hand a bag of seeds could contain a great variety of different kinds of seeds showing resistance against many insect pests, but that all those seeds would have a genome designed precisely for the product quality to be sold after harvest. Genomic research offers a bright future and will greatly speed up modern breeding and add considerably to its precision. Here we also reintroduce some old concepts that will drive modern agriculture closer to the promotion of biodiversity.

In the future, organic farming needs go hand in hand with modern breeding methods including genetic engineering. This is an absolute need, but it will also be very difficult to achieve, since first-generation transgenic crops are either not suitable for organic farming or even worse, they work against such visionary strategies. Indeed, perhaps we need some novel products which fit to terms such as *organo-transgenic crops* and *organic precision biotechnology*.

This vision would of course break up the harsh, present day debate on the PA, and we would at last have the possibility to develop a balanced approach to difficult PA decisions, which needs as a basis a balanced approach to risk assessment, including different kinds of knowledge just as described above. Under these auspices, we will have at least a chance to make a breakthrough in the present-day PA debate – but if we continue to fight about factual knowledge alone, there is little hope of solving these problems, problems which have an international impact and need to be treated according to the latest insights in management and the systems approach.

If we really want to make progress, we have to abstain from the Western model of risk, which is always calculated on a formula with an intriguing logic: *Risk = hazard × probability.*

A lot of people do not realize that this leaves the evaluation completely on the negative side. The worst effect of the overestimation of the PA promoted as a *principle* is its absolute focus on the negative aspects of biotechnology. This has been summarized by Elizabeth M. Whelan, president of the American Council on Science and Health, and aptly sums up the shortcomings of the precautionary principle [24].

- First, it always assumes worst-case scenarios.
- Second, it distracts consumers and policy makers alike from the known and proven threats to human health.
- Third, it assumes no health detriment from the proposed regulations and restrictions, i.e. the PA overlooks the possibility that real public health risks can be associated with the elimination of minuscule, hypothetical risks.

This is why we should all advocate the Chinese meaning of the word *risk*, which comprises two risk elements, namely hazard and chance. With the discursive approach, following the systems approach of Rittel [14] and his long time companion Frank West Churchman (see foreword in Ref [25]), we have a chance to work in the complex environment to evaluate risk *and* chance with professional methods.

Bibliography

[1] A. PRAKASH, K. L. KOLLMAN, *Int. Studies Quart.* **2003**, *47* (4), 617–641.

[2] P. MOORE, *IPA Rev.* **2004**, *56* (1) 10–13.

[3] H. MILLER, *Chem. & Ind.* **1996**, December issue, 1000.

[4] United Nations Convention on Biological Diversity, **1992**. http://www.biodiv.org/doc/publications/guide.asp

[5] United Nations Montreal Protocol Biotechnology, **2000**. http://www.jus.uio.no/lm/biosafety.montreal.protocol.2000/doc

[6] United nations Cartagena Protocol on Biosafety, **2003**. http://www.biodiv.org/doc/publications/bs-brochure-03-en.pdf

[7] R. BATE, *IPA Rev.* **2004**, *56* (1) 14–15.

[8] R. TREN, R. BATE, *Malaria and the DDT Story*, The Institute of Economic Affairs, Profile Books, London, **2001**.

[9] S. BRIER, *Syst. Res. Behav. Sci.* **2000**, *17* (5), 433–458.

[10] A. J. N. JUDGE, *Knowl. Organ.* **1995**, *22* (2), 82–88.

[11] C. MITCHAM, R. FRODEMAN, *Sci. Technol. Hum. Val.* **2003**, *28* (1), 180–183.

[12] J. MITTELSTRASS, *Chem. Ing. Tech.* **1994**, *66* (3), 309–315.

[13] H. RITTEL, In: *Developments in Design Methodology* (N. Cross, ed), John Wiley & Sons, New York, **1984**, pp 317–327.

[14] H. RITTEL, *Planen, Entwerfen, Design, Ausgewählte Schriften.* Kohlhammer, Stuttgart, **1992**.

[15] H. RITTEL, M. WEBER, *Policy Sci.* **1973**, *4* (1), 155–169.

[16] M. SCHWANINGER, *Kybernetes* **2001**, *30* (9–10), 1209–1222.

[17] M. VAN MANEN, *Qual. Health Res.* **2001**, *11* (6), 850–852.

[18] K. AMMANN, *The Impact of Agricultural Biotechnology on Biodiversity – A Review*, **2004**. *http://www.botanischergarten.ch/Biotech-Biodiv/Report-Biodiv-Biotech12.pdf*

[19] J. CONKLIN, *Wicked Problems and Fragmentation*, CogNexus Institute, **2003**. *http://www.cognexus.org/id29.htm*

[20] R. JUNGK, N. MÜLLERT, *Future Workshops: How to Create Desirable Futures.* Institute for Social Inventions, London, UK, **1987**.

[21] G. FISCHER, P. EHN, Y. ENGESTRÖM et al., *Symmetry of Ignorance and Informed Participation.* Proceedings of the

Participatory Design Conference (PDC '02) (J. G. T. Binder, I. Wagner, eds), **2002**, pp 426–428.

[22] R.A. ROGERS, *Western J. Comm.* **1998**, *62* (3), 244–272.

[23] A.S. MUULA, J.M. MFUTSO-BENGO, *Croat. Med. J.* **2003**, *44* (1), 102–106.

[24] H. MILLER, G. CONCO, *Precautionary Principle Stalls Advances in Food Technol-* ogy. Competitive Enterprise Institute, **2000**. http://www.cei.org/utils/printer. cfm?AID=1758

[25] N. VERMA, *Similarities, Connections, and Systems: The Search for a New Rationality for Planning and Management.* Lexington Books, Lanham, MD, **1998**.

Subject Index